T0336135

VOLUME ONE HUNDRED AND TWENTY FOUR

ADVANCES IN
COMPUTERS

Power-Efficient Network-on-Chips:
Design and Evaluation

VOLUME ONE HUNDRED AND TWENTY FOUR

ADVANCES IN
COMPUTERS
Power-Efficient Network-on-Chips:
Design and Evaluation

Edited by

ALI R. HURSON
Missouri University of Science and Technology,
Rolla, MO, United States

HAMID SARBAZI-AZAD
Department of Computer Engineering,
Sharif University of Technology;
School of Computer Science,
Institute for Research in Fundamental Sciences (IPM),
Tehran, Iran

ACADEMIC PRESS
An imprint of Elsevier

ELSEVIER

Academic Press is an imprint of Elsevier
50 Hampshire Street, 5th Floor, Cambridge, MA 02139, United States
525 B Street, Suite 1650, San Diego, CA 92101, United States
The Boulevard, Langford Lane, Kidlington, Oxford OX5 1GB, United Kingdom
125 London Wall, London, EC2Y 5AS, United Kingdom

First edition 2022

ISBN: 978-0-323-85688-1
ISSN: 0065-2458

For information on all Academic Press publications
visit our website at https://www.elsevier.com/books-and-journals

Publisher: Zoe Kruze
Developmental Editor: Cindy Angelita Gardose
Production Project Manager: James Selvam
Cover Designer: Greg Harris

Typeset by STRAIVE, India

Working together
to grow libraries in
developing countries

www.elsevier.com • www.bookaid.org

Contents

7. Power-efficient network-on-chip design by partial topology reconfiguration   217

Mehdi Modarressi and S. Hossein SeyyedAghaei Rezaei

8. The design of an energy-efficient deflection-based on-chip network   257

Rachata Ausavarungnirun and Onur Mutlu

9. Power-gating in NoCs   319

Hossein Farrokhbakht, Shaahin Hessabi, and N. Enright Jerger

Contributors

Homa Aghilinasab
School of Computer Science, Institute for Research in Fundamental Sciences (IPM), Tehran, Iran

Negar Akbarzadeh
Department of Computer Engineering, Sharif University of Technology, Tehran, Iran

Fawaz Alazemi
Kuwait University, Kuwait, Kuwait

Rachata Ausavarungnirun
King Mongkut's University of Technology North Bangkok, Bangkok, Thailand

Bella Bose
Oregon State University, Corvallis, OR, United States

Kun-Chih (Jimmy) Chen
National Sun Yat-sen University, Kaohsiung, Taiwan

Lizhong Chen
Oregon State University, Corvallis, OR, United States

Masoumeh Ebrahimi
Royal Institute of Technology, Stockholm, Sweden

Hossein Farrokhbakht
University of Toronto, Toronto, ON, Canada

Shaahin Hessabi
Sharif University of Technology, Tehran, Iran

N. Enright Jerger
University of Toronto, Toronto, ON, Canada

Amirhossein Mirhosseini
Department of Electrical Engineering and Computer Science, University of Michigan, Ann Arbor, MI, United States

Mehdi Modarressi
School of Electrical and Computer Engineering, Faculty of Engineering, University of Tehran; School of Computer Science, Institute for Research in Fundamental Sciences (IPM), Tehran, Iran

Onur Mutlu
SAFARI Research Group, ETH Zürich, Switzerland

Mohammad Sadrosadati
School of Computer Science, Institute for Research in Fundamental Sciences (IPM), Tehran, Iran

Hamid Sarbazi-Azad
Department of Computer Engineering, Sharif University of Technology; School of Computer Science, Institute for Research in Fundamental Sciences (IPM), Tehran, Iran

S. Hossein SeyyedAghaei Rezaei
School of Electrical and Computer Engineering, Faculty of Engineering, University of Tehran, Tehran, Iran

Behnaz Soltani
Department of Computer Engineering, Sharif University of Technology, Tehran, Iran

Ling Wang
School of Telecommunication Engineering, Xidian University, Xi'an, The People's Republic of China

Xiaohang Wang
School of Software Engineering, South China University of Technology, Guangzhou, The People's Republic of China

Preface

Traditionally, *Advances in Computers*, the oldest series to chronicle the rapid evolution of computing, annually publishes several volumes, each one typically comprising four to eight chapters, describing new developments in the theory and applications of computing.

The 124th volume is a thematic volume titled "Power-efficient network-on-chips: Design and evaluation," inspired by recent advances in power management for network-on-chip platforms. Different components of a system-on-chip (SoC), cores, and memory blocks in a chip multiprocessor (CMP) communicate through an on-chip interconnection fabric called the on-chip network or the network-on-chip (NoC). NoCs are one of the most important parts contributing to the whole SoC or CMP performance and power consumption. The efficient design of NoCs to offer higher performance with lower power and energy consumption has attracted much attention from researchers and engineers. This volume is a collection of nine chapters and each chapter explores innovative solutions to tackle major challenges in the design and implementation of NoCs.

Sadrosadati et al., in Chapter 1 titled "An efficient DVS scheme for on-chip networks," propose an efficient dynamic voltage scaling (DVS) scheme to reduce both static and dynamic power of NoCs. When the supply voltage is scaled down (i.e., lower clock frequency), NoCs suffer from significant performance overhead. This chapter tackles this issue and proposes a new reconfigurable arbitration logic to enable multiple latencies with different slack times, i.e., a DVS-based NoC scheme without any performance overhead. In this method, slack times increase to employ lower supply voltages for routers, hence decreasing the power consumption. It is shown that the proposed method can reduce power consumption up to 45.7% over the baseline architecture without any performance degradation.

Chapter 2 introduces a traffic-load-aware virtual channel power-gating scheme for NoCs. The NoC employs several virtual channels per physical channel to mitigate the head-of-line blocking issue in transmitting network packets. Unfortunately, virtual channels (VCs) are power-hungry resources. Sadrosadati et al. observe that even for heavy traffic loads, a number of virtual channels are idle while using significant static power. As a result, one can employ the power-gating technique to switch off such idle VCs. In this chapter, a new VC power-gating mechanism that is effective for various traffic loads is introduced. It should be noted that most prior works are suitable for light traffic loads. The proposed scheme dynamically activates virtual channels according to the traffic of the NoC. Results indicate that the proposed technique reduces the average static power consumption of the NoC by 40%.

CPU–GPU-integrated systems are emerging as a high-performance and easily programmable heterogeneous platform to facilitate development of data-parallel programs. In Chapter 3, titled "A power-performance balanced network-on-chip for mixed CPU-GPU systems," Mirhosseini et al. introduce the balanced and compromised between power consumption and performance in a heterogeneous platform comprised of CPUs and GPUs. Memory-intensive GPU workloads create heavy on-chip traffic, badly producing traffic congestion around the nodes hosting the last-level cache (LLC) banks; such traffic congestions drastically harm CPU performance. A congestion-optimized NoC design can mitigate this problem using large channel resources, but when there is little or no GPU traffic, the NoC becomes suboptimal, showing larger unloaded packet latencies due to the long critical path latencies. To address such shortcomings, a reconfigurable voltage-scalable on-chip network for a CPU–GPU heterogeneous on-chip system called the bimodal NoC architecture for CPU–GPU heterogeneous systems (BiNoCHS) is introduced. For CPU-dominated light traffic loads, the proposed NoC operates at nominal voltage with high clock frequency using a topology optimized for low hop counts and a simple routing algorithm, hence maximizing CPU performance. Under high-intensity GPU traffic loads, BiNoCHS behaves as a near-threshold NoC, activating additional resources and using adaptive routing to resolve congestion. Evaluation results show that the average CPU/GPU performance is improved by 57.3%/33.6% over a latency-optimized network under congestion.

Router-based NoCs offer superior scalability over bus-based interconnects. However, they incur significant power and area overhead due to complex router structures. In Chapter 4 titled "Routerless networks-on-chip," Alazemi et al., as the title suggests, propose a routerless NoC in which routers are completely eliminated. A novel design that utilizes on-chip wiring resources smartly to achieve comparable hop count and scalability as a router-based NoC is introduced. Compared with a conventional mesh NoC, the model achieves 9.5× reduction in power consumption, 7.2× reduction in area, 2.5× reduction in zero-load packet latency, and 1.7× increase in throughput.

Chen and Ebrahimi discuss designing router algorithms for power- and temperature-aware NoCs in Chapter 5. As technology development advances, the complexity of the on-chip interconnection grows dramatically in contemporary large-scale multiprocessor SoC (MPSoC) systems. The high integration density of such systems operating at high operation frequency leads to larger power density and serious temperature problems. Higher temperature limits the chip performance and also causes higher leakage power and lower reliability. In this chapter, titled "Routing algorithm design for power- and temperature-aware NoCs," the authors investigate the correlation between power and temperature in the NoC and introduce a new thermal model. Using the proposed thermal model, a novel routing design methodology for power- and temperature-aware NoCs is introduced.

Approximate computing is a viable solution when relaxing the accuracy constraint of applications in emerging data-intensive applications, image/video processing, machine learning, and big data domains. In Chapter 6, "Approximate communication for energy-efficient network-on-chip," Wang considers approximate communication for designing energy-efficient NoCs. In NoCs with a large number of connected components, approximate communication can be used to leverage relaxed accuracy for an energy-efficient platform. Such an approach improves network performance and reduces power consumption by reducing network load, optimizing data transmission, and optimizing network architecture design.

Network topology greatly affects energy consumption, performance, cost, and design time of the NoC employed in today's systems-on-chip (SoCs). Partial topology reconfiguration can be used as a remedy to tackle these issues. Chapter 7, titled "Power-efficient network-on-chip design by partial topology reconfiguration" by Modarressi and Rezaei, articulates a new architecture capable of establishing virtual adaptive links between nonadjacent network nodes of a regular mesh NoC in order to shorten the distance. The proposed architecture benefits from the regularity, desirable design time, and scalability of a regular mesh design while offering the same superior power/performance of customized irregular topologies. Experimental results show significant improvements over the state-of-the-art NoC designs in terms of flexibility and energy efficiency, with a negligible area overhead.

Chapter 8 addresses the design of a deflection-based energy-efficient NoC. As the number of cores scales up to tens and hundreds, the energy consumption of routers across various types of NoCs increases significantly. A major source of the NoC's energy consumption comes from the input buffers of routers. To mitigate this high energy cost, a bufferless router design has been proposed that utilizes deflection routing to resolve port contention. Although this approach results in good performance in comparison to buffered NoCs at low traffic, it badly suffers from performance degradation under high network load. In order to maintain high performance and energy efficiency under both low and high network loads, in this chapter, Ausavarungnirun and Mutlu discuss critical drawbacks of traditional bufferless designs and focus on recent research on two major modifications in an attempt to improve the overall performance of the traditional bufferless network-on-chip design. Design tradeoffs of these two modifications are addressed, and experimental results based on common chip multiprocessor (CMP) configurations with various network topologies are presented.

Finally, in Chapter 9, Farrokhbakht, Hessabi, and Jerger discuss power gating as a method to reduce the static power of NoCs. Power gating is a widely used technique to power off idle blocks for reducing static power with shortcomings such as unwanted latency overhead and additional power consumption when applied to NoCs. An early wake-up technique is a common solution to alleviate the latency overhead of the power-gating technique. This chapter presents and analyzes several early wake-up techniques as advanced in the literature: TooT, SMART, SPONGE, and Muffin.

We hope that readers find this volume of interest, and useful for teaching, research, and other professional activities. We welcome feedback on the volume as well as suggestions for topics of future volumes.

ALI R. HURSON
Missouri University of Science and Technology,
Rolla, MO, United States

HAMID SARBAZI-AZAD
Department of Computer Engineering, Sharif University
of Technology (SUT), Tehran, Iran
School of Computer Science, Institute for Research
in Fundamental Sciences (IPM), Tehran, Iran

CHAPTER ONE

Traffic-load-aware virtual channel power-gating in network-on-chips

Mohammad Sadrosadati[a], Amirhossein Mirhosseini[b],
Negar Akbarzadeh[c], Mehdi Modarressi[a,d], and Hamid Sarbazi-Azad[a,c]

[a]School of Computer Science, Institute for Research in Fundamental Sciences (IPM), Tehran, Iran
[b]Department of Electrical Engineering and Computer Science, University of Michigan,
Ann Arbor, MI, United States
[c]Department of Computer Engineering, Sharif University of Technology, Tehran, Iran
[d]School of Electrical and Computer Engineering, Faculty of Engineering, University of Tehran, Tehran, Iran

Contents

Abstract

Network-on-Chips (NoCs) employ several virtual channels per input port to mitigate head-of-line blocking issue in transmitting network packets. Unfortunately, these virtual channels are power-hungry resources that significantly contribute to the total power consumption of NoCs. In particular, we make the key observation that even in high load traffic, a number of virtual channels are idle, imposing significant static power overhead. Prior works use power-gating technique to switch off idle VCs and reduce the static power consumption. However, we observe that prior works are mostly suitable for low traffic loads and are ineffective in high traffic loads. In this chapter, we aim to propose a new power-gating mechanism for virtual channels that is effective for various traffic loads. Our proposal dynamically activates virtual channels according to the traffic of the NOC. Virtual channels enable reaching higher throughput in case of heavy traffic

Advances in Computers, Volume 124
ISSN 0065-2458
https://doi.org/10.1016/bs.adcom.2021.09.001

loads. Therefore to avoid performance drop, we adjust the number of virtual channels for each port dynamically according to the requirements of each workload. Results indicate that our proposed technique reduces static power consumption by 40%, on average, with negligible performance overhead.

1. Introduction

Due to the scaling of technology size and inability to increase the frequency of chips proportionally, Systems on Chip (SOCs) and chip Multiprocessors with relatively high number of cores have been introduced [1–3]. The main challenge of CMP is designing an efficient scalable communication interface for cores. Nowadays Network-on-Chip (NoC) is used rather than traditional buses or point to point connections [1,2].

NoCs consume high portion of the total power of CMP system. This claiming is more challenging for static power which is growing significantly with technology size scaling [4]. Moreover, performance of NoC has direct impact on total CMP execution time since the latency of packet affects the total execution time of CMP system [5]. To this end, utilizing power reduction techniques for NoCs must contemplate the performance and power trade offs carefully to maintain the normal performance of the system [5,6].

With power-gating technique, idle units can be turned off until they are needed. Power-gating is widely adopted in digital systems since it can reduce leakage power extremely [4,5,7–12]. Power-gating technique has two main overheads. First, it imposes some dynamic energy overhead to turn on a power-gated unit. To mitigate this overhead, a unit should be in the power-gating state long enough (i.e., breakeven time) to compensate the energy overhead of switching on. For the cases in which a unit is turned off and on frequently, power-gating would incur high power overhead due to the power line switching overhead. Therefore, power gating should be applied for units which have low utilization, i.e., high potential of turning off and non-fragmented idleness to decrease power line switching overhead. Second, a power-gated unit cannot become active immediately and we need to wait for amount of time (i.e., wake-up latency) to have this unit ready to use. This wake-up latency can potentially impose some performance overhead. To mitigate this overhead, power-gating technique should intelligently anticipate future needs for power-gated units and start waking up process soon enough.

Routers of NoCs are the main contributor in NoC power consumption, and can be a candidate for applying power-gating technique. However, previous study has observed that their idleness is so fragmented such that there is not enough time for power-gating [5]. It seems that employing power-gating for finer-grained units such as Virtual Channels (VCs) that have low utilization with long and continuous idle time would be more applicable [10–12]. The power/performance trade-off for VCs is also a big challenge since VCs determine the throughput of the NoC. Several pieces of related work propose different power-gating techniques for VCs. However, prior works achieve power reduction at the price of some (often considerable) performance overhead, especially in high traffic loads.

In this chapter, we propose a new power gating technique for VCs in which we dynamically adjust the number of active VCs of the routers depending on the traffic load. This technique reduces power significantly while retaining performance in high traffic loads. Simulation results show that our technique reduces static power up to 40% compared to the baseline architecture with negligible performance overhead.

The rest of this chapter is organized as follows: In Section 2, some previous works related to VCs leakage power reduction are briefly discussed. In Section 3, the proposed method is described in details. Section 4 evaluates our proposal by various metrics, and finally Section 5 concludes this chapter.

2. Related work

Several methods have been proposed to reduce the power consumption of NoCs [4,5,8–30]. Most of the previous methods have either turned off some unused elements by power-gating techniques which are only useful for the static power consumption [4,5,7–12,17,22,23,31–33] or have reduced the voltage of some elements in order to reduce both dynamic and static power consumptions [11,13,21,23,34–41]. Some recent works have tested and analyzed DVFS approaches on real platforms [42,43]. Finally, there are many designs that have used emerging technologies like SOI photonics [18,26], Carbon Nanotubes [44], and Wireless RF [45–48] to improve both performance and power consumption of NoCs.

We focus specially on techniques that use power-gating for VCs to compare them with our work.

The main focus of Matsutani et al. [11], is to reduce both dynamic and static power of NoCs. However, since we are working on optimizing the static power of NoCs, similarly, we just cover the part of this paper that pays attention to static power of NoCs. This paper uses two policies for turning off and on VCs which are described as follows:

- Turn off policy: Router detects VCs that are idle longer than $T_{idle-detect}$ (4 cycles in their method) and turns them off immediately.
- Turn on policy: In this policy, router turns on a VC when it cannot find a free VC at the next-hop router to send the packet. Therefore, adjacent router should send the wake up request via a flit. This request can be granted or denied according to the $T_{break-even}$, which is the minimum required time for compensating the energy overhead of power gating technique. Regarding to T_{wakeup} which determines the time to make the power-gated VC fully activated, the wake up request is sent speculatively in the routing stage of the router's pipeline.

Furthermore, Matsutani et al. [10] also proposed a more fine-grained method dividing the router architecture into several power domains; namely, VC buffer, Output latches, CBMUX, and VCMUX and activates/deactivates them in different phases. They suggested several mechanisms for early wakeup control using lookahead policies. Basically, every flit looks two hops ahead to send the wakeup request signal. These two hops help the architecture to mitigate the wakeup latency.

Yin et al. [12] propose a VC utilization factor for turning on and off the VCs. They define two thresholds: low and high thresholds. The authors turn on VCs when their utilization exceeds the high threshold and turn them off when it falls below the low threshold. They adjust these thresholds statically regarding to workload characteristics.

3. Proposed method

3.1 Motivation

The latency of NoC have direct impact on the performance of the whole NoC-based CMP system. Therefore, utilizing power-optimization techniques for NoCs would not be beneficial without maintaining the performance of the whole system. Our observations indicate that state-of-the-art techniques that employ power gating techniques for VCs would result in high performance overhead as they increase packet latency and thus reduce the maximum throughput of NoC relatively. Unfortunately, this performance overhead can also result in higher energy consumption.

Previous studies apply power gating to VCs based on the number of idle cycles. However, this is not an appropriate metric since: (1) If we overestimate the threshold for deciding when to power gate a VC, we would sacrifice the achievable power savings. To this end, previous studies have selected a relatively low threshold (e.g., 4 cycles in Ref. [11]). This approach would result in high performance overhead as some VCs have been turned off when they are needed. (2) This metric evaluates each VC individually while different VCs in a port are usually not different from each other and it would be much more efficient to control each port rather than each VC. To further elaborate upon this problem, two example situations could be considered. One of them is where the idleness is distributed on different VCs of a port and thus, the idleness period of each VC is not long enough and cannot be detected. The other situation is that a VC of a port is idle while the other VCs are totally busy, hence this idleness does not necessarily mean that the idle VC should be turned off. Another study [12] chooses VC utilization as a threshold metric which suffers from the same shortcoming. If the period based on which the utilization has been calculated is set to be short, the estimated utilization would be arbitrary and unreliable. On the contrary, long periods would prevent detecting traffic changes and thus, would result to decide based on obsolete data and statistics.

Most of the state-of-the-art studies [10,11] send a wake up signal to the next-hop router in cases which there is no idle VC at the receiver. In this situation, stall happens until the VC of the receiver gets activated which can degrade performance significantly. To avoid this overhead, some works propose to send wake up requests a few cycles before the packet receives at the aimed router whose VC has been turned off and does not have any other free VC [11] or use look ahead methods to generate wake up signals for the router which is two hops away [10]. Our observations indicate that these techniques cannot fully mitigate the performance overhead of power gating since sending wake up package is completely dependent on the traffic flow. This means that powering off a VC not only makes the network congested due to the fewer active VCs but also makes the wake up process hard as wake up flits cannot flow fluently through the network. This positive feedback loop can avoid turning on the other power-gated VCs. Similarly, the same scenario exists for the other work which relies on the utilization factor of VCs [12].

According to previous discussions, a proper metric for power-gating and waking up the VCs should be devised which at least has the following characteristics:

- being independent of idle cycles of VCs
- regarding the whole available VCs of a port rather than each VC by itself
- relying on current status of network with considering run-time traffic changes
- being independent of traffic flow of the network

In the next part, we will propose a power gating method which satisfies these conditions.

3.2 The method

This section describes our proposed method for power-gating and waking up the VCs. To avoid aforementioned shortcomings of the previous methods, we introduce a new metric for each port, i.e., the ratio of wins to the losses of VC allocation requests from the upstream router. Not only this metric is independent of the idle cycles of the VC, but also it refers to the whole port rather than each VC. In addition, since both the number of wins and losses of allocation requests change similarly with the traffic flow, this metric is also independent of the traffic flow.

We add two counters to each output port to count the number of wins and losses of VC allocation (VCA) requests. To evaluate the obtained ratio, we add a comparator to compare the number of losses (multiplied by a threshold) with the number of wins. Since we set the threshold value such that to be a power of 2, only a shift operation is required for implementing the multiplication. Simulation results indicate that we need 8 and 32 threshold values to wake up and power gate VCs, respectively. Accordingly, we turn a VC off and on when the ratio goes beyond 32 and falls below 8, respectively. An upstream router sends wake up and power gate signals to the downstream router to turn on and off its VCs. The downstream router wakes up a power gated VC if it is power gated for more than $T_{\text{break–even}}$ and power gates it when it is idle and not already allocated. To ensure the correctness of the obtained ratio, we include the following options:

- we connect overflow signal of one counter to the reset signal of the other.
- we reset both the number of wins and losses when the state of a port changes, i.e., when we turn on/off a VC.
- we evaluate the results at least 100 cycles after the state of a port changes.

We can power gate all the VCs of an input port without worrying about the functionality and connectivity of the whole network since we can turn them on whenever needed. Nevertheless, to mitigate the notable cost and performance overhead of waking up an input port which is totally power gated, we do not allow the last active VC of a port to be power gated normally if all other VCs are already power gated. We only allow an input port to be completely power gated when it is not going to be used by the routing algorithm. This condition happens when a port does not receive any VC allocation request for a long time for example 1000 cycles. There are some cases in which more than one active VC is required at the input port like some special protocols or applications that reserve some VCs for a particular data stream. In these cases, we define virtualized VCs to support the required number of VCs if there are not enough active VCs in the port.

We do not apply the same configuration to all routers since they experience different loads depending on their locations in the network. We use link utilization to distinguish between different routers. We observe that link utilization follows almost a similar pattern in most of the synthetic and real traffic workloads. This pattern indicates that central parts of the network experience high link utilization despite marginal parts which experience much lower values of link utilization. The distribution of link utilization for an 8×8 mesh under uniform traffic is shown in Fig. 1 as an example. Accordingly as indicated in Fig. 2, we partition routers into hot, warm, and cold regions. We choose twice and half of the normal waking up and power gating thresholds i.e., 16 and 64 for hot and 4 and 16 for cold routers, respectively.

Partitioning routers into hot, warm, and cold regions is not possible when the traffic load is not high enough. Since, in low traffic loads all the routers should be put in cold region however our baseline pattern would mark some of them as warm or hot, mistakenly. Similarly, some traffic loads like hotspot are not compatible with the baseline pattern and therefore, may incur high performance and power overheads [23,49]. To address the possible mismatch between real traffic load and baseline pattern, we dynamically change the state of a router, i.e., hot, warm and cold, using two counters at each port with regard to the following conditions: to a hotter state.

- *counter1* is incremented for VCs that have been idle through $T_{break-even}$ successive cycles which makes them candidate of being power gated.
- *counter2* is incremented when a wake up signal is received at a router to turn on one of its ports whose VCs are power gated for less than $T_{break-even}$.

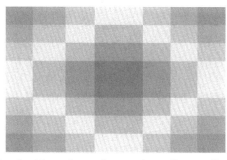

Fig. 1 Link utilization for 2D mesh topology under uniform traffic pattern.

Fig. 2 Cold, warm, and hot routers distribution in 2-D mesh NoC, green color depicts cold, yellow color depicts warm, and red color depicts hot routers.

- When *counter1* exceeds 31, both counters would be reset to zero and the intended router would be configured to a colder state.
- when *counter2* exceeds 7, both counters would be reset to zero and the intended router would be configured.

4. Evaluation

Section 4.1 includes description of simulation environment, including network configuration, traffic pattern and comparison points. Evaluation results of power consumption, performance, overheads, and scalability of the proposed method are discussed in Sections 4.2–4.5, respectively.

4.1 Simulation environment

Booksim 2.0 [50] is used for evaluating performance of the proposed design, including throughput and latency metrics. The topology of the evaluated networks is 2D mesh since it is regular and compatible with manufacturing process compared to other well-known topologies. Other details of design configurations is included in Table 1. As described in the previous sections,

Table 1 Network simulation parameters.

Parameter	For synthetic	For real workload
Network size	8×8 or 10×10 or 12×12	8×8
Number of VC per input port	4	2
Routing function	XY	
Flow control	Wormhole	
Allocation policy	Separable input first [51]	
VC length (flits)	4	2
Packet length (flits)	4	...
Flit length (bits)	128	
$T_{break-even}$ (cycles)	15	
T_{wakeup} (cycles)	4	

we introduce $T_{break-even}$ as the required cycles to compensate the wake up overhead of a VC. If we misadjust the value of these parameter, dynamic power of the proposed power gating method would be increased. We adjust $T_{break-even}$ at 15 cycles with regard to [8] and the obtained results for static and dynamic power of VCs in 45 nm technology with 1.2 GHz working frequency. The wake up latency (T_{wakeup}) is also set to be 4 cycles according to [10]. To study static power consumption, dynamic power consumption, and area of the proposed design, we implement a typical VC at transistor level in HSPICE simulation environment using 45 nm technology.

The baseline network is composed of 2-cycle virtual channel routers like the architecture described in Ref. [51]. We compare the proposed method with the SSVC proposed by Matsutani *et al.* [11], since it works exactly on power gating VCs rather than other works that combine multiple techniques and approaches like [10,12]. We compare our design with the baseline and SSVC designs in terms of performance (average network latency and maximum network throughput) and power consumption (static power and energy consumption). To investigate area and dynamic power overheads, we compare both SSVC and the proposed design with the baseline architecture.

We evaluate our design with both synthetic and real traffic patterns. Synthetic traffic patterns are extracted from *bit complement, bit reversal, shuffle, tornado, transpose, uniform*. To evaluate our proposal using real workloads, we use Netrace traces that model the Parsec suite traffic on an 8×8 network [52].

4.2 Power analysis

Fig. 3 indicates the reduction in leakage power for various synthetic traffic patterns including the overhead that will be discussed in Section 4.5. It is obvious that the proposed method efficiently manages the traffic loads with different traffic patterns by selecting the most suitable number of VCs. We observe that the number of active VCs in our design is almost similar to the real number of VCs that is used by the network to deliver packets efficiently. In fact, our design responds to the traffic load changes adaptively such that in each case, the maximum possible number of VCs are power gated while maintaining the performance.

Fig. 3 indicate that our design reduces leakage power by 15% and 40% for heavy and light traffic loads compared to the baseline, respectively. Comparing the proposed design with SSVC shows that in case of low traffic loads our design consumes lower static power, however, for high traffic

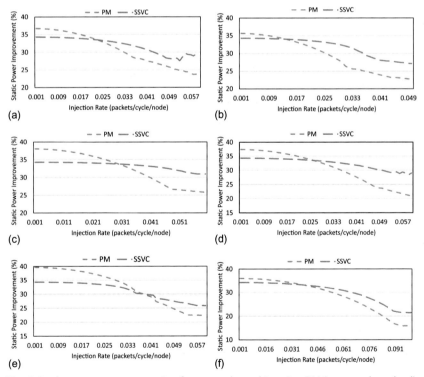

(a)   (b)   (c)   (d)   (e)   (f)

Fig. 3 Leakage power consumption for network on chip using PM (proposed method), SSVC, and baseline, (A) bit complement, (B) bit reversal, (C) shuffle, (D) tornado, (E) transpose, and (F) uniform.

loads, SSVC has lower leakage power consumption results. However, this is not a failure as performance deterioration of the SSVC method in high traffic region has been diminished. In other words, in high traffic loads, our method only activate those VCs that are needed to maintain the NoC performance level.

4.3 Performance analysis

Figs. 4 and 5 indicates the performance evaluation of our design including the maximum network throughput with different synthetic traffic loads and average network latency for Netrace workloads. According to the obtained results, our proposed method degrades total network throughput only by 0.3% on average. This is while the network throughput of SSVC falls down

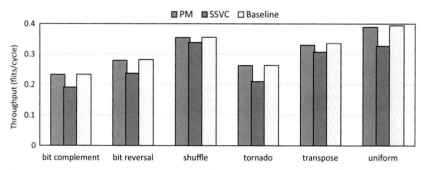

Fig. 4 Network throughput analysis using PM (proposed method), SSVC, and baseline.

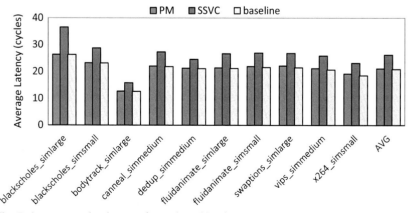

Fig. 5 Average packet latency for real workloads using PM (proposed method), SSVC, and baseline.

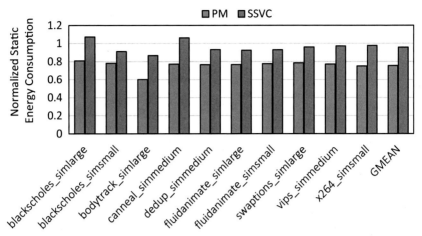

Fig. 6 Normalized static energy consumption for PM and SSVC under real workloads.

to 14.2%. Similarly, our proposed design increases average network latency by 1.3% on average for real workloads, however, SSVC incurs 25.5% higher latency compared to the baseline architecture.

To provide a better comparison between leakage power and performance, we measure the static energy factor or the leakage power delay product whose results are reported in Fig. 6 normalized to the baseline architecture. According to the figure, our method outperforms SSVC in terms of static energy factor for all workloads.

4.4 Scalability

Fig. 7 and Table 2 depict leakage power and maximum network throughput of our proposed design for different network sizes, respectively compared to the baseline architecture which can provide better understanding of the scalability. According to Fig. 7, static power do not change with the network size scaling and remains almost constant for different sizes of network before reaching to the saturation region. However, this amount of improvement degrades at saturation point and beyond for all network sizes to keep performance overhead as less as possible. According to the results depicted in Table 2, performance overhead of the proposed design is so low that is negligible for all network sizes.

4.5 Area overhead and power consumption

To enable power gating for VCs, we need to add some switches between VDD and VC which impose higher 5.8% static and higher 3.6% dynamic

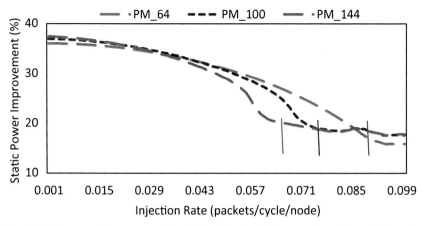

Fig. 7 Static power improvement for different network sizes under uniform traffic pattern (vertical lines show the saturation points).

Table 2 Throughput for different network sizes under uniform traffic pattern.

Mechanism	64 Nodes	100 Nodes	144 Nodes
PM	0.38 flits/cycle	0.32 flits/cycle	0.27 flits/cycle
Baseline	0.39 flits/cycle	0.33 flits/cycle	0.28 flits/cycle

power compared to the baseline VC. Moreover, the additional units that build up our design increase dynamic power up to 2.4%. The modified VCs and added units impose an area overhead of 3.1% and 0.9%, respectively.

5. Conclusion

Static power consumption is becoming a serious challenge in deep nanometer technologies. A viable solution for mitigating higher overhead of leakage power is power-gating which is widely used in large multiprocessors which employ on-chip networks for their inter-core communications. We observe that virtual channels not only consume significant amount of total network leakage power but also are idle most of the time during the activity of the chip. Therefore, based on our observations, we proposed an efficient power gating technique for VCs which can turn them on and off on demand. Moreover, the proposed method is adaptive and dynamically changes with the current traffic load. Simulation results indicate that the proposed technique improves leakage power significantly without sacrificing the performance. Comparing the proposed design with the baseline architecture in terms of static power, maximum network throughput and average network latency indicated that we can save leakage power by up to 40% with only 0.3% performance degradation.

References

[1] L. Benini, G. De Micheli, Networks on chips: a new soc paradigm, Computer 35 (1) (2002) 70–78.

[2] P. Guerrier, A. Greiner, A generic architecture for on-chip packet-switched interconnections, in: Proceedings of the Conference on Design, Automation and Test in Europe, Ser. DATE'00, 2000, pp. 250–256.

[3] J.D. Owens, W.J. Dally, R. Ho, D. Jayasimha, S.W. Keckler, L.-S. Peh, Research challenges for on-chip interconnection networks, IEEE Micro 27 (5) (2007) 96.

[4] A. Samih, R. Wang, A. Krishna, C. Maciocco, C. Tai, Y. Solihin, Energy-efficient interconnect via router parking, in: 2013 IEEE 19th International Symposium on High Performance Computer Architecture (HPCA), IEEE, 2013, pp. 508–519.

[5] L. Chen, L. Zhao, R. Wang, T.M. Pinkston, Mp3: minimizing performance penalty for power-gating of clos network-on-chip, in: Proceedings of the IEEE Symposium on High-Performance Computer Architecture, IEEE, 2014, pp. 296–307.

[6] A. Mishra, R. Das, S. Eachempati, R. Iyer, N. Vijaykrishnan, C. Das, A case for dynamic frequency tuning in on-chip networks, in: 42nd Annual IEEE/ACM International Symposium on Microarchitecture, 2009. MICRO-42, December, 2009, pp. 292–303.

[7] L. Chen, T.M. Pinkston, Nord: node-router decoupling for effective power-gating of on-chip routers, in: 2012 45th Annual IEEE/ACM International Symposium on Microarchitecture, IEEE, 2012, pp. 270–281.

[8] Z. Hu, A. Buyuktosunoglu, V. Srinivasan, V. Zyuban, H. Jacobson, P. Bose, Microarchitectural techniques for power gating of execution units, in: Proceedings of the 2004 International Symposium on Low Power Electronics and Design, Ser. ISLPED'04, 2004, pp. 32–37.

[9] G. Kim, J. Kim, S. Yoo, Flexibuffer: reducing leakage power in on-chip network routers, in: Design Automation Conference (DAC), 2011 48th ACM/EDAC/IEEE, June, 2011, pp. 936–941.

[10] H. Matsutani, M. Koibuchi, D. Ikebuchi, K. Usami, H. Nakamura, H. Amano, Ultra fine-grained run-time power gating of on-chip routers for cmps, in: 2010 Fourth ACM/IEEE International Symposium on Networks-on-Chip (NOCS), May, 2010, pp. 61–68.

[11] H. Matsutani, M. Koibuchi, D. Wang, H. Amano, Adding slow-silent virtual channels for low-power on-chip networks, in: Proceedings of the Second ACM/IEEE International Symposium on Networks-on-Chip, Ser. NOCS'08, IEEE Computer Society, 2008, pp. 23–32.

[12] J. Yin, P. Zhou, S.S. Sapatnekar, A. Zhai, Energy-efficient timedivision multiplexed hybrid-switched NoC for heterogeneous multicore systems, in: Proceedings of the IEEE International Symposium on Parallel and Distributed Processing (IPDPS), IEEE, 2014, pp. 293–303.

[13] A. Ansari, A. Mishra, J. Xu, J. Torrellas, Tangle: route-oriented dynamic voltage minimization for variation-afflicted, energy-efficient on-chip networks, in: 2014 IEEE 20th International Symposium on High Performance Computer Architecture (HPCA), IEEE, 2014, pp. 440–451.

[14] M. Arjomand, H. Sarbazi-Azad, A comprehensive power-performance model for nocs with multi-flit channel buffers, in: Proceedings of the 23rd International Conference on Supercomputing, 2009, pp. 470–478.

[15] M. Arjomand, H. Sarbazi-Azad, Power-performance analysis of networks-on-chip with arbitrary buffer allocation schemes, IEEE Trans. Comput. Aid. Des. Integ. Circuits Syst. 29 (10) (2010) 1558–1571.

[16] M. Arjomand, H. Sarbazi-Azad, Voltage-frequency planning for thermal-aware, low-power de sign of regular 3-d nocs, in: 2010 23rd International Conference on VLSI Design, IEEE, 2010, pp. 57–62.

[17] L. Chen, D. Zhu, M. Pedram, and T. M. Pinkston, "Power punch: towards non-blocking power-gating of noc routers," in 2015 IEEE 21st International Symposium on High Performance Computer Architecture (HPCA). IEEE, 2015, pp. 378–389.

[18] I. Fujiwara, M. Koibuchi, T. Ozaki, H. Matsutani, H. Casanova, Augmenting low-latency hpc network with free-space optical links, in: 2015 IEEE 21st International Symposium on High Performance Computer Architecture (HPCA), IEEE, 2015, pp. 390–401.

[19] M. Jalili, J. Bourgeois, H. Sarbazi-Azad, Power-efficient partially-adaptive routing in on-chip mesh networks, in: 2016 International SoC Design Conference (ISOCC), IEEE, 2016, pp. 65–66.

[20] P. Mehrvarzy, M. Modarressi, H. Sarbazi-Azad, Power-and performance-efficient cluster-based network-on-chip with reconfigurable topology, Microprocess. Microsyst. 46 (2016) 122–135.

[21] A. Mirhosseini, M. Sadrosadati, F. Aghamohammadi, M. Modarressi, H. Sarbazi-Azad, Baran: bimodal adaptive reconfigurable-allocator network-on-chip, ACM Trans. Parallel Comput. (TOPC) 5 (3) (2019) 1–29.

[22] A. Mirhosseini, M. Sadrosadati, A. Fakhrzadehgan, M. Modarressi, H. Sarbazi-Azad, An energy-efficient virtual channel powergating mechanism for on-chip networks, in: Proceedings of the 2015 Design, Automation & Test in Europe Conference & Exhibition, EDA Consortium, 2015, pp. 1527–1532.

[23] A. Mirhosseini, M. Sadrosadati, B. Soltani, H. Sarbazi-Azad, T.F. Wenisch, Binochs: bimodal network-on-chip for cpu-gpu heterogeneous systems, in: Proceedings of the Eleventh IEEE/ACM International Symposium on Networks-on-Chip, Ser. NOCS'17, ACM, 2017, pp. 7:1–7:8.

[24] M. Modarressi, H. Sarbazi-Azad, Power-aware mapping for reconfigurable noc architectures, in: 2007 25th International Conference on Computer Design, IEEE, 2007, pp. 417–422.

[25] M. Modarressi, H. Sarbazi-Azad, A. Tavakkol, Performance and power efficient on-chip communication using adaptive virtual point-to-point connections, in: 2006 3rd ACM/IEEE International Symposium on Networks-on-Chip, IEEE, 2009, pp. 203–212.

[26] J. Pang, C. Dwyer, A.R. Lebeck, More is less, less is more: molecular-scale photonic noc power topologies, in: Proceedings of the Twentieth International Conference on Architectural Support for Programming Languages and Operating Systems, 2015, pp. 283–296.

[27] D. Rahmati, H. Sarbazi-Azad, S. Hessabi, A.E. Kiasari, Power-efficient deterministic and adaptive routing in torus networks-on-chip, Microprocess. Microsyst. 36 (7) (2012) 571–585.

[28] R. Sabbaghi-Nadooshan, M. Modarressi, H. Sarbazi-Azad, A novel high-performance and low-power mesh-based noc, in: 2008 IEEE International Symposium on Parallel and Distributed Processing, IEEE, 2008, pp. 1–7.

[29] R. Sabbaghi-Nadooshan, M. Modarressi, H. Sarbazi-Azad, The 2d sem: a novel high-performance and low-power mesh-based topology for networks-on-chip, Int. J. Parallel Emerg. Distrib. Syst. 25 (4) (2010) 331–344.

[30] N. Teimouri, M. Modarressi, H. Sarbazi-Azad, Power and performance efficient partial circuits in packet-switched networks-on-chip, in: 2013 21st Euromicro International Conference on Parallel, Distributed, and Network-Based Processing, IEEE, 2013, pp. 509–513.

[31] H. Aghilinasab, M. Sadrosadati, M.H. Samavatian, H. Sarbazi-Azad, Reducing power consumption of gpgpus through instruction reordering, in: Proceedings of the 2016 International Symposium on Low Power Electronics and Design, Association for Computing Machinery, 2016, pp. 356–361.

[32] M. Sadrosadati, S.B. Ehsani, H. Falahati, R. Ausavarungnirun, A. Tavakkol, M. Abaee, L. Orosa, Y. Wang, H. Sarbazi-Azad, O. Mutlu, Itap: idle-time-aware power management for gpu execution units, ACM Trans. Arch. Code Optim. (TACO) 16 (1) (2019) 1–26.

[33] M. Sadrosadati, A. Mirhosseini, S.B. Ehsani, H. Sarbazi-Azad, M. Drumond, B. Falsafi, R. Ausavarungnirun, O. Mutlu, Ltrf: enabling high-capacity register files for gpus via hardware/software cooperative register prefetching, in: Proceedings of the Twenty-Third International Conference on Architectural Support for Programming Languages and Operating Systems, Association for Computing Machinery, 2018, pp. 489–502.

[34] C.-H.O. Chen, S. Park, T. Krishna, L.-S. Peh, A low-swing crossbar and link generator for low-power networks-on-chip, in: Proceedings of the International Conference on Computer-Aided Design, Ser. ICCAD'11, 2011, pp. 779–786.

[35] T. Krishna, J. Postman, C. Edmonds, L.-S. Peh, P. Chiang, Swift: a swing-reduced interconnect for a token-based network-on-chip in 90 nm cmos, in: 2010 IEEE International Conference on Computer Design, IEEE, 2010, pp. 439–446.

[36] H. Matsutani, Y. Hirata, M. Koibuchi, K. Usami, H. Nakamura, H. Amano, A multi-vdd dynamic variable-pipeline on-chip router for cmps, in: Design Automation Conference (ASP-DAC), 2012 17th Asia and South Pacific, January, 2012, pp. 407–412.

[37] T. Nakamura, H. Matsutani, M. Koibuchi, K. Usami, H. Amano, Fine-grained power control using a multi-voltage variable pipeline router, in: 2012 IEEE 6th International Symposium on Embedded Multicore Socs (MCSoC), IEEE, 2012, pp. 59–66.

[38] M. Sadrosadati, A. Mirhosseini, H. Aghilinasab, H. Sarbazi-Azad, An efficient dvs scheme for on-chip networks using reconfigurable virtual channel allocators, in: 2015 IEEE/ACM International Symposium on Low Power Electronics and Design (ISLPED), 2015.

[39] L. Shang, L.-S. Peh, N.K. Jha, Dynamic voltage scaling with links for power optimization of interconnection networks, in: High Performance Computer Architecture, 2003. HPCA-9, 2003. Proceedings. The Ninth International Symposium on IEEE, 2003, pp. 91–102.

[40] F. Worm, P. Ienne, P. Thiran, G. De Micheli, An adaptive low-power transmission scheme for on-chip networks, in: Proceedings of the 15th International Symposium on System Synthesis. ACM, 2002, pp. 92–100.

[41] J. Yin, P. Zhou, A. Holey, S.S. Sapatnekar, A. Zhai, Energy-efficient non-minimal path on-chip interconnection network for heterogeneous systems, in: Proceedings of the 2012 ACM/IEEE International Symposium on Low Power Electronics and Design, ACM, 2012, pp. 57–62.

[42] P. Bogdan, Mathematical modeling and control of multifractal workloads for data-center-on-a-chip optimization, in: Proceedings of the 9th International Symposium on Networks-on-Chip, 2015, pp. 1–8.

[43] R. David, P. Bogdan, R. Marculescu, Dynamic power management for multicores: case study using the intel scc, in: 2012 IEEE/IFIP 20th International Conference on VLSI and System-on-Chip (VLSI-SoC), IEEE, 2012, pp. 147–152.
[44] S. Fujita, K. Nomura, K. Abe, T.H. Lee, 3-d nanoarchitectures with carbon nanotube mechanical switches for future on-chip network beyond cmos architecture, IEEE Trans. Circuits Syst. I 54 (11) (2007) 2472–2479.
[45] S. Abadal, E. Alarcon, A. Cabellos-Aparicio, J. Torrellas, WiSync: an architecture for fast on-chip synchronization through wireless-enabled global communication, in: ACM SIGARCH Computer Architecture News, ACM, 2016.
[46] S. Abadal, M. Nemirovsky, E. Alarcón, A. Cabellos-Aparicio, Networking challenges and prospective impact of broadcast-oriented wireless networks-on-chip, in: Proceedings of the 9th International Symposium on Networks-on-Chip, 2015, pp. 1–8.
[47] K. Chang, S. Deb, A. Ganguly, X. Yu, S.P. Sah, P.P. Pande, B. Belzer, D. Heo, Performance evaluation and design trade-offs for wireless network-on-chip architectures, ACM J. Emerg. Technol. Comput. Syst. (JETC) 8 (3) (2012) 1–25.
[48] K. Duraisamy, Y. Xue, P. Bogdan, P.P. Pande, Multicast-aware high-performance wireless network-on-chip architectures, IEEE Trans. Very Large Scale Integr. (VLSI) Syst. 25 (3) (2017) 1126–1139.
[49] M. Sadrosadati, A. Mirhosseini, S. Roozkhosh, H. Bakhishi, H. Sarbazi-Azad, Effective cache bank placement for gpus, in: Design, Automation Test in Europe Conference Exhibition (DATE), 2017.
[50] N. Jiang, D.U. Becker, G. Michelogiannakis, J. Balfour, B. Towles, D.E. Shaw, J. Kim, W.J. Dally, A detailed and flexible cycle-accurate network-on-chip simulator, in: 2013 IEEE International Symposium on Performance Analysis of Systems and Software (ISPASS), IEEE, 2013, pp. 86–96.
[51] W. Dally, B. Towles, Principles and Practices of Interconnection Networks, Morgan Kaufmann Publishers Inc., 2003.
[52] J. Hestness, B. Grot, S.W. Keckler, Netrace: dependency-driven trace-based network-on-chip simulation, in: Proceedings of the Third International Workshop on Network on Chip Architectures, ACM, 2010, pp. 31–36.

About the authors

Mohammad Sadrosadati received his PhD degree in Computer Engineering from Sharif University of Technology, Tehran, Iran, in 2019. He is currently a senior researcher in the SAFARI research group at ETH Zurich. His research interests are in the areas of heterogeneous computing, processing-in-memory, memory systems, and interconnection networks. Due to his achievements and impact on improving the energy efficiency of GPUs, he won Khwarizmi Youth Award, one of the most prestigious awards, as the first laureate in 2020, to honor and embolden him to keep taking even bigger steps in his research career.

Amirhossein Mirhosseini received his PhD in Computer Science and Engineering from the University of Michigan in 2021. His research interests center around datacenters and cloud computing, with a particular focus on microarchitecture enhancements and accelerator design, as well as queueing and scheduling optimizations for latency sensitive microservices. He did his undergraduate studies in Sharif University of Technology, where he participated in a series of research studies on network-on-chips and GPU architectures.

Negar Akbarzadeh received her BSc and MSc in electrical engineering from Shahid-Beheshti University in 2014 and 2016, respectively. She started her PhD in Computer Engineering at Sharif University of Technology in 2017 and is currently a researcher at HPCAN lab of the department of Computer Engineering. Her research interests include computer architecture, memory systems, NoCs and GPUs.

Mehdi Modarressi received the BSc degree in computer engineering from Amirkabir University of Technology, Tehran, Iran, in 2003, and the MSc and PhD degrees in computer engineering from Sharif University of Technology, Tehran, Iran, in 2005 and 2010, respectively. He is currently an assistant professor with the Department of Electrical and Computer Engineering, University of Tehran, Tehran, Iran. His research focuses on different aspects of high-performance computer architecture and parallel processing with a particular emphasis on networks-on-chip and neuromorphic architectures.

Hamid Sarbazi-Azad received his BSc in electrical and computer engineering from Shahid-Beheshti University, Tehran, Iran, in 1992, his MSc in computer engineering from Sharif University of Technology, Tehran, Iran, in 1994, and his PhD in computing science from the University of Glasgow, Glasgow, UK, in 2002. He is currently professor of computer engineering at Sharif University of Technology and School of Computer Science, Institute for Research in Fundamental Sciences (IPM), Tehran, Iran. His research interests include high-performance computer/memory architectures, NoCs and SoCs, parallel and distributed systems, performance modeling/evaluation, and storage systems, on which he has published about 400 refereed conference and journal papers. He received Khwarizmi International Award in 2006, TWAS Young Scientist Award in engineering sciences in 2007, and Sharif University Distinguished Researcher awards in 2004, 2007, 2008, 2010, and 2013. He is now an associate editor of ACM Computing Surveys, Elsevier Computers and Electrical Engineering, and CSI Journal on Computer Science and Engineering.

CHAPTER TWO

An efficient DVS scheme for on-chip networks

Mohammad Sadrosadati[a], Amirhossein Mirhosseini[b], Negar Akbarzadeh[c], Homa Aghilinasab[a], and Hamid Sarbazi-Azad[a,c]

[a]School of Computer Science, Institute for Research in Fundamental Sciences (IPM), Tehran, Iran
[b]Department of Electrical Engineering and Computer Science, University of Michigan, Ann Arbor, MI, United States
[c]Department of Computer Engineering, Sharif University of Technology, Tehran, Iran

Contents

Abstract

Network-on-Chips (NoCs) consume a significant portion of multiprocessors' total power. Dynamic Voltage Scaling (DVS) which can reduce both static and dynamic power consumption is widely applied to NoCs. However, prior DVS schemes usually impose significant performance overhead to NoCs as NoCs need to work with lower clock frequencies when the supply voltage is scaled down. In this chapter, we propose a novel DVS scheme for NoCs with no performance overhead. We reduce power consumption when there is few Virtual Channels (VCs) that have active allocation requests at each cycle compared to the total number of available VCs. To enable multiple latencies with different slack times, we propose a new reconfigurable arbitration logic. In this method, we increase slack times to employ lower supply voltages for routers and accordingly, decrease the power consumption. We show that the proposed method can reduce power up to 45.7% over the baseline architecture without any performance degradation.

Advances in Computers, Volume 124
ISSN 0065-2458
https://doi.org/10.1016/bs.adcom.2021.09.002
21

1. Introduction

Networks-on-Chip (NoCs) provide a scalable communication infra-structure to connect different cores of a Chip Multi-Processor (CMP) or multi-core and many-core Systems-on-Chip (SoCs) [1]. Since communication plays a crucial role in both performance and power consumption of a CMP system, it is very important that the NoC works in an optimum and efficient manner [2].

Recent studies investigate that NoCs consume about 9% to 36% of the total power consumption [3–6]. Static power plays a relatively important role as dynamic power in deep sub-micron era in which the technology size shrink to less than 45 nm. To this end, an efficient power saving method is required which can reduce not only the dynamic power but also the static power of a chip [6,7]. The proposed power saving method have to maintain NoC latency as it has direct impact on the total performance of a parallel CMP system [2].

With Dynamic Voltage Scaling (DVS), both dynamic and static power consumption can be reduced [8]. However, by scaling the supply voltage, working frequency also needs to be scaled due to timing violations. Since downscaling clock frequency leads to performance loss in digital systems, we need to increase slack time while downscaling the supply voltage to avoid timing violations. With this approach, there is no need to decrease frequency when decreasing the supply voltage and accordingly the total power consumption.

We propose a novel method on top of DVS which dynamically scales slack time through the pipeline stage of the router. Since the Virtual Channel Allocation (VCA) stage is the critical stage and determines the working fre-quency of the router (see Section 2 for more details), we try to reduce the delay of this stage when allocation requests of active VCs are less than the total number of input VCs at each allocation cycle. We observe that even with high traffic loads the allocation requests for each output VC is extremely less than the number of input VCs at each cycle. By decreasing the delay and accord-ingly increasing the slack time of the VCA stage, we are able to reduce the supply voltage without performance degradation or timing violations.

To decrease the cost of the proposed method, we cluster routers and scale the voltage of the whole nodes at the cluster even though the proposed method can be applied to each single router individually. We can scale the voltage of the whole NoC at once if the cluster size is set to the NoC size.

We will investigate different tradeoffs to decide on the granularity size of the voltage scaling approach in this chapter. We implement the proposed method on two baseline router architectures; one with ripple carry arbiters and the other one with fast logarithmic delay arbiters for both VC and switch allocation units. We evaluate the achievable power saving in these two baseline under real and synthetic traffic loads. Obtained results indicate that by applying the proposed method, we can save power by up to 45.7% without any performance degradation. This chapter makes the following contributions:

- We present a new approach that enables scaling the slack time of pipeline stage for each router even under high traffic loads.
- By increasing the slack time, we are able to reduce supply voltage without incurring any performance overhead or timing violation.
- We decrease cost of the proposed approach by applying DVS to clusters of routers rather than each router individually.

The rest of this chapter is organized as follows: in Section 2, some background information is presented as a motivation for the proposed method. In Section 3, the details of the proposed method are discussed. Section 4 evaluates the method and reports the results. In Section 5, a brief report of related work is presented and finally, Section 6 concludes this chapter.

2. Motivation and background

VCs help to improve performance of NoCs since they work as an independent buffers for input ports of a router. VCs not only enable traffic classification and deadlock freedom, but also resolve Head-of-Line (HoL) blocking problem which helps to improve throughput and resource utilization of the NoC [9]. We study the impact of number of VCs on NoC throughput using Booksim 2.0 simulator. Table 1 reports the results. According to the table, by increasing the number of VCs per input port from 1 to 2 and 2 to 4, the network saturation rate improves by 43% and 14% on average, respectively. The obtained results confirms the key role of VCs in determining the throughput of NoC.

Table 1 Network saturation rate for different number of VCs (flits/cycle).

Traffic pattern	1-VC	2-VC	4-VC
Uniform	0.216	0.336	0.392
Bit complement	0.152	0.204	0.228

Despite the aforementioned advantages of VCs, they have some power and latency overheads. To enable VCs, we need to add extra VCA stage to the pipeline path of a wormhole switched router. Since according to the most of state-of-the-art implementations, VCA stage has the highest delay among the pipeline stages [10], we need to modify clock period of the whole NoC. Timing details of different pipeline stages of a router, including Route Computation (RC), VC Allocation (VCA), Switch Allocation (SA), and Switch Traversal (ST), are shown in Table 2 when the number of VCs increase from 1 to 4. Results are generated through implementing and synthesizing RTL model of router in 45 nm technology through Synopsys design flow tool. Two-stage round-robin arbiters are used to implement VC allocator. The first stage arbitrates among the candidate output VCs and the second stage arbiter arbitrates among the allocation requests to the output VC which were selected by the first stage. These arbiters are implemented according to the ripple carry architecture [9]. Simulation results indicate that in all cases the latency of traversing one-millimeter links is less than latency of VCA.

Table 3 reports timing details for different pipeline stages of a four-VC router when supply voltage is scaled. According to the reported results by reducing the supply voltage, only the VCA stage violates the timing slack constraint. In fact, since VCA stage has the highest delay among the pipeline stages, it is the critical stage and determines the clock frequency of the NoC.

Table 2 Delay (ps) of router pipeline stages for different number of VC.

Router pipeline stages	1-VC	2-VC	3-VC	4-VC
Route computation	63	63	63	63
Virtual channel allocation	–	520	730	960
Switch allocation	275	345	380	415
Switch traversal	138	158	176	176

Table 3 Delay (ps) of 1-GHz router pipeline stages with 4 VCs using different supply voltages.

Router pipeline stages	1.0	0.9	0.8	0.7
Route computation	63	67	81	109
Virtual channel allocation	960	1021	1235	1675
Switch allocation	415	441	534	724
Switch traversal	176	187	226	307

Therefore, to avoid performance overhead and timing violation when reducing supply voltage, we need to increase slack time of VCA stage.

We observe that under low traffic loads, most of the VCs have no active allocation requests and correspondingly any flit to send to the network. According to Fig. 1, this fact also validates in case of high traffic loads as our observations show that the average number of VCs with active allocation requests under high traffic loads still is much less than the total number of VCs. This is because of two reasons: first, VCA (along with RC) is only performed for the head flit of each packet. Therefore, when other flits of the packet are being transferred to the downstream router, no RC or VCA is needed to be done for them. As a result, if we assume the average number of flits per packet to be N, arbitration will only be performed, at most, once in each N cycles. Second, arbitration is performed for each output port separately in order to allocate VCs in that port. Therefore, since different input VCs of a router are usually routed to different output ports, even if all of the input VCs have head flits in a cycle, VCA does not need to be done among all of them for each output port as they are headed towards different directions. Moreover, local congestion, which happens frequently in modern NoC workloads and traffics (e.g., GPU [4,11,12]), causes the utilization of both the VCs and the arbitration slots of the VC allocators to vary a lot across different routers [13]. Fig. 1 shows the percentage of VCs with active allocation requests under different traffic loads with different numbers of VCs and packet sizes. As can be seen in this figure, arbitration slots of the VC allocator units are still underutilized even in highest traffic loads.

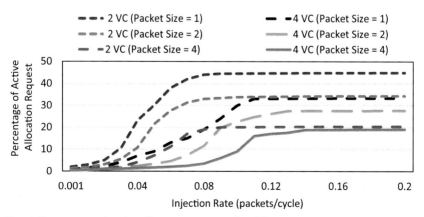

Fig. 1 Percentage of active allocation requests for different injection rates under uniform traffic in an 8×8 mesh using XY routing.

We show that the latency of VCA stage can be reduced when some VCs do not have any active allocation requests and therefore can be avoided at arbitration process. In fact, number of input VCs with active allocation request is used to adaptively set the delay of VCA stage of a wormhole-switched router. By reducing the latency of the VCA stage which is on the critical path, we are able to reduce supply voltage and thus, power consumption without incurring performance overhead.

3. Proposed method

All of input VCs usually do not have allocation requests at each cycle. In this case, a large arbiter to support all VCs is totally inefficient and prolongs the critical path of the whole router. This section proposes our methodology which dynamically scales the slack time of VCA unit according to the maximum number of VCs that have active allocation requests. This approach provides the opportunity of reducing supply voltage for the routers that have few VCs with active allocation requests.

3.1 Dynamic VCA datapath

This section investigates details of VCA unit architecture and presents the required changes to enable our proposed method. The three main components of our proposed architecture are Request Measurement (RM), Request Scheduling (RS) and Variable Delay Arbitration (VDA) units which are shown in Fig. 2A.

We discussed earlier that if we decrease the delay of VCA stage, we can increase its slack time accordingly which provides the opportunity of applying DVS to the router, i.e., reducing its supply voltage. Since modifying the supply voltage through DVS takes some time, for being more accurate, we introduce epoch which models this latency. Previous study [10] has shown that an epoch with 50 ns, i.e., 50 cycles for 1 GHz router is sufficient to ensure correct operation of DVS. RM unit calculates the maximum number of active allocation requests at each cycle which is used in the next epoch to determine the number of VCs.

RS unit is responsible for discovering active allocation requests and allocating them in the few request lines that are available. For instance, 10 request lines exists at each output virtual channel of a 2D mesh with 2 virtual channels at each port. Accordingly, requests are in the format of 10-bit vector in which a request from an input channel turns the corresponding bit to "1." Fig. 2B shows an example. Assuming that we have allowed to have 4 active requests at

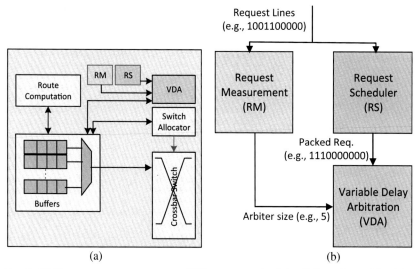

Fig. 2 (A) The request measurement (RM), request scheduler (RS), and variable delay arbitration (VDA) within a router's datapath and (B) an example request vector flowing through BARAN units to illustrate RM and RS operation.

each cycle, for a 10,011,010 vector request, RS unit maps the requests to the four inputs of the arbiter and assigns the grant signal to the corresponding requester. In conditions in which more than 4 requests are received at the VCA unit, RS unit fairly picks out only 4 of the request and the unselected requests should have been resended at the incoming cycles just like the case of arbitration loss.

VDA unit is the alternative for an allocation unit of a conventional router. The *size* input port of VDA unit captures the number of active VCs to enable scaling latency and slack time with regard to the current status of the NoC. Fig. 3 indicates the possibly simplest architecture for the VDA unit which is composed of multiple independent arbiters that have various number of input requests with different latencies which arbitrates according to the Size input. This architecture fully satisfies the requirements of our proposed design, however, it is not the best design in terms of power and area metrics. To this end, we will propose a more efficient architecture called Reduced Logic Arbiter (RLA) later in this chapter.

3.2 Implementation details

This section describes architectural details of the modules utilized in the proposed router architecture.

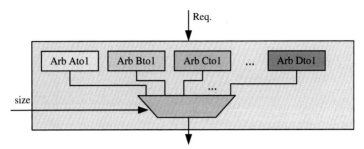

Fig. 3 High-level (conceptual) view of VDA unit.

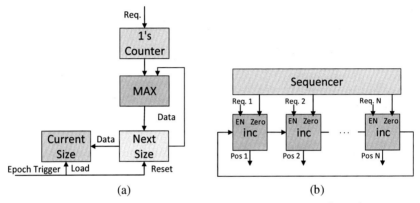

Fig. 4 (A) Microarchitecture of RM unit and (B) microarchitecture of RS unit.

(1) *Request Measurement Unit:* We mentioned that RM unit has to calculate the maximum number of active allocation requests at each epoch. As indicated in Fig. 4A, this unit consists of one combinational logic to count the allocation requests at each epoch, one combinational logic for calculating the maximum value and 2 registers for storing maximum values of the previous and current epochs. The VC allocator only handles the number of input VCs that are set by the router.

(2) *Request Scheduler Unit:* RS unit is responsible for mapping current active requests to fewer request lines accepted by the VDA unit. Furthermore, to assign arbitration result (grant signal) to the corresponding requester, it has to remember the mapping. To architect this unit, we design an incrementer logic which activates incrementation and resets output through enable and zero signals, respectively. Fig. 4B indicates the architecture of this unit. This logic gets the position of the previous active request and increments it if the current position also contains an active request. For the cases in which we have larger active requests than the allowed value, we start counting from different positions at each allocation cycle by resetting the output through zero signal to ensure fairness.

RS unit has to complete its task at RC stage after route computation as VDA unit works at VCA stage. As an example, the delay of RS for 4 VCs is 304 ps however, the slack time for RC stage is much larger. Thus, since the slack time of RC stage is much larger than the slack time of VCA stage and knowing that its slack time grows proportionally with the number of VCs at each port according to Table 1), we can finalize both routing and request scheduling stages within the same cycle.

(3) *Variable Delay Arbitration Unit:* As discussed earlier, VDA unit consists of multiple arbiters each having different delays and consequently slack times. This section proposes Reduced Logic Arbiter (RLA) which contains a more power and area efficient implementation of this unit. Fig. 5A indicates the architecture of RLA which is composed of various programmable priority arbiters cascaded through their carry in/out ports. To determine depth and latency of arbiter a multiplexer with a feedback line is added. Fig. 5A shows a baseline 20:1 RLA that can be configured as 5:1, 10:1, 15:1, and 20:1 arbiters. The 5:1 arbiters are programmable priority arbiters which grant a request if they have given the priority or the priority will be passed to the next request. Fig. 5B indicates the basic architecture of a 5:1 sliced ripple carry arbiter including 5 cascaded 1-bit-slice blocks as depicted in Fig. 5C. It should be noted that this general architecture can be utilized for other types of arbiters that have logarithmic delay or multi-cycle architecture. The main bottleneck of designing fast arbiters with logarithmic delay is that the linear latency of RC unit would not fit in the RC stage when the RLA delay is logarithmic. However, we observe that this condition rarely happens in recent designs since the number of VCs per port is normally less than 10.

3.3 DVS granularity

We mentioned before that the aim of our proposed method is reducing the latency of VCA stage according to the number of active allocation requests to enable increasing its slack time. This way we can decrease the supply voltage of the router through applying DVS to save power. As it is possible to have different supply voltages for the adjacent routers in our design, we need to add some level shifters between them to convert the voltage when needed.

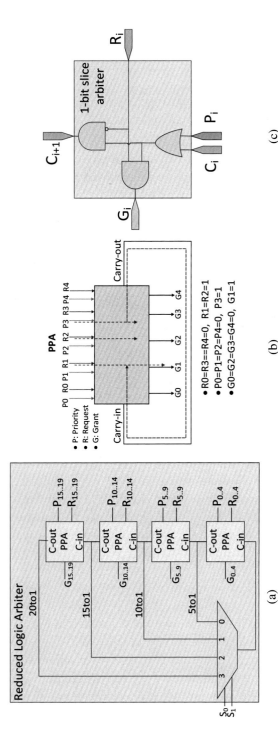

Fig. 5 (A) The design of RLA in a 2D Mesh topology with 4 VCs per each input port. The RLA is composed of four 5:1 Programmable Priority Arbiters (PPA) where each PPA module takes 5 bits of Request (R) and 5 bits of Priority (P) as input signals and generates 5 bits of Grant (G) signals as output. The Carry-Out signal generated by each module is fed to the next module as the Carry-In input to form a 20:1 PPA. The multiplexer selects the correct feedback line in order to determine the arbiter's depth and latency, (B) a high level conceptual view of a PPA functionality; It searches from the priority position and wraps around if needed to find the first request and sets its grant signal and (C) 1-bit slice arbiter.

Although this level shifters are needed only when the upstream router has a lower supply voltage than the downstream one, these units should be placed between each two routers since the voltage levels are determined dynamically during the run time and they can be easily changed during the next epoch. We observe that since the latency of these level shifters completely fit in the corresponding slack time, they do not affect the total latency of the system which is determined by the VCA stage.

Though aforementioned level shifters do not impose performance overhead to the system, they incur high area and power overheads. To alleviate this problem, we cluster routers such that only one DVS unit is utilized for each cluster and thus, level shifters are placed only between the links that connect the routers of the adjacent clusters. Nevertheless this technique sacrifices accuracy, since one voltage is applied to the whole cluster according to the global maximum number of VCs with active allocation requests across all the routers in the cluster. Per cluster calculation of the maximum number of VCs with active allocation requests can be lengthy, but a root node at each cluster can be selected to receive maximum values from all cluster nodes and then apply the results to DVS logic to set the supply voltage of the entire cluster. The maximum value of each node can be sent either on the data network or dedicated narrow links between the cluster nodes and the root.

4. Evaluation

This section discusses the simulation environment and presents the obtained results of power consumption, area, and performance for the proposed architecture.

4.1 Simulation environment

We design our proposed NoC based on a router model presented in [9] and implement it using Booksim 2.0 simulator [14]. Moreover, we implement the RTL model of the proposed NoC in Synopsys Design Flow tool and extract its corresponding SPICE netlist. For evaluating power consumption, we use HSIM that enables transistor level simulation.

2D mesh topology is selected for our evaluation since it is regular and simple to fabricate for on-chip implementation [15]. Other details about the proposed NoC is included in Table 4. These values are selected according to the most of today's common implementation of NoC designs. To ensure the correctness of RTL implementation, we calculate throughput and average packet latency for both synthetic and real workloads and compare them with

Table 4 Network simulation parameters.

Parameter	Value
Network size	8×8
Routing function	XY
Flow control	Wormhole
Number of VC per input port	4, 2
VC buffer length	4
Epoch (ns)	50
Cluster size	4, 16, 64

Table 5 Clock frequencies of the both baseline architectures with different number of virtual channels.

VCs	BL	IBL
2	1.9 GHz	2.4 GHz
4	1 GHz	1.6 GHz

the results obtained from Booksim 2.0. Results prove the correctness and accuracy of the RTL model.

We have explained in Section 3.3 that the architecture of VDA unit is general and does not belong to any specific arbiter. We implement our proposed architecture (PM) on top of Baseline (BL) using ripple carry arbiters and the Improved BaseLine (IBL) architecture using fast arbiters that have logarithmic delay for both VC and switch allocation units [16]. Clock frequencies of the both baselines with regard to the number of virtual channels are included in Table 5. Transistor level implementation of our proposed method including the VDA unit is done in 45 nm technology size with HSPICE and the slack times versus different VDA sizes are extracted. The minimum supply voltage with regard to the slack time that ensures the correctness of the design is also reported in Table 6.

We evaluate our design using both synthetic (uniform random traffic) and real traffic patterns. Real traffic workloads include Netrace traces that model the Parsec suite traffic on an 8×8 networked CMP (low traffic loads) [17] and some memory-intensive GPU workloads (high traffic loads) extracted from GPGPU-sim v3 [18]. Micro-architectural details of these traffic loads is included in Table 7. We place one DVS unit per each 4-router cluster to decrease cost.

Table 6 Selected supply voltages (volt) for different configurations of RLA implemented on BL and IBL with different number of VCs.

2 VCs			4 VCs		
VDA size	BL VDD	IBL VDD	VDA size	BL VDD	IBL VDD
4:1	0.75	0.75	5:1	0.7	0.75
6:1	0.85	0.85	10:1	0.75	0.75
8:1	0.95	0.95	15:1	0.85	0.85
10:1	1		20:1	1	1

Table 7 CMP and GPGPU microarchitectures.

System unit	CMP	GPGPU
Cores	64 on-chip, in-order, Alpha ISA, 2 GHz	56 clusters, 1 SM/cluster, 32 SPs/SM 700 MHz
L1 Cache	32 KB Instruction/32 KB data, 4-way associative, 64B lines, MESI coherence	2 KB instruction/16 KB data, 4-way associative, 128B line
L2 Cache	64 banks fully shared S-NUCA, 16 MB, 64B line, 8-way associative	8 banks shared, 384 KB, 256B line, 8-way associative
Memory	150 cycle access time, 8 on-chip memory controllers	100 cycle access time, 8 on-chip memory controllers

4.2 Power analysis

Fig. 6 indicates power savings achieved via the proposed method over both the baseline and improved baseline architectures. We can see that by increasing the injection rate of network from zero-load to after-saturation-rate, we can gain up to 37.1–33.1% power reduction on average, respectively. Similarly, Fig. 7 also, indicates power improvements of the proposed design for real workloads. It can be seen that our method can save power up to 43.7% for the real workloads. Obtained results prove the efficiency of our method in reducing power under both low and high traffic loads (Table 8).

4.3 Impact of cluster size

Fig. 8 indicates power savings obtained under uniform traffic pattern for various cluster sizes. We observe that larger cluster sizes result in better power saving results in low traffic loads. This is mainly due to the fact that the power overhead of larger clusters is much lower than the small ones, even

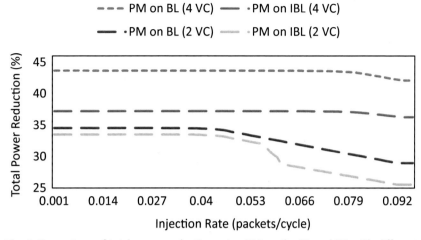

Fig. 6 Percentage of total power reduction using PM on the BL and IBL with different number of VCs.

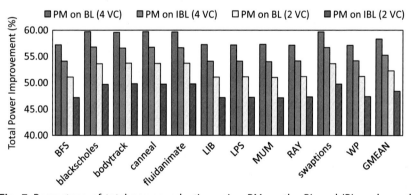

Fig. 7 Percentage of total power reduction using PM on the BL and IBL under real workloads.

Table 8 Area and power breakdown for 4 VCs.

Unit	Area	Power
Buffers	39.4%	56.5%
Routing function	0.7%	0.3%
Allocators	4.4%	1.9%
Crossbar	50.2%	36.9%
Our added units	5.3%	4.4%

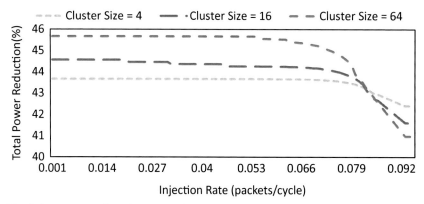

Fig. 8 Percentage of total power reduction using PM on the BL with different cluster sizes under uniform traffic pattern.

Table 9 Area overhead for different cluster sizes.

Size = 4	Size = 16	Size = 64
5.3%	3.2%	0.6%

though for larger cluster sizes we have less accuracy in determining the suitable supply voltage. However, in high traffic loads, the inaccuracy of determining the suitable supply voltage is more important. We observe that the improvement results for large clusters under high traffic loads is smaller compared to the small cluster sizes. Fig. 8 shows that under low and high traffic loads, cluster with the size of the whole network and four leads to the best result in each case, respectively. Table 9 includes the area overheads of various cluster sizes. According to the table, area overhead of the cluster with the whole network size is below 1.5% which is the best configuration for our method.

4.4 The overheads

Since our proposed method does not change clock frequency and pipeline depth of the router, it maintains performance to a good extent. The only problem that is associated with our method and other previous methods that use DVS [8,19–26] is that when the supply voltage is downscaled, the noise sensitivity increases and the likelihood of incorrect packet arrival increases as well. If a packet reaches the destination with errors, a retransmission phase is usually needed that has a bad effect on performance. In this chapter, we do not consider this problem as it is cited in many related works [8,25]. The proposed method also incurs some area and power overheads which are included in Table 8 for a cluster with 4 routers. All these numbers have been taken into account in our power reduction results.

5. Related work

Several methods have been proposed to reduce the power consumption of NoCs [2,6–8,14,19–23,25–44]. Most of the previous methods have either turned off some unused elements by power–gating techniques which are only useful for the static power consumption [2,6,7,21,30,32,33,36,37] or have reduced the voltage of some elements in order to reduce both dynamic and static power consumptions [8,19–23,25,26,35,37,43]. Some recent works have tested and analyzed DVFS approaches on real platforms [45,46]. Finally, there are many designs that have used emerging technologies like SOI photonics [47,48], Carbon Nanotubes [49], and Wireless RF [50–53] to improve both performance and power consumption of NoCs.

MVCG-VP proposed by Matsutani et al. reduces power of NoCs by dynamically scaling the pipeline of router from 2-cycle to 3-cycle and vice versa. In the 3-cycle configuration, the slack time of pipeline stages can be increased due to the reduction of the critical path delay. In this case, voltage can be decreased from 1.2 to 0.83 which improves both static and dynamic power consumptions. Nakamura et al. [22] proposed MVFG-VP on top of the previous MVCG-VP method which takes advantage of the fact that different ports of the router have different link utilizations. They pair input and output ports and modify their supply voltage and pipeline structure, separately. Though both of these works [20,22] improve power significantly in low traffic loads, they have not much effect in high traffic loads since they use the 2-cycle configuration to maintain performance in high traffic loads.

Yin et al. [26] utilize Dynamic Voltage and Frequency Scaling (DVFS) method according to the utilization of the routers per epoch and introduce two thresholds, i.e., $Tkreskold_{high}$ and $Threshold_{low}$. In this approach, when the utilization of a router goes beyond the $Tkreskold_{high}$ through an epoch, its voltage and frequency are scaled up. In an opposite way, voltage and frequency are scaled down when the utilization is lower than $Tkreskold_{low}$. Load balancing is established through a non-minimal routing function as various routers have different utilizations. Though this method increases packet latency, it does not downgrade the performance since, this method is designed specifically for GPU workloads which are sensitive to throughput of network rather than its latency [12,54].

We propose a novel method for reducing power of NoCs which can be applied both to latency-intensive and throughput-intensive workloads without incurring any performance overhead. Despite previous methods that were designed for specific traffic loads or type of workloads, our technique can be employed for both low and high traffic loads with various patterns when running different types of workloads.

6. Conclusion

One of the most efficient methods for reducing both static and dynamic power consumption of NoCs is DVS. Allocation process of VCs has the highest latency among the pipeline stages of a wormhole-switched router and thus, determines the pipeline frequency. We propose a method to reduce latency and accordingly increase slack time of this stage when there are fewer VCs with active allocation requests compared to the maximum available VCs. Our proposed method maps the many allocation request to few active request lines and reduces power supply of the routers utilizing an arbitration unit which its length and latency can be configured. Evaluation results for different metrics indicate that our method reduces power efficiently under different traffic loads. Simulation results indicate that the proposed method can save power up to 45.7% without incurring any performance overhead.

References

[1] L. Benini, G. De Micheli, Networks on chips: a new soc paradigm, Computer 35 (1) (2002) 70–78.
[2] L. Chen, L. Zhao, R. Wang, T.M. Pinkston, Mp3: minimizing performance penalty for power-gating of clos network-on-chip, in: 2014 IEEE 20th International Symposium on High Performance Computer Architecture (HPCA), IEEE, 2014, pp. 296–307.
[3] J. Leng, T. Hetherington, A. ElTantawy, S. Gilani, N.S. Kim, T.M. Aamodt, V.J. Reddi, GPUWattch: enabling energy optimizations in GPGPUs, ACM SIGARCH Computer Architecture News, ACM, 2013, pp. 487–498.
[4] M. Sadrosadati, A. Mirhosseini, S. Roozkhosh, H. Bakhishi, H. Sarbazi-Azad, Effective cache bank placement for gpus, in: Design, Automation Test in Europe Conference Exhibition (DATE), 2017.
[5] M. Sadrosadati, S.B. Ehsani, H. Falahati, R. Ausavarungnirun, A. Tavakkol, M. Abaee, L. Orosa, Y. Wang, H. Sarbazi-Azad, O. Mutlu, Itap: idle-time-aware power management for gpu execution units, ACM Trans. Arch. Code Optim. (TACO) 16 (1) (2019) 1–26.
[6] A. Samih, R. Wang, A. Krishna, C. Maciocco, C. Tai, Y. Solihin, Energy-efficient interconnect via router parking, in: 2013 IEEE 19th International Symposium on High Performance Computer Architecture (HPCA), IEEE, 2013, pp. 508–519.
[7] L. Chen, T.M. Pinkston, Nord: node-router decoupling for effective power-gating of on-chip routers, in: Proceedings of the 2012 45th Annual IEEE/ACM International Symposium on Microarchitecture, IEEE Computer Society, 2012, pp. 270–281.

[8] A. Ansari, A. Mishra, J. Xu, J. Torrellas, Tangle: route-oriented dynamic voltage min-imization for variation-afflicted, energy-efficient on-chip networks, in: 2014 IEEE 20th International Symposium on High Performance Computer Architecture (HPCA), IEEE, 2014, pp. 440–451.

[9] W.J. Dally, B.P. Towles, Principles and Practices of Interconnection Networks, Elsevier, 2004.

[10] A.K. Mishra, R. Das, S. Eachempati, R. Iyer, N. Vijaykrishnan, C.R. Das, A case for dynamic frequency tuning in on-chip networks, in: 2009 42nd Annual IEEE/ACM International Symposium on Microarchitecture (MICRO), IEEE, 2009, pp. 292–303.

[11] A. Bakhoda, J. Kim, T.M. Aamodt, Designing on-chip networks for throughput accel-erators, ACM Trans. Arch. Code Optim. (TACO) 10 (3) (2013) 21.

[12] M. Sadrosadati, A. Mirhosseini, S.B. Ehsani, H. Sarbazi-Azad, M. Dru-mond, B. Falsafi, R. Ausavarungnirun, O. Mutlu, Ltrf: enabling high-capacity register files for gpus via hardware/software cooperative register prefetching, in: Proceedings of the Twenty-Third International Conference on Architectural Support for Programming Languages and Operating Systems, Association for Computing Machinery, 2018, pp. 489–502.

[13] P. Bogdan, R. Marculescu, Workload characterization and its impact on multicore platform design, in: Proceedings of the Eighth IEEE/ACM/IFIP International Conference on Hardware/Software Code-Sign and System Synthesis, ACM, 2010, pp. 231–240.

[14] N. Jiang, D.U. Becker, G. Michelogiannakis, J. Balfour, B. Towles, D.E. Shaw, J. Kim, W.J. Dally, A detailed and flexible cycle-accurate network-on-chip simulator, in: Performance Analysis of Systems and Software (ISPASS), 2013 IEEE International Symposium on IEEE, 2013, pp. 86–96.

[15] W.J. Dally, Performance analysis of k-ary n-cube interconnection networks, IEEE Trans. Comput. 39 (6) (1990) 775–785.

[16] G. Dimitrakopoulos, N. Chrysos, K. Galanopoulos, Fast arbiters for on-chip network switches, in: Computer Design, 2008. ICCD 2008. IEEE International Conference on IEEE, 2008, pp. 664–670.

[17] J. Hestness, B. Grot, S.W. Keckler, Netrace: dependency-driven trace-based network-on-chip simulation, in: Proceedings of the Third International Workshop on Network on Chip Architectures, ACM, 2010, pp. 31–36.

[18] A. Bakhoda, G.L. Yuan, W.W. Fung, H. Wong, T.M. Aamodt, Analyzing cuda work-loads using a detailed gpu simulator, in: 2009 IEEE International Symposium on Performance Analysis of Systems and Software, IEEE, 2009, pp. 163–174.

[19] T. Krishna, J. Postman, C. Edmonds, L.-S. Peh, P. Chiang, Swift: a swing-reduced interconnect for a token-based network-on-chip in 90nm cmos, in: Computer Design (ICCD), 2010 IEEE International Conference on IEEE, 2010, pp. 439–446.

[20] H. Matsutani, Y. Hirata, M. Koibuchi, K. Usami, H. Nakamura, and H. Amano, "A multi-vdd dynamic variable-pipeline on-chip router for cmps," in 17th Asia and South Pacific Design Automation Conference. IEEE, pp. 407–412.

[21] H. Matsutani, M. Koibuchi, D. Wang, H. Amano, Adding slow-silent virtual channels for low-power on-chip networks, in: Proceedings of the Second ACM/IEEE International Symposium on Networks-on-Chip, IEEE Computer Society, 2008, pp. 23–32.

[22] T. Nakamura, H. Matsutani, M. Koibuchi, K. Usami, H. Amano, Fine-grained power control using a multi-voltage variable pipeline router, in: Embedded Multicore Socs (MCSoC) 2012 IEEE 6th International Symposium on IEEE, 2012, pp. 59–66.

[23] L. Shang, L.-S. Peh, N.K. Jha, Dynamic voltage scaling with links for power optimi-zation of interconnection networks, in: High Performance Computer Architecture, 2003. HPCA-9, 2003. Proceedings. The Ninth International Symposium on. IEEE, 2003, pp. 91–102.

[24] K. Siozios, I. Anagnostopoulos, D. Soudris, Multiple vdd on 3d noc architectures, in: Electronics, Circuits, and Systems (ICECS), 2010 17th IEEE International Conference on IEEE, 2010, pp. 831–834.

[25] F. Worm, P. Ienne, P. Thiran, G. De Micheli, An adaptive low-power transmission scheme for on-chip networks, in: Proceedings of the 15th International Symposium on System Synthesis, ACM, 2002, pp. 92–100.

[26] J. Yin, P. Zhou, A. Holey, S.S. Sapatnekar, A. Zhai, Energy-efficient non-minimal path on-chip interconnection network for heterogeneous systems, in: Proceedings of the 2012 ACM/IEEE International Symposium on Low Power Electronics and Design, ACM, 2012, pp. 57–62.

[27] M. Arjomand, H. Sarbazi-Azad, A comprehensive power performance model for nocs with multi-flit channel buffers, in: Proceedings of the 23rd International Conference on Supercomputing, 2009, pp. 470–478.

[28] M. Arjomand, H. Sarbazi-Azad, Power-performance analysis of networks-on-chip with arbitrary buffer allocation schemes, IEEE Trans. Comput. Aid. Des. Integ. Circuits Syst. 29 (10) (2010) 1558–1571.

[29] M. Arjomand, H. Sarbazi-Azad, Voltage-frequency planning for thermal-aware, low-power de-sign of regular 3-d nocs, in: 2010 23rd International Conference on VLSI Design, IEEE, 2010, pp. 57–62.

[30] L. Chen, D. Zhu, M. Pedram, T.M. Pinkston, Power punch: towards non-blocking power-gating of noc routers, in: 2015 IEEE 21st International Symposium on High Performance Computer Architecture (HPCA), IEEE, 2015, pp. 378–389.

[31] M. Jalili, J. Bourgeois, H. Sarbazi-Azad, Power-efficient partially-adaptive routing in on-chip mesh networks, in: 2016 International SoC Design Conference (ISOCC), IEEE, 2016, pp. 65–66.

[32] G. Kim, J. Kim, S. Yoo, Flexibuffer: reducing leakage power in on-chip network routers, in: 2011 48th ACM/EDAC/IEEE Design Automation Conference (DAC), IEEE, 2011, pp. 936–941.

[33] H. Matsutani, M. Koibuchi, D. Ikebuchi, K. Usami, H. Nakamura, H. Amano, Ultra fine-grained run-time power gating of on-chip routers for cmps, in: 2010 Fourth ACM/IEEE International Symposium on Networks-on-Chip, IEEE, 2010, pp. 61–68.

[34] P. Mehrvarzy, M. Modarressi, H. Sarbazi-Azad, Power-and performance-efficient cluster-based network-on-chip with reconfigurable topology, Microprocess. Microsyst. 46 (2016), 122135.

[35] A. Mirhosseini, M. Sadrosadati, F. Aghamohammadi, M. Modarressi, H. Sarbazi-Azad, Baran: bimodal adaptive reconfigurable-allocator network-on-chip, ACM Trans. Parallel Comput. (TOPC) 5 (3) (2019) 1–29.

[36] A. Mirhosseini, M. Sadrosadati, A. Fakhrzadehgan, M. Modarressi, H. Sarbazi-Azad, An energy-efficient virtual channel powergating mechanism for on-chip networks, in: Proceedings of the 2015 Design, Automation & Test in Europe Conference & Exhibition, EDA Consortium, 2015, pp. 1527–1532.

[37] A. Mirhosseini, M. Sadrosadati, B. Soltani, H. Sarbazi-Azad, T.F. Wenisch, Binochs: bimodal network-on-chip for cpu-gpu heterogeneous systems, in: Proceedings of the Eleventh IEEE/ACM International Symposium on Networks-on-Chip, Ser. NOCS'17, ACM, 2017, pp. 7:1–7:8.

[38] M. Modarressi, H. Sarbazi-Azad, Power-aware mapping for reconfigurable noc archi-tectures, in: 2007 25th International Conference on Computer Design, IEEE, 2007, pp. 417–422.

[39] M. Modarressi, H. Sarbazi-Azad, A. Tavakkol, Performance and power efficient on-chip communication using adaptive virtual point-to-point connections, in: 2009 3rd ACM/IEEE International Symposium on Networks-on-Chip, IEEE, 2009, pp. 203–212.

[40] D. Rahmati, H. Sarbazi-Azad, S. Hessabi, A.E. Kiasari, Power-efficient deterministic and adaptive routing in torus networks-on-chip, Microprocess. Microsyst. 36 (7) (2012) 571–585.

[41] R. Sabbaghi-Nadooshan, M. Modarressi, H. Sarbazi-Azad, A novel high-performance and low-power mesh-based noc, in: 2008 IEEE International Symposium on Parallel and Distributed Processing, IEEE, 2008, pp. 1–7.

[42] R. Sabbaghi-Nadooshan, M. Modarressi, H. Sarbazi-Azad, The 2d sem: a novel high-performance and low-power mesh-based topology for networks-on-chip, Int. J. Parallel Emerg. Distrib. Syst. 25 (4) (2010) 331–344.

[43] M. Sadrosadati, A. Mirhosseini, H. Aghilinasab, H. Sarbazi-Azad, An efficient dvs scheme for on-chip networks using reconfigurable virtual channel allocators, in: 2015 IEEE/ACM International Symposium on Low Power Electronics and Design (ISLPED), 2015.

[44] N. Teimouri, M. Modarressi, H. Sarbazi-Azad, Power and performance efficient partial circuits in packet-switched networks-on-chip, in: 2013 21st Euromicro International Conference on Parallel, Distributed, and Network-Based Processing, IEEE, 2013, pp. 509–513.

[45] P. Bogdan, Mathematical modeling and control of multifractal work-loads for data-center-on-a-chip optimization, in: Proceedings of the 9th International Symposium on Networks-on-Chip, 2015, pp. 1–8.

[46] R. David, P. Bogdan, R. Marculescu, Dynamic power management for multicores: case study using the intel scc, in: 2012 IEEE/IFIP 20th International Conference on VLSI and System-on-Chip (VLSI-SoC), IEEE, 2012, pp. 147–152.

[47] I. Fujiwara, M. Koibuchi, T. Ozaki, H. Matsutani, H. Casanova, Augmenting low-latency hpc network with free-space optical links, in: 2015 IEEE 21st International Symposium on High Performance Computer Architecture (HPCA), IEEE, 2015, pp. 390–401.

[48] J. Pang, C. Dwyer, A.R. Lebeck, More is less, less is more: molecular-scale photonic noc power topologies, in: Proceedings of the Twentieth International Conference on Architectural Support for Programming Languages and Operating Systems, 2015, pp. 283–296.

[49] S. Fujita, K. Nomura, K. Abe, T.H. Lee, 3-d Nanoarchitectures with carbon nanotube mechanical switches for future on-chip network beyond cmos architecture, IEEE Trans. Circuits Syst. I 54 (11) (2007) 2472–2479.

[50] S. Abadal, E. Alarcon, A. Cabellos-Aparicio, J. Torrellas, WiSync: an architecture for fast on-chip synchronization through wireless-enabled global communication, ACM SIGARCH Computer Architecture News, ACM, 2016.

[51] S. Abadal, M. Nemirovsky, E. Alarcón, A. Cabellos-Aparicio, Networking challenges and prospective impact of broadcast-oriented wireless networks-on-chip, in: Proceedings of the 9th International Symposium on Networks-on-Chip, 2015, pp. 1–8.

[52] K. Chang, S. Deb, A. Ganguly, X. Yu, S.P. Sah, P.P. Pande, B. Belzer, D. Heo, Performance evaluation and design trade-offs for wireless network-on-chip architectures, ACM J. Emerg. Technol. Comput. Syst. (JETC) 8 (3) (2012) 1–25.

[53] K. Duraisamy, Y. Xue, P. Bogdan, P.P. Pande, Multicast-aware high-performance wireless network-on-chip architectures, IEEE Trans. Very Large Scale Integ. (VLSI) Syst. 25 (3) (2017) 1126–1139.

[54] H. Aghilinasab, M. Sadrosadati, M.H. Samavatian, H. Sarbazi-Azad, Reducing power consumption of gpgpus through instruction reordering, in: Proceedings of the 2016 International Symposium on Low Power Electronics and Design, Association for Computing Machinery, 2016, pp. 356–361.

About the authors

Mohammad Sadrosadati received his Ph.D. degree in Computer Engineering from Sharif University of Technology, Tehran, Iran, in 2019. He is currently a senior researcher in the SAFARI research group at ETH Zurich. His research interests are in the areas of heterogeneous computing, processing-in-memory, memory systems, and interconnection networks. Due to his achievements and impact on improving the energy efficiency of GPUs, he won Khwarizmi Youth Award, one of the most prestigious awards, as the first laureate in 2020, to honor and embolden him to keep taking even bigger steps in his research career.

Amirhossein Mirhosseini received his PhD in Computer Science and Engineering from the University of Michigan in 2021. His research interests center around datacenters and cloud computing, with a particular focus on microarchitecture enhancements and accelerator design, as well as queueing and scheduling optimizations for latency sensitive microservices. He did his undergraduate studies in Sharif University of Technology, where he participated in a series of research studies on Network-on-Chips and GPU architectures.

Negar Akbarzadeh received her BSc and MSc in Electrical Engineering from Shahid-Beheshti University in 2014 and 2016, respectively. She started her Ph.D. in Computer Engineering at Sharif University of Technology in 2017 and is currently a researcher at HPCAN lab of the department of Computer Engineering. Her research interests include computer architecture, memory systems, NoCs and GPUs.

Homa Aghilinasab received her first M.Sc. degree from Sharif University of Technology, Tehran, Iran, in 2015 and her second M.Sc. degree from University of Waterloo, Waterloo, ON, Canada, in 2020. She is currently a Senior Software Engineer at AMD, Markham, ON, Canada. Her role is focused on Multimedia driver development.

 **Hamid Sarbazi-Azad** received his BSc in electrical and computer engineering from Shahid-Beheshti University, Tehran, Iran, in 1992, his MSc in computer engineering from Sharif University of Technology, Tehran, Iran, in 1994, and his Ph.D. in computing science from the University of Glasgow, Glasgow, UK, in 2002. He is currently professor of computer engineering at Sharif University of Technology and School of Computer Science, Institute for Research in Fundamental Sciences (IPM), Tehran, Iran. His research interests include high-performance computer/memory architectures, NoCs and SoCs, parallel and distributed systems, performance modeling/evaluation, and storage systems, on which he has published about 400 refereed conference and journal papers. He received Khwarizmi International Award in 2006, TWAS Young Scientist Award in engineering sciences in 2007, and Sharif University Distinguished Researcher awards in 2004, 2007, 2008, 2010 and 2013. He is now an associate editor of ACM Computing Surveys, Elsevier Computers and Electrical Engineering, and CSI Journal on Computer Science and Engineering.

CHAPTER THREE

A power-performance balanced network-on-chip for mixed CPU-GPU systems ☆

Amirhossein Mirhosseini[a], Mohammad Sadrosadati[b], Behnaz Soltani[c], and Hamid Sarbazi-Azad[b,c]

[a]Department of Electrical Engineering and Computer Science, University of Michigan, Ann Arbor, MI, United States
[b]School of Computer Science, Institute for Research in Fundamental Sciences (IPM), Tehran, Iran
[c]Department of Computer Engineering, Sharif University of Technology, Tehran, Iran

Contents

Abstract

CPU-GPU integrated systems are emerging as a high-performance and easily-programmable heterogeneous platform to facilitate development of data-parallel software. Network-intensive GPU workloads generate high on-chip traffic, producing local

☆This chapter presents an extended version of BiNoCHS [1].

Copyright © 2022 Elsevier Inc.
All rights reserved.
45

congestion near hot Last Level Cache (LLC) banks, drastically harming CPU performance. Congestion-optimized on-chip network designs can mitigate this problem through their large virtual and physical channel resources. However, when there is little or no GPU traffic, such networks become suboptimal, as they exhibit higher unloaded packet latencies due to their longer critical path delays. In this chapter, we introduce BiNoCHS, a reconfigurable voltage-scalable on-chip network for CPU-GPU heterogeneous systems. Under CPU-dominated low-traffic scenarios, BiNoCHS operates at nominal-voltage and high clock frequency with a topology optimized for low hop count and simple routing strategy, maximizing CPU performance. Under high-intensity GPU/mixed workloads, it transitions to a near-threshold mode, activating additional routers/channels and adaptive routing to resolve congestion. Our evaluation results demonstrate that BiNoCHS improves CPU/GPU performance by an average of 57.3%/33.6% over a latency-optimized network under congestion, while improving CPU performance by 32.8% over high-bandwidth design in unloaded scenarios.

1. Introduction

Gpus have become the mainstream accelerator for throughput applications by offering massive thread-level parallelism (TLP) [2]. Historically, CPUs and GPUs were connected via a PCIe bus and CPUs offloaded data-parallel kernels to the GPU by explicitly copying data to GPU memory. However, such explicit data movement imposes significant performance overheads that, in some cases, outweigh the potential gains of GPU acceleration. To reduce communication costs and improve programmability, recent designs, such as Intel's Skylake or NVIDIA's Denver, have integrated GPU/CPU cores on the same die, sharing Last-Level Cache (LLC) banks and memory controllers through a Network-on-Chip (NoC), as in Fig. 1A.

Though integration simplifies the programming model, it creates challenges in designing an on-chip network that is well suited for the differing memory access patterns of CPU and GPU applications [3]. CPUs typically issue memory accesses individually or in small bursts and benefit greatly from optimizations that reduce latency, but rarely saturate NoC bandwidth or queueing buffers [4]. In contrast, high-TLP GPU workloads with working sets that overflow local caches generate intense NoC traffic to access LLC banks [5,6]. Moreover, their traffic patterns are typically not uniform. Instead, temporal locality within and across warps tends to create hot spots that traverse the LLC banks, yielding transient local congestion in the on-chip network [7].

When an on-chip network is highly congested, head-of-line blocking and queuing delays grow to dominate overall latency. Above some critical load threshold, packet latencies increase exponentially and the latency distribution becomes heavy-tailed [8]. Although GPUs can often hide

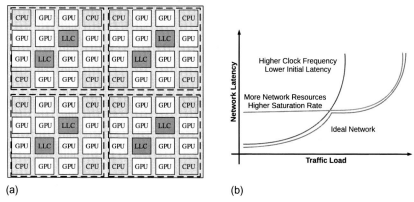

(a)   (b)

Fig. 1 (A) A single-chip CPU-GPU system. Memory controllers are co-located with LLC banks. (B) Ideal NoC behavior for heterogeneous systems.

moderately high latencies through TLP, the high (and high variance) latencies to LLC banks under congested network scenarios cannot be effectively hidden, leading to frequent stalls [9]. In application conditions where CPU and GPU cores operate concurrently, heavy congestion induced by GPU traffic can severely degrade CPU performance [10]. Therefore, many recent research studies have proposed techniques, such as resource partitioning and prioritizing CPU traffic over GPU traffic, in order to reduce CPU/GPU interference in the on-chip network.

Congestion can be reduced in a straight-forward way by adding network resources, such as additional virtual and physical channels [6,11–22]. These resources increase the network's saturation bandwidth and help to resolve local congestion and head-of-line blocking, rapidly reducing queueing delays. However, adding additional virtual and physical channels necessarily adds delay to router critical paths. These delays increase the minimum packet traversal latency when the network is unloaded, sacrificing performance for workloads with low traffic. Moreover, adding additional channels increases the on-chip network power requirement; to maintain the same power budget, the network voltage and frequency must be scaled down, further increasing the minimum traversal latency [23–25].

As such, statically optimizing an NoC for a heterogeneous system presents the trade-off illustrated in Fig. 1B. If one optimizes the on-chip network for latency-sensitive CPU traffic—minimizing the number of virtual and physical channels and maximizing clock frequency—one arrives at the latency-vs-load profile indicated by the blue line ("Higher Clock Frequency; Lower Initial Latency") in the figure. Conversely, if one statically provisions additional channels at the cost of clock frequency, one arrives at the green trade-off curve

("More Network Resources; Higher Saturation Rate"), which is well-suited for GPU workloads but penalizes latency in low-load scenarios.

In this chapter, we introduce BiNoCHS [1]—a Bimodal NoC architecture for CPU-GPU Heterogeneous Systems. BiNoCHS is an adaptive NoC design for heterogeneous systems that uses dynamic concentration and link reconfiguration to switch between low-latency and congestion-optimized operating modes, enabling a latency-vs-load profile as seen in the red line ("Ideal Network") in Fig. 1B. BiNoCHS adapts at coarse granularity between modes when the average on-chip network load crosses the critical threshold illustrated in the figure.

BiNoCHS adapts its topology, routing, physical/virtual channel resources, and operating frequency/voltage to optimize for each traffic scenario. Under low-traffic scenarios, BiNoCHS operates at nominal voltage with a low-diameter concentrated mesh where multiple cores (e.g., 4) share a single router, accessing it through a concentrator. This mode enables the highest clock frequency and lowest hop count, minimizing the on-chip network traversal latency, which is a critical factor in CPU performance. However, when traffic is high (e.g., during an NoC-intensive GPU kernel), BiNoCHS transitions to a near-threshold operating mode where it activates additional routers and links, deactivating the concentrator and instead connecting each core to a dedicated router. Furthermore, it enables additional virtual channels in each router. The larger topology spreads load and provides greater routing flexibility, while the additional virtual channels reduce head-of-line blocking. BiNoCHS uses adaptive routing in this mode to exploit the higher path diversity and additional virtual channels to further flatten the traffic and route around hot-spots. Under high load, the reduced queuing delays more than compensate for the higher hop-count and reduced link frequency, improving both latency and throughput. The intuition is similar to the trade-off between M/M/1 and M/M/k queues at high load; having more servers can provide greater average benefit than reducing service time when queuing delays are considerably larger than service times.

BiNoCHS employs near-threshold design techniques [26] to reduce the voltage as much as possible in its congestion-optimized operating mode. The reduced operating voltage frees the power budget to activate the additional hardware resources and channels. In the rest of this chapter, we describe BiNoCHS operation in detail and discuss how it leverages near-threshold design. Our evaluation demonstrates that BiNoCHS congestion-optimized mode improves CPU and GPU performance over a baseline latency-optimized network under NoC-intensive GPU workloads by an average of 57.3%, and 33.6%, respectively, within the baseline NoC power budget.

Moreover, the low-latency mode, which operates at nominal voltage, improves CPU performance by up to 32.8% over several bandwidth-optimized baseline NoCs under low-traffic CPU workloads.

2. Background and motivation

2.1 CPU-GPU heterogeneous systems

Tight CPU/GPU integration is an emerging trend in heterogeneous system architectures and thus has been increasingly studied in the recent literature. CPU/GPU SoC integration opens up many new challenges, such as resource sharing and interference. GPU applications, due to their high degree of data-parallelism, tend to overwhelm shared hardware resources, such as memory and network bandwidth. Many recent studies propose mechanisms, such as partitioning and prioritization, to minimize the interference of CPU and GPU on-chip traffic [3,5,10]. Another way of addressing this problem is by provisioning different resources, such as networks, for CPU and GPU units [27]. However, static resource provisioning is suboptimal and leads to overall bandwidth underutilization because GPU resource demands vary drastically across workloads. For example, as shown in Fig. 2, average network load can vary up to 20× among different workloads in three widely-used GPGPU benchmark suites. Moreover, various phases of the same GPU application may have disparate levels of intensity on different resources. Fig. 3 shows network load of a Computational Fluid Dynamics workload (from the Rodinia Benchmark suite [28]) as an example. As this figure illustrates, the network load in different phases of this workload can vary by up to 2.5×, which illustrates why static sub-network provisioning is not desirable.

In this chapter, we argue that, due to the varying bandwidth demands of GPU workloads and latency-sensitivity of CPU workloads, a bandwidth-optimized network is not always the best option for heterogeneous systems. While partitioning [3], throttling [10], and other interference management techniques can limit the impact of GPU traffic on the on-chip network, when on-chip network interference is low (e.g., if the GPU is idle or its working set fits within L1), a latency-optimized network is a better option for CPU-GPU systems. Thus, we introduce a bimodal on-chip networks for CPU-GPU heterogeneous systems, which can switch between a latency-optimized and a congestion-optimized mode depending on the load of the GPU workload. Interference management techniques are orthogonal to our approach and can also be employed in our design's congestion-optimized mode (when interference is high) to further improve performance.

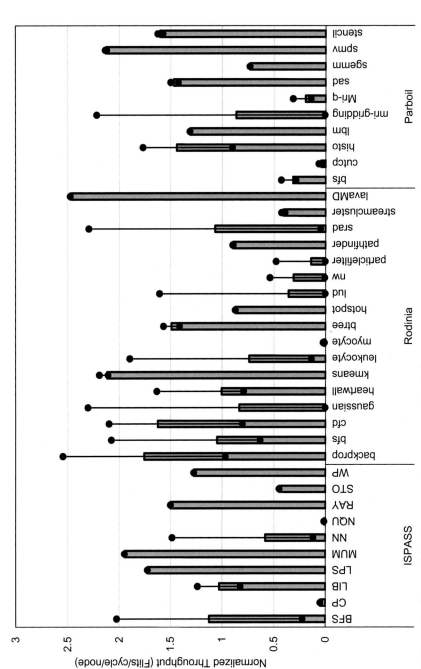

Fig. 2 Normalized network loads across various GPU workloads form ISPASS, Rodinia, and Parboil benchmark suites.

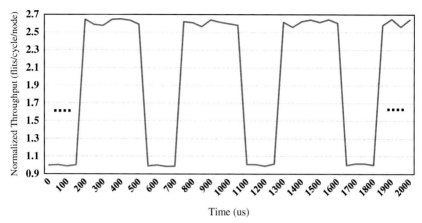

Fig. 3 Normalized network loads within different phases of a CFD GPU workload in a 2 ms window.

2.2 Near-threshold design

For NoC-intensive workload phases, to reduce contention, BiNoCHS reduces the NoC voltage and frequency to create head room in the power budget to activate additional network resources. A key challenge of near-threshold design is the differing scaling behavior of logic and SRAM structures. When the voltage is decreased from nominal to near-threshold, both logic and storage structures slow drastically. However, the slowdown of storage elements is steeper, and storage becomes relatively slower as compared to logic; this effect gets worse as the feature size becomes smaller [29]. As a consequence, aggressive pipeline architectures that are tuned for nominal voltage operation are sub-optimal in the near-threshold regime—their clock frequency is dominated by the access latency of buffers.

Moreover, storage structures become failure-prone when the voltage is lowered below a critical threshold, typically referred to as V_{min} [26]. V_{min} can be reduced by using more and larger transistors (or more fins in FinFET designs) [29]. However, larger transistors imply significant area and power overheads, especially in systems designed for efficient nominal-voltage operation. As such, recent near-threshold designs [26,29] choose different near-threshold operating voltages for logic and storage elements and use dual-rail voltage supplies to optimize overall energy efficiency while ensuring data integrity. We similarly use dual supply voltages with optimized near-threshold operating points for logic and storage structures in our dual-operating-mode routers.

3. BiNoCHS architecture

As we have noted, there is a trade-off between latency-optimized and bandwidth-optimized on-chip networks. Bandwidth-optimized networks employ additional resources, such as virtual and physical channels, to provide extra bandwidth. However, these extra resources usually come at the cost of lower clock frequencies for two reasons: first, extra channels necessarily prolong the critical path of the routers, and second, to maintain the power budget of the network, its voltage and frequency are usually down-scaled when adding extra network resources. As a result, under high traffic scenarios, additional resources play a crucial role in resolving local congestion and mitigating head-of-line blocking—providing higher effective bandwidth. Nonetheless, when the network is unloaded, the additional resources can actually hurt performance by increasing the minimum traversal latency.

In pure-GPU systems, bandwidth optimized networks are usually preferred over latency-optimized networks as GPUs are able to hide network latencies through TLP. However, since most CPU workloads are highly sensitive to network latency, a bandwidth-optimized network is only preferred for a CPU-GPU heterogeneous system when the traffic load is high and queuing delays dominate traversal latencies. To address this requirement, we introduce BiNoCHS. BiNoCHS is a voltage-scalable reconfigurable concentrated mesh (Cmesh) NoC design (targeting 22 nm CMOS technology) for CPU-GPU heterogeneous systems. BiNoCHS provides two modes, operating at nominal and near-threshold voltages, which are designed for low-traffic CPU-dominated and GPU-heavy NoC-intensive workloads, respectively.

BiNoCHS's nominal-voltage mode targets low-traffic conditions where GPUs are idle or do not issue much NoC traffic (e.g., working sets fit in local caches). This mode seeks to minimize network latency by maximizing clock frequency and reducing hop count. NoC-intensive GPU workloads, on the other hand, impose far higher load on the network, inducing transient local congestion at hot LLC banks. BiNoCHS activates additional network resources at high load to increase effective bandwidth and resolve local congestion caused by high load. Under congested scenarios, where latency is dominated by queuing delays and head-of-line blocking, adding routers and channels provides greater benefit than higher frequency links. This is analogous to the trade-off between M/M/1 and M/M/k queues at high load; increasing k provides higher average benefit than reducing the service time. As a result, BiNoCHS's near-threshold mode minimizes NoC voltage and frequency, reapportioning the freed power budget to activate additional routers and channels.

To this end, BiNoCHS provisions separate supply voltage lines for router logic and storage elements, allowing optimized logic and storage supply voltage in near-threshold mode. Router logic elements comprise route computation units, allocators, crossbars, and controllers, whereas storage elements comprise the buffers assigned for each virtual channel. In our design, we use a logic supply voltage (V_{logic}) of 0.5 V. Prior work [29] shows that voltages below 0.5 V impose increasingly challenging performance and leakage overheads. In contrast, the near-threshold voltage used for storage elements ($V_{storage}$) is the V_{min} of 6 T SRAM in 22 nm technology—0.6 V [29,30].

At a high level, BiNoCHS provides higher effective bandwidth and resolves local congestion in its near-threshold mode through three techniques: provisioning more virtual channels, reducing network concentration, and using non-minimal adaptive routing. The following subsections describe each of techniques in greater detail as well as the methodology for switching between the two operating modes.

3.1 More virtual channels

Provisioning more virtual channels (VCs) at each router input port alleviates local congestion by reducing head-of-line (HoL) blocking. HoL blocking arises when a packet that is blocked due to link congestion precludes another packet traveling in the same channel (but which requires a different outbound link) from making forward progress. HoL blocking can form a chain of blocked packets spanning several routers, causing congestion on a single link to spread throughout the network. Additional virtual channels allow packets with ready outbound links to route past blocked packets, preventing congestion from spreading.

BiNoCHS uses two different supply voltages ($V_{storage} = 0.6$ V and $V_{logic} = 0.5$ V) for router buffers and logic under near-threshold operation. Hence, buffers are relatively faster than logic in near-threshold mode as compared to nominal operation (though both storage and logic are slower in absolute terms). The disparity between storage and logic delay allows us to activate additional buffers on each input port to increase the number of VCs. Our SPICE simulations show that, in near-threshold mode, we can double the number of VCs (from two to four)—the shorter buffer access times compensate for the additional decoding/multiplexing time to route to the additional channels.

However, additional VCs increase arbitration delay in the Virtual-Channel Allocation (VCA) unit. When multiple head flits contend for a single output port, the VCA unit must arbitrate among them and choose which output VC must be allocated to each packet. The VCA's arbiters must be timed for a worst-case condition where all input channels contend for the same output VC. Under nominal operation (two VCs per input), the arbiter must be able to accept at most 10 requests (up to five input ports, including the local ingress, with two VCs each) in each allocation cycle. In near-threshold mode, when the number of VCs doubles to four, the arbiter must now scale to accept 20 requests in a single allocation cycle.

Since BiNoCHS adapts between nominal and near-threshold modes, its VCA must support both 10-input and 20-input arbitration. However, as the VCA is often on the router critical path, we do not want to penalize nominal operation with a 20-input arbiter. Instead, we use the reconfigurable Reduced Logic Arbiter (RLA) introduced in [23]. RLA builds a single high-radix arbiter from two cascaded arbiters. Fig. 4 illustrates BiNoCHS's arbiter, which comprises two 10-input programmable priority arbiters and a multiplexer that selects the correct feedback line. Under nominal-voltage operation, when only 10 inputs are needed, the second arbiter is disabled and the multiplexer is pre-steered to select the output of the first, minimizing impact on the critical path. We use the fast carry-look-ahead arbiters proposed in prior work [31].

By using RLA, our approach avoids significant critical-path impact at nominal voltage. However, RLA prolongs the clock period under near-threshold mode by about 46% (relative to a plain 20-input look-ahead arbiter operating at V_{logic}), which translates to significant performance overheads. We address this delay using two insights: First, in near-threshold operation, tiny voltage increases can enable substantial frequency gains. Second, although arbiters have long critical paths, they are small and do not consume much power (relative, e.g., to crossbars). Hence, we can reduce delay impact by up-sizing transistors or over-driving the arbitration circuit with a small voltage increase. We select the latter approach in BiNoCHS and connect the arbitration circuit supply to $V_{storage}$ rather than V_{logic}. The higher supply voltage allows RLA to operate with a clock frequency only 6% slower than a design with a plain 20-input arbiter at V_{logic}.

Furthermore, our SPICE simulations show that increasing $V_{storage}$ by only 0.02v can close the remaining 6% performance gap with a negligible (less than 1%) power overhead. We illustrate the final BiNoCHS reconfigurable voltage-scalable router design in Fig. 5A. Two supply voltages,

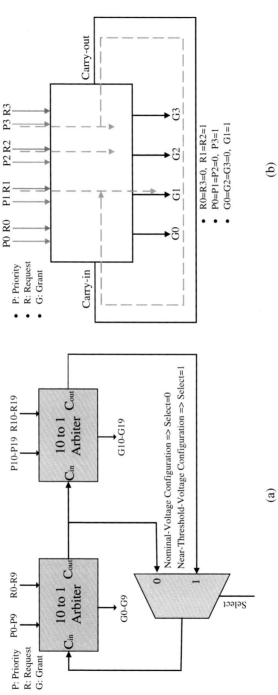

Fig. 4 (A) RLA used in BiNoCHS. (B) A high-level view of programmable priority arbiter functionality. The arbiter searches from the last priority position and wraps around if needed to find the first request, setting the corresponding grant signal.

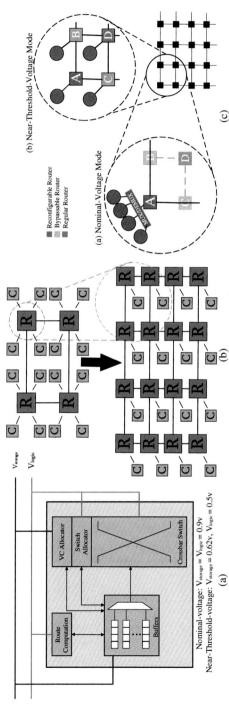

Fig. 5 (A) BiNoCHS reconfigurable voltage-scalable router, (B) reducing C factor, and (C) BiNoCHS reconfigurable topology.

(a)

Nominal-voltage: $V_{storage} = V_{logic} = 0.9v$
Near-Threshold-voltage: $V_{storage} = 0.62v$, $V_{logic} = 0.5v$

$V_{storage}$
V_{logic}

Route Computation
VC Allocator
Switch Allocator
Crossbar Switch
Buffers

(b)

C
C
C
C
C
C
C
C
C
C
R
R
R
R
C
C
C
C

R
R
R
R
R
R
R
R
R
R
R
R
R
R
R
R
C
C
C
C
C
C
C
C
C
C
C
C

(b) Near-Threshold-Voltage Mode

A
B
C
D

(a) Nominal-Voltage Mode

A
B
C
D

Reconfigurable Router
Bypassable Router
Regular Router

(c)

V_{logic} and $V_{storage}$, are available in each router. $V_{storage}$ supplies the buffers and the arbiters used in VC allocation whereas V_{logic} supplies all other elements. In near-threshold mode, $V_{storage}$ and V_{logic} are 0.62v and 0.5v, respectively, while both are set to 0.9v in the nominal-voltage mode. Note that while the near-threshold mode of a BiNoCHS router has four VCs, BiNoCHS only provides two logical lanes that can be used by application-level protocols (e.g., cache coherence) to avoid deadlocks. Our system does not require additional lanes; a system that requires more than two lanes can employ other mechanisms, such as flit reservation along with VC remapping/time-multiplexing, to virtualize more lanes on fewer VCs per port [32].

3.2 Reducing network concentration

Network concentration (or "C factor") refers to the number of nodes connected to each router. Reducing network concentration brings down the injection load at each router. Hence, the router can provide higher effective bandwidth to its remaining node(s). BiNoCHS reduces the network concentration in its near-threshold mode by activating additional routers and links, reassigning nodes to the new routers as shown in Fig. 5B. The expanded network provides higher bandwidth and greater path diversity. These features mitigate congestion under high traffic from NoC-intensive GPU workloads.

Our simulations show that, in near-threshold mode, the power consumption of BiNoCHS's reconfigurable router is around 25% of a baseline (non-configurable) router's power consumption if its C factor is also reduced by four. Under high traffic loads, dynamic power dominates the router power breakdown and, hence, by reducing voltage, frequency, and concentration, BiNoCHS can compensate for the power consumption of its extra VCs.

To reduce the network concentration by a factor of four in the near-threshold mode, we double the number of routers in each mesh dimension and deactivate the concentrator. The additional routers must be bypassed (i.e., behave as wires) in nominal-voltage operation. To this end, we introduce another type of router, which we call a *bypassable router*, shown in Fig. 6. A bypassable router offers two power states: nominal operation and bypass. In nominal operation, it behaves as a conventional router with one (non-concentrated) ingress port. In bypass operation, all buffering, routing, and arbitration logic is power-gated, and the crossbar is bypassed, instead directly connecting either east–west or north–south links. Because of BiNoCHs topology, a bypassable router need provide only one of horizontal or vertical bypass, depending on its position in the network.

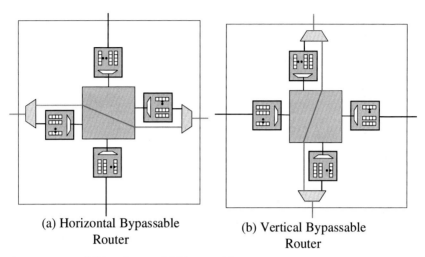

(a) Horizontal Bypassable
Router

(b) Vertical Bypassable
Router

Fig. 6 Horizontal (A) and vertical (B) bypassable routers.

Table 1 Characteristics of different routers in a cluster.

Router	Reconfigurable?	Bypassable?	#VC	VCA arbitersize	$V_{storage}$ (v)	V_{logic} (v)
A	Yes	No	2(NV), 4(NT)	RLA {10(NV), 20(NT)}	0.9(NV), 0.62(NT)	0.9(NV), 0.5(NT)
B	No	Horizontal	4	20 inputs	0.62	0.5
C	No	Vertical	4	20 inputs	0.62	0.5
D	No	No	4	20 inputs	0.62	0.5

We build our reconfigurable network as illustrated in Fig. 5C. The network consists of clusters of four routers with each cluster comprising a reconfigurable voltage-scalable router, two bypassable near-threshold routers, and one regular near-threshold router. We describe the characteristics of each type of router in Table 1 and illustrate the design of a cluster in Fig. 5C (where A–D refer to the four types of routers as described in the table). When the on-chip network is configured for near-threshold mode, all of the routers are activated and the reconfigurable router (A) is configured in its near-threshold mode. Under nominal-voltage operation, all near-threshold routers (B, C, and D) are power-gated (the bypassable ones (B and C) are configured for bypass) and the reconfigurable router (A) is configured for nominal-voltage mode.

3.3 Using adaptive routing

Adaptive routing algorithms allow packets to take multiple paths to avoid congestion and flatten the on-chip network traffic distribution. Fully adaptive algorithms, which permit non-minimal routes, can further improve NoC efficiency as they allow packets to turn around to avoid becoming blocked in congested regions. In fact, prior works [33] have shown that non-minimal adaptive routing performs best when the traffic has a hot-spot pattern and local congestion is frequent (e.g., as in NoC-intensive GPU workloads). However, when the traffic intensity is low and the network is unloaded, non-minimal adaptive routing can lead to considerably higher packet latencies as it can increase the number of hops a packet may take. Therefore, adaptively routed designs are most desirable for highly congested high-load networks.

In BiNoCHS, we exploit this dichotomy and use non-minimal adaptive routing in the near-threshold congestion-optimized operating mode, where adaptive routing provides maximal benefit. However, in the nominal-voltage mode, adaptive routing provides no benefit—packets are rarely blocked and can simply follow minimal routes. So, BiNoCHS disables the adaptive Route Computation (RC) units in its latency-optimized nominal-voltage mode. Instead, the NoC uses the deterministic dimension-ordered routing algorithm.

Adaptive routing is synergistic with BiNoCHS additional VCs and greater path diversity in the congestion-optimized mode. Although BiNoCHS, does not directly increase bisection bandwidth, adaptive routing nevertheless can utilize the extra routes and VCs more effectively (i.e., utilize the available bandwidth better), avoiding head-of-line blocking and providing a higher saturation throughput.

Adaptive routing is prone to deadlocks. In BiNoCHS, we use the Odd-Even turn model to provide deadlock-freedom [34]. Odd-Even turn model guarantees deadlock-freedom by disallowing some paths depending on the location of each router. To select between the different available routes allowed by the turn model, we follow the procedure described in [35]. At each step, the packet must take a productive channel (i.e., toward its destination) if possible. If no legal productive channel is free, a legal non-productive channel might be taken. We use local buffer occupancy information to select between different available productive/non-productive channels.

Previous works have shown that, under uniform traffic patterns, local selection functions may in fact harm network performance as they distribute the load unevenly [33,35]. However, in CPU-GPU heterogeneous systems, the traffic pattern is already non-uniform and includes many locally congested hot-spots. As such, even random path selections can improve network performance by flattening the traffic [36]. That said, there are selection functions that take into account global network status to further improve performance [33]; we did not use these to avoid additional complexities of such mechanisms (propagating information, cost functions, ...).

While our routing scheme is deadlock-free, live-locks are still possible, as there is no guarantee that a packet will eventually reach its destination. To avoid live-locks, we impose a limit on the number of non-productive links (misroutes) a packet may take. Once a packet reaches the misroute threshold, it may still be adaptively routed but may select only productive channels at each hop. Unlike some complex adaptive routing algorithms that prolong the router critical path, our routing scheme has a simple implementation; it only involves a few parallel comparisons for buffer occupancy and channel-productivity checks at each hop. Therefore, BiNoCHS adaptive router can work within the same clock frequency as the dimension-order routing.

3.4 Switching modes

Prior work has shown that GPU applications tend to have coarse-grain phase behavior at the granularity of kernels that last for 10 s of microseconds [37,38]. Even for a 10us kernel, we can optimize for simplicity rather than switching latency and still incur a negligible performance penalty. BiNoCHS switches modes by draining all packets from the network and reconfiguring routers/links while no packets are in flight. Draining the network avoids numerous complexities that arise when attempting to reconfigure routes with packets in transit. Draining the network involves a few corner-to-corner network round-trip latencies. In our experiments, draining never required more than 200 ns. We assume another 100 ns for voltage/frequency scaling and reconfiguring the network topology, using voltage regulation techniques described in [39,40], resulting in an overall mode switch latency of 300 ns. Even if mode switches occur every 10 μs, the network would be unavailable only 3% of the time. The performance overhead is typically lower as (1) in most cases, workloads do not change phases at this frequency, and (2) CPU/GPU cores can continue execution while the network is switching, partially hiding the mode switch latency.

Whereas we assume phase changes occur at a coarse granularity, we need a mechanism to detect and react to them quickly. Most related works that classify GPU kernels based on their memory intensity use the number of stall cycles observed in each streaming multiprocessor to do so. However, we care particularly about network intensity and some workloads, such as the "Back propagation" kernel from Rodinia, have few memory stalls despite imposing a high load on the network, due to high memory-level parallelism. Therefore, we use total GPU L1 Misses-per-Kilo-Cycle (MPKC) [41] as a metric for detecting phase changes. At each LLC bank, we count the number of arriving requests in a 4096-cycle epoch. (We have experimentally determined that shorter detection epochs lead to instability.) At the end of each epoch, we aggregate the MPKC counts through a chain of messages sent among the LLC banks and the last bank triggers a mode switch if the average crosses a pre-selected threshold. We tuned the transition threshold empirically and use the same threshold for all applications. After the mode switch decision has been made, routers do not accept any new packets until the network has fully drained and reconfigured. A global wired-or logic signal that spans all routers is used to signal when the network is fully drained.

3.5 Discussion

To summate the implementation details of BiNoCHS, we briefly discuss its overheads. Requiring dual voltage rails complicates layout. Recent work [29] reports that dual voltage rails impose about a 5% area overhead on the whole design. A second source of overhead is the level converters required in the router when signals transition between voltage domains (e.g., at the input port of the arbiters). Our SPICE simulations indicate that both the area and delay overheads of the level converters are negligible in our design (less than 1%). The biggest overheads of BiNoCHS arise from the added logic for reconfigurability and phase change detection, such as the concentrator MUXs needed to connect the routers with cores, the switches needed in the bypassable routers, the adaptive RC units, and the MPKC counters in LLC banks. Our results indicate that these logic elements impose 11% area, 6% performance, and 2% power overheads, which we include in our reported performance and energy results. Note that many of the MUXs are inserted into control structures and are not on the design's critical path.

4. Evaluation

4.1 Methodology

We use GPGPU-Sim V3.2.2 [42] which employs Booksim 2.0 [43] as a detailed cycle-accurate on-chip network simulator to analyze the performance of BiNoCHS. We have augmented this infrastructure with a trace-driven x86 CMP simulator to model the joint traffic from both CPU and GPU applications. The microarchitectural details of our GPU and CPU architectures are listed in Table 4. We also implemented an RTL model of our reconfigurable router and synthesized it in 22 nm technology using a Synopsys design flow to obtain power estimates. We extract netlists from synthesis and use SPICE simulations to obtain peak frequency estimates at each operating voltage. We model a 64 node CPU-GPU heterogeneous system with a tile-based layout as in Fig. 1A where each 4×4 tile is composed of 10 GPU cores, 4 CPU cores, and 2 LLC banks. We assume a one-to-one coupling between LLC banks and memory controllers as in prior works [5,10]. Whereas memory controllers have conventionally been placed at the edges of a mesh, with modern flip-chip packages and 3D stacked chips, this is no longer a requirement; we achieve better performance by placing them in the center of the mesh in our baseline configuration. We have chosen the GPU/CPU core ratio to be 2.5/1 as the area of different 22 nm Intel Haswell cores is roughly 2–3 × larger than the area of an NVIDIA Maxwell GPU core scaled to 22 nm technology. The NoC transitions between a 16×16 mesh and a 4×4 concentrated mesh topology with a C factor of four (each router is attached to four nodes) in its near-threshold/nominal-voltage modes. The key simulation parameters of the NoC are reported in Table 2.

To contrast BiNoCHS with the following baselines:

1. C4VC2 and C4VC4: 4×4 Cmesh topology (C = 4) with two and four VCs per input port.

Table 2 Network simulation parameters.

Parameter	value
Number of nodes	64
Topology	2D Mesh/Cmesh
Routing	Dimension-order(XY)/adaptive
Flow control	Wormhole
Number of VC per input port	4/2
VC buffer length (flits)	4
Flit size (bits)	256
Router microarchitecture	2-cycle speculative pipeline

2.   C1VC4: 8×8 mesh topology (Cmesh with $C = 1$) with four VCs per input port.

3.   C4VC4AD and C1VC4AD: C4VC4 and C1VC4 baselines augmented with non-minimal adaptive routing (same as in BiNoCHS).

We again use our RTL model to obtain frequency estimates for each baseline; voltage and maximum frequency for each design are shown in Table 3. We used maximum network throughput, average packet latency, power consumption, and energy per flit as comparison metrics (Fig. 7).

We study three classes of GPGPU workloads from the Rodinia [28] benchmark suite; purely compute-intensive applications with moderate network loads, purely NoC-intensive applications with heavy traffic, and applications with phasechanging behaviors where BiNoCHS transitions modes depending on the network load in different phases. We also consider cases where no GPU kernel is launched and the network accommodates only CPU traffic. On the CPUs in all conditions, we run multiprogrammed workloads from different mixes of SPEC CPU 2006 applications. We have considered three CPU workload mixes with different ranges of Misses-Per-Kilo-Instructions (MPKI) to model different CPU traffic loads and different sensitivities to network latency. Applications used in each workload mix are reported in Table 5.

4.2 Application results

Fig. 8 shows normalized average packet latency, CPU IPC, GPU IPC, network energy per flit, and network energy-delay-product (EDP) under each pair of CPU-GPU workloads.

Under low load, clock frequency and hop count have direct effects on packet latency while the extra VCs, routers, and/or links provide no benefit, since there is no congestion or head-of-line blocking. Non-minimal routing can also hurt performance due to higher hop counts. Hence, when the GPU workload is purely compute-intensive or the GPU nodes are idle, we see that all C4VC4, C4VC4AD, C1VC4, and C1VC4AD hurt network latency by up to 28.1%, 52.9%, 115.2%, and 141.5% relative to C4VC2, respectively, due to their lower clock frequencies (and higher hop count in C1VC4 and C1VC4AD). These latencies translate to up to 8.2%, 17.3%, 24.9%, and 27.7% reduction in CPU performance. Moreover, baselines with non-minimal adaptive routing perform worse than their counterpart with deterministic routing. However, thanks to BiNoCHS reconfigurablity mechanisms, it manages to minimize performance overhead, with network

Table 3 Supply voltage and working frequency for different on-chip network designs.

Parameter	C4VC2	C4VC4	CV4VC4AD	C1VC4	C1VC4AD	BiNoCHS (NT)	BiNoCHS (NV)
Supply voltage (v)	0.9	0.9	0.9	0.9	0.9	$V_{storage}=0.62,\ V_{logic}=0.5$	$V_{storage}=V_{logic}=0.9$
Frequency (GHz)	2.5	2.0	2.0	2.0	2.0	0.9	2.3

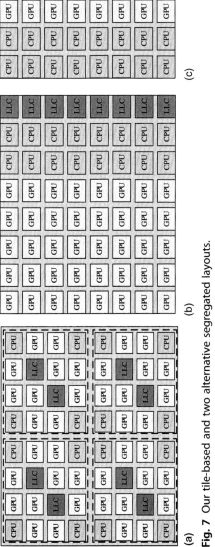

(a)

(b)

(c)

Fig. 7 Our tile-based and two alternative segregated layouts.

Table 4 CPU and GPU microarchitectures.

System unit	CPU	GPU
Cores	16 × 86 cores, 3-wide OoO, 64-entry ROB, 2.0 GHz	40 clusters, 1 SM/cluster, 32 SPs/SM, 1.4 GHz
Private caches	32 KB L1, 256 KB L2, 4-way, 128B line size	16 KB 4-way, 128B line size
Shared LLC	8 MB 16-way, 128B Line Size, 8 banks	
Main memory	GDDR5, 8 MCs, FR-FCFS, 800 MHz, 8 DRAM-banks/MC	

Table 5 CPU workload mixes.

H	soplex, mcf, milc, omnetpp, soplex, mcf, milc, omnetpp, soplex, mcf, milc, omnetpp, soplex, mcf, milc, omnetpp
M	milc, omnetpp, lbm, leslie, soplex, mcf, milc, omnetpp, lbm, leslie, soplex, mcf, milc, omnetpp, lbm, leslie
L	leslie, lbm, libq, sphinx, astar, hmmer, gems, bzip2, go, deal, sjeng, gromacs, soplex, mcf, omnetpp, milc

Table 6 Two different designs with separate CPU/GPU sub-networks.

	GPU-optimized		Balanced	
Parameter	CPU subnet	GPU subnet	CPU subnet	GPU subnet
Voltage	0.8v	0.6v	0.9v	0.6v
Frequency	1800 MHz	800 MHz	2100 MHz	700 MHz
C factor	4	1	4	1
VC count	1	3	2	2
Routing	XY	Adaptive	XY	Adaptive

latency and CPU performance at most 7.3% and 1.5% worse than of C4VC2, respectively. These results imply that, even with an unlimited NoC power budget, a single non-reconfigurable design with many resources is not well-suited for heterogeneous systems, especially with respect to CPU performance. GPU performance, on the other hand, does not vary drastically across designs, since GPUs are able to hide reasonable latencies through TLP. Energy efficiency tracks network and CPU performance; all C4VC4, C4VC4AD, C1VC4, and C1VC4AD designs consume more energy than C4VC2 and are less cost effective in terms of EDP. Their energy overhead arises because

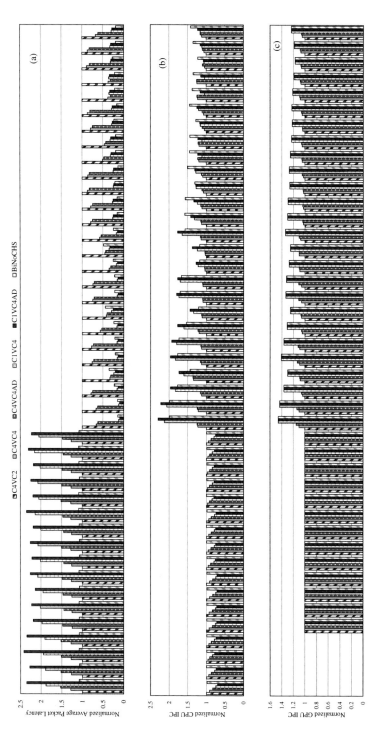

Fig. 8 Normalized (A) average packet latency, (B) CPU IPC, (C) GPU IPC, (D) energy per flit, and (E) EDP per flit under different CPU-GPU workload pairs.
(Continued)

Fig. 8—Cont'd

of the static energy consumed by the additional resources in these designs. In contrast, BiNoCHS power-gates unneeded resources, avoiding energy overhead.

With NoC-intensive GPU workloads, C4VC2, C4VC4, and C4VC4AD NoC designs become saturated and incur large NoC latencies and CPU stalls. Even GPUs cannot hide and sustain such long network latencies. BiNoCHS, on the other hand, activates additional resources to resolve local congestion and uses non-minimal adaptive routing to avoid congested regions, improving performance remarkably. As Fig. 8 shows, BiNoCHS improves network, CPU, and GPU performance by an average of 75.5%, 57.3%, and 33.6%, respectively, compared to C4VC2 as a latency-optimized baseline. It also improves network energy efficiency and EDP by 45.6% and 86.7%, on average, respectively. Note that neither extra virtual channels nor adaptive routing used in BiNoCHS are sufficient alone; neither C4VC4 nor C4VC4AD improves network performance considerably in such congested conditions. C1VC4 and C1VC4AD, on the other hand, outperform BiNoCHS in terms of network and CPU performance when the traffic load is high (GPU performance is almost the same). However, these baselines consume 3.2× and 2.9× more energy-per-flit than BiNoCHS and their EDPs are also 2.2× and 1.8× more, respectively. EDP indicates cost-effectiveness of a design and shows that C1VC4 and C4VC4AD do not offer enough performance for the extra energy they consume. Moreover, assuming the network power budget is not limited, BiNoCHS can simply work under higher voltage and frequency and offer almost the same performance as C1VC4AD when the network traffic load is high.

Finally, under GPU workloads with phase-changing behavior, BiNoCHS outperforms all other designs, in terms of network and CPU performance, as it can reconfigure between low-latency and congestion-optimized modes as the traffic load of the network changes. In particular, BiNoCHS improves CPU/network performance compared to C4VC2 and C1VC4AD, the most latency-optimized and bandwidth-optimized baselines, by 42.2%/82.5% and 15.8%/37.8% on average, respectively. Energy and EDP results track network performance. Nonetheless, in terms of GPU performance, it is almost the same as C4VC2 and C1VC4AD and only outperforms other baselines. BiNoCHS offers the highest CPU performance improvements for workload pairs with CFD and SRAD as the GPU workload, due to drastic changes in the network load of CFD and SRAD workloads (up to 5X) within their different execution phases. The Guassian Elimination workload, on the other hand, achieves the smallest CPU performance improvements from BiNoCHS because it is a heavily synchronized workload comprising many short kernels that are frequently shorter than 10us. Therefore, this workload transitions modes frequently, losing performance due to modeswitch overhead. Nevertheless, even for this workload, BiNoCHS still performs better than all other baselines offering 27.6% and 8.2% CPU performance improvements over C4VC2 and C1VC4AD, respectively.

4.3 Comparison with dual-network designs

BiNoCHS seeks to address the challenge that CPU and GPU traffic have different latency and bandwidth demands. Whereas CPU applications are highly latency-sensitive, GPUs are latency tolerant and benefit from bandwidth-optimized networks. Therefore, unlike GPU-only systems, CPU-GPU heterogeneous systems only benefit from a bandwidth-optimized network when the GPU kernel is network-intensive. In such congested network scenarios, a latency-optimized network saturates and packet latencies increase rapidly. As suggested by prior work [27], an alternative solution to address differing CPU-GPU traffic demands is to partition CPU and GPU traffic onto disjoint physical sub-networks wherein the CPU sub-network is latency-optimized and the GPU sub-network is bandwidth-optimized. However, this solution is suboptimal as the sub-networks must be statically provisioned while the relative CPU/GPU traffic demands vary considerably across workloads. Hence, partitioned networks result in overall bandwidth under-utilization and impede CPU/GPU performance. We have considered two dual-network designs *GPU-optimized* and *Balanced* wherein there are two statically provisioned sub-networks for CPU and GPU. In both, the CPU sub-network is latency-optimized while the GPU sub-network is bandwidth-optimized. Sub-networks have been provisioned such that both consume, in the aggregate, almost the same amount of power as BiNoCHS. Details of both designs are reported in Table 6. As this table shows, GPU-opt design dedicates more hardware resources and power budget to the GPU network while balanced design allocates roughly the same hardware and power budget to each.

Fig. 9 compares CPU/GPU performance of GPU-optimized and balanced designs with BiNoCHS. As shown in this figure, both of these designs can impede both CPU and GPU performance due to overall bandwidth under-utilization in many cases. In particular, both of these designs always lead to lower GPU performance. This result is consistent with prior reports [3,10] that show GPUs do not benefit from network resource partitioning. Moreover, CPU performance is only better than BiNoCHS in the balanced

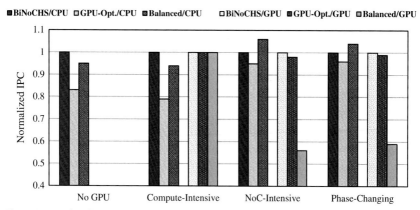

Fig. 9 Normalized CPU/GPU performance with BiNoCHS and two dual-network designs.

design under high traffic (by 5.4%), where GPU performance is degraded by 42.5%. Our result indicates that having two different sub-networks is only able to mitigate GPU interference on CPU traffic in congested conditions. Traffic interference is an orthogonal problem that can be solved using various techniques, such as throttling, dynamic VC partitioning, and traffic prioritization [3,5,10]. These techniques are generally orthogonal to BiNoCHS and can be applied in its congestion-optimized mode as well.

4.4 Effect of alternative layouts

We compare CPU/GPU performance results gained by BiNoCHS under different chip layouts depicted in Fig. 7. We consider two alternative layouts, denoted GCM and CGM in Fig. 7. In both layouts, CPU cores, GPU cores, and LLC banks are segregated, which is representative of some prior commercial floorplans. Such segregated layouts may be easier to implement if CPU, GPU, and LLC tiles differ markedly in geometry. Fig. 10 shows the relative CPU/GPU performance results of CGM and GCM compared with our integrated layout under varying CPU-GPU workload pairs. Under the GCM layout, CPU nodes are placed close to LLC banks as CPU applications tend to be more latency-sensitive. As our results show, this layout yields slightly better CPU performance under low-traffic GPU workloads and idle-GPU conditions due to smaller hop counts for CPU requests. However, this layout sacrifices CPU performance drastically under congested scenarios; placing CPU nodes close to LLC banks causes all CPU requests to traverse congested regions of the network, considerably increasing latencies. As prior work has shown [44,45], under congested network

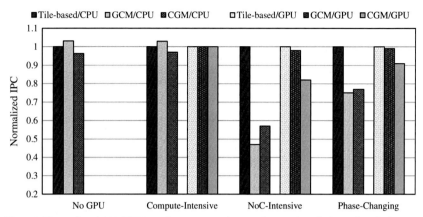

Fig. 10 Normalized CPU/GPU performance with BiNoCHS under tile-based and alternative layouts.

scenarios, there is at best a weak correlation between hop counts and packet latencies as some packets might be blocked for a long time.

Under the CGM layout, we have swapped the locations of the CPU and GPU cores. As illustrated in Fig. 10, this layout yields worse performance than BiNoCHS for both CPU and GPU in almost all conditions. This degradation arises mostly because this layout increases the average hop count for CPU packets and reduces the effective bandwidth for GPU packets, providing the worst of both worlds. CPU performance degradations are more apparent in unloaded network scenarios where hop counts correspond to packet latencies whereas GPU performance degradations are more apparent in congested scenarios where bandwidth demands are higher.

5. Related work

CPU/GPU Interference: Many recent works have sought to minimize CPU/GPU interference on shared resources in heterogeneous systems. Lee and Kim [46] propose cache partitioning to reduce interference at LLC banks. Ausavarungnirun et al. [41], Jeong et al. [47], and Jog et al. [48] propose memory scheduling techniques that reduce interference in memory controllers by exploiting various forms of isolation and prioritization in each type of memory request. Several works [3,5] have explored VC partitioning and traffic prioritization techniques to reduce interference in the NoC. Other works [49–52] have proposed varying forms of throttling in GPU cores to manage contention in both network and memory hierarchy. Kayiran *et al.* [10] deploy throttling techniques to mitigate interference in CPU-GPU heterogeneous systems. BiNoCHS, on the other hand, observes that not only is interference minimal when the GPU workload is mostly compute-intensive, but even a bandwidth-optimized network that targets only NoC-intensive GPU workloads can hurt CPU performance when the network load is low. To this end, it proposes a bimodal design that transitions between latency-optimized and congestion-optimized modes depending on the GPU traffic load. Almost all prior interference management techniques can be orthogonally employed in combination with BiNoCHS congestion-optimized mode.

Application aware and heterogeneous NoCs: Mishra *et al.* [27] propose asymmetric resource allocation to address varying application demands. Das *et al.* [53] propose several prioritization techniques to address different slack times of multiple applications. Kumar *et al.* [54] propose express channels for latency-critical applications. Nicopoulos *et al.* [55] propose a reconfigurable NoC design that regulates the number and the organization of VCs at run-time to adapt to the traffic load. Yin *et al.* [24] characterize such slack times for GPU packets and propose to route them on non-minimal paths. Choi *et al.* [56] propose a hybrid wired-wireless NoC design to service different latency/bandwidth demands in heterogeneous systems. Abadal *et al.* [57] propose a similar hybrid design to differentiate accesses to latency-sensitive synchronization variables from other memory accesses in a parallel program. Various works [45] have also proposed placement strategies for different memory controllers and LLC banks to distribute the load in homogeneous and heterogeneous multicores. To the best of our knowledge, BiNoCHS is the first reconfigurable design that takes a holistic approach to optimize various characteristics of the network based on the traffic characteristics in CPU-GPU heterogeneous systems.

Dual-voltage near-threshold architectures: Using multiple supply voltages for logic and storage structures in near-threshold regimes has been explored in a variety of contexts other than NoCs [29,58–62]. Miller *et al.* [59] use multiple voltage lanes to mitigate process variation effects in slower units. Dreslinski *et al.* [60] proposed using multiple voltages for processor cores and caches to make caches relatively faster and hence, make the whole system more energy efficient. A similar approach is used in the Intel Claremont [62] design. Gopireddy *et al.* [29] go one step further by using multiple supply voltages inside a core for logic-intensive and storage-intensive structures to reduce the power consumption of the core as much as possible. Therefore, they can turn on more cores during the parallel phases of program execution and increase energy efficiency by improving performance within the same power budget. Moreover, since they use higher voltage levels for storage elements, they can make these elements larger and gain some performance and energy improvements. BiNoCHS deploys a similar strategy to hide the latency of more VCs and larger VC allocators to minimize the reconfigurability costs of the on-chip network.

6. Conclusion

Memory-intensive GPU workloads impose high loads and cause time-varying local congestion near hot LLC banks. While additional network resources, such as virtual and physical channels, can improve saturation bandwidth and resolve such local congestion, they negatively impact the clock frequency and, thus, unloaded packet latencies. Higher unloaded latency drastically hurts performance in CPU-dominated low-traffic conditions as, unlike GPUs, CPUs are unable to hide latencies longer than a few clock cycles. In this chapter, we introduced BiNoCHS as a reconfigurable voltage-scalable NoC design for CPU-GPU heterogeneous systems. In nominal-voltage mode, it offers high clock frequency and low hop counts to maximize CPU performance. In near-threshold mode, it trades off clock frequency for network resources and employs adaptive routing to resolve congestion and postpone network saturation. Our experimental results demonstrated that BiNoCHS improves CPU/GPU performance under a congested network by 57.3%/33.6%, on average, over a latency-optimized NoC and CPU performance in an unloaded environment by 32.8% over a bandwidth-optimized NoC.

References

[1] A. Mirhosseini, M. Sadrosadati, B. Soltani, H. Sarbazi-Azad, T.F. Wenisch, Binochs: bimodal network-on-chip for cpu-gpu heterogeneous systems, in: Proceedings of the Eleventh IEEE/ACM International Symposium on Networks-on-Chip, 2017, pp. 1–8.
[2] M. Sadrosadati, A. Mirhosseini, S.B. Ehsani, H. Sarbazi-Azad, M. Drumond, B. Falsafi, R. Ausavarungnirun, O. Mutlu, Ltrf: enabling highcapacity register files for gpus via hardware/software cooperative register prefetching, in: Proceedings of the Twenty-Third International Conference on Architectural Support for Programming Languages and Operating Systems, Association for Computing Machinery, 2018, pp. 489–502.
[3] J. Lee, S. Li, H. Kim, S. Yalamanchili, Adaptive virtual channel partitioning for network-on-chip in heterogeneous architectures, ACM Trans. Des. Autom. Electron. Syst. (TODAES) 18 (4) (2013) 1–28.

[4] A. Mirhosseini, M. Sadrosadati, M. Zare, H. Sarbazi-Azad, Quantifying the difference in resource demand among classic and modern noc workloads, in: 2016 IEEE 34th International Conference on Computer Design (ICCD), IEEE, 2016, pp. 404–407.

[5] J. Zhan, O. Kayiran, G.H. Loh, C.R. Das, Y. Xie, Oscar: Orchestrating stt-ram cache traffic for heterogeneous cpu-gpu architectures, in: 2016 49th Annual IEEE/ACM International Symposium on Microarchitecture (MICRO), IEEE, 2016, pp. 1–13.

[6] A. Mirhosseini, M. Sadrosadati, F. Aghamohammadi, M. Modarressi, H. Sarbazi-Azad, Baran: bimodal adaptive reconfigurable-allocator network-on-chip, ACM Trans. Parallel Comput. 5 (3) (2019).

[7] H. Jang, J. Kim, P. Gratz, K.H. Yum, E.J. Kim, Bandwidth-efficient on-chip interconnect designs for gpgpus, in: Proceedings of the 52nd Annual Design Automation Conference, 2015, pp. 1–6.

[8] P. Bogdan, R. Marculescu, Workload characterization and its impact on multicore platform design, in: 2010 IEEE/ACM/IFIP International Conference on Hardware/ Software Codesign and System Synthesis (CODES + ISSS), IEEE, 2010, pp. 231–240.

[9] N. Vijaykumar, G. Pekhimenko, A. Jog, A. Bhowmick, R. Ausavarungnirun, C. Das, M. Kandemir, T.C. Mowry, O. Mutlu, A case for core-assisted bottleneck acceleration in gpus: enabling flexible data compression with assist warps, in: Proceedings of the 42nd Annual International Symposium on Computer Architecture, 2015, pp. 41–53.

[10] O. Kayiran, N.C. Nachiappan, A. Jog, R. Ausavarungnirun, M.T. Kandemir, G.H. Loh, O. Mutlu, C.R. Das, Managing gpu concurrency in heterogeneous architectures, in: 2014 47th Annual IEEE/ACM International Symposium on Microarchitecture, IEEE, 2014, pp. 114–126.

[11] A. Mirhosseini, M. Sadrosadati, A. Fakhrzadehgan, M. Modarressi, and H. Sarbazi-Azad, "An energy-efficient virtual channel power-gating mechanism for on-chip networks," in Proceedings of the 2015 Design, Automation & Test in Europe Conference & Exhibition. EDA Consortium, 2015, pp. 1527–1532.

[12] P. Mehrvarzy, M. Modarressi, H. Sarbazi-Azad, Power-and performance-efficient cluster-based network-on-chip with reconfigurable topology, Microprocess. Microsyst. 46 (2016) 122–135.

[13] M. Jalili, J. Bourgeois, H. Sarbazi-Azad, Power-efficient partially-adaptive routing in on-chip mesh networks, in: 2016 International SoC Design Conference (ISOCC), IEEE, 2016, pp. 65–66.

[14] N. Teimouri, M. Modarressi, H. Sarbazi-Azad, Power and performance efficient partial circuits in packet-switched networks-on-chip, in: 2013 21st Euromicro International Conference on Parallel, Distributed, and Network-Based Processing, IEEE, 2013, pp. 509–513.

[15] D. Rahmati, H. Sarbazi-Azad, S. Hessabi, A.E. Kiasari, Power-efficient deterministic and adaptive routing in torus networks-on-chip, Microprocess. Microsyst. 36 (7) (2012) 571–585.

[16] R. Sabbaghi-Nadooshan, M. Modarressi, H. Sarbazi-Azad, The 2d sem: a novel high-performance and low-power mesh-based topology for networks-on-chip, Int. J. Parallel Emerg. Distrib. Syst. 25 (4) (2010) 331–344.

[17] M. Arjomand, H. Sarbazi-Azad, Power-performance analysis of networks-on-chip with arbitrary buffer allocation schemes, IEEE Trans. Comput. Aid. Des. Integ. Circuits Syst. 29 (10) (2010) 1558–1571.

[18] M. Arjomand, H. Sarbazi-Azad, Voltage-frequency planning for thermal-aware, low-power design of regular 3-d nocs, in: 2010 23rd International Conference on VLSI Design, IEEE, 2010, pp. 57–62.

[19] M. Arjomand, H. Sarbazi-Azad, A comprehensive power-performance model for nocs with multi-flit channel buffers, in: Proceedings of the 23rd International Conference on Supercomputing, 2009, pp. 470–478.

[20] M. Modarressi, H. Sarbazi-Azad, A. Tavakkol, Performance and power efficient on-chip communication using adaptive virtual point-to-point connections, in: 2009 3rd ACM/IEEE International Symposium on Networks-on-Chip, IEEE, 2009, pp. 203–212.

[21] R. Sabbaghi-Nadooshan, M. Modarressi, H. Sarbazi-Azad, A novel high-performance and low-power mesh-based noc, in: 2008 IEEE International Symposium on Parallel and Distributed Processing, IEEE, 2008, pp. 1–7.

[22] M. Modarressi, H. Sarbazi-Azad, Power-aware mapping for reconfigurable noc architectures, in: 2007 25th International Conference on Computer Design, IEEE, 2007, pp. 417–422.

[23] M. Sadrosadati, A. Mirhosseini, H. Aghilinasab, H. Sarbazi-Azad, An efficient dvs scheme for on-chip networks using reconfigurable virtual channel allocators, in: 2015 IEEE/ACM International Symposium on Low Power-Electronics and Design (ISLPED), IEEE, 2015, pp. 249–254.

[24] J. Yin, P. Zhou, A. Holey, S.S. Sapatnekar, A. Zhai, Energy-efficient non-minimal path on-chip interconnection network for heterogeneous systems, in: Proceedings of the 2012 ACM/IEEE International Symposium on Low Power Electronics and Design, 2012, pp. 57–62.

[25] H. Matsutani, Y. Hirata, M. Koibuchi, K. Usami, H. Nakamura, H. Amano, A multi-vdd dynamic variable-pipeline on-chip router for cmps, in: 17th Asia and South Pacific Design Automation Conference, IEEE, 2012, pp. 407–412.

[26] R.G. Dreslinski, M. Wieckowski, D. Blaauw, D. Sylvester, T. Mudge, Near-threshold computing: reclaiming moore's law through energy efficient integrated circuits, Proc. IEEE 98 (2) (2010) 253–266.

[27] A.K. Mishra, O. Mutlu, C.R. Das, A heterogeneous multiple network-on-chip design: an application-aware approach, in: 2013 50th ACM/EDAC/IEEE Design Automation Conference (DAC), IEEE, 2013, pp. 1–10.

[28] S. Che, M. Boyer, J. Meng, D. Tarjan, J.W. Sheaffer, S.-H. Lee, K. Skadron, Rodinia: a benchmark suite for heterogeneous computing, in: 2009 IEEE International Symposium on Workload Characterization (IISWC), IEEE, 2009, pp. 44–54.

[29] B. Gopireddy, C. Song, J. Torrellas, N.S. Kim, A. Agrawal, A. Mishra, Scalcore: designing a core for voltage scalability, in: 2016 IEEE International Symposium on High Performance Computer Architecture (HPCA), IEEE, 2016, pp. 681–693.

[30] C.-H. Jan, U. Bhattacharya, R. Brain, S.-J. Choi, G. Curello, G. Gupta, W. Hafez, M. Jang, M. Kang, K. Komeyli, et al., A 22 nm soc platform technology featuring 3-d tri-gate and high-k/metal gate, optimized for ultra low power, high performance and high density soc applications, in: 2012 International Electron Devices Meeting, IEEE, 2012, pp. 1–3.

[31] G. Dimitrakopoulos, N. Chrysos, K. Galanopoulos, Fast arbiters for on-chip network switches, in: 2008 IEEE International Conference on Computer Design, IEEE, 2008, pp. 664–670.

[32] M. Evripidou, C. Nicopoulos, V. Soteriou, J. Kim, Virtualizing virtual channels for increased network-on-chip robustness and upgradeability, in: 2012 IEEE Computer Society Annual Symposium on VLSI, IEEE, 2012, pp. 21–26.

[33] P. Gratz, B. Grot, S.W. Keckler, Regional congestion awareness for load balance in networks-on-chip, in: 2008 IEEE 14th International Symposium on High Performance Computer Architecture, IEEE, 2008, pp. 203–214.

[34] G.-M. Chiu, The odd-even turn model for adaptive routing, IEEE Trans. Parallel Distrib. Syst. 11 (7) (2000) 729–738.

[35] W.J. Dally, H. Aoki, Deadlock-free adaptive routing in multicomputer networks using virtual channels, IEEE Trans. Parallel Distrib. Syst. 4 (4) (1993) 466–475.

[36] L.G. Valiant, A scheme for fast parallel communication, SIAM J. Comput. 11 (2) (1982) 350–361.

[37] J. Leng, T. Hetherington, A. ElTantawy, S. Gilani, N.S. Kim, T.M. Aamodt, V.J. Reddi, Gpuwattch: enabling energy optimizations in gpgpus, in: Proceedings of the 40th Annual International Symposium on Computer Architecture, 2013, pp. 487–498.

[38] A. Sethia, S. Mahlke, Equalizer: dynamic tuning of gpu resources for efficient execution, in: 2014 47th Annual IEEE/ACM International Symposium on Microarchitecture, IEEE, 2014, pp. 647–658.

[39] W. Kim, M.S. Gupta, G.-Y. Wei, D. Brooks, System level analysis of fast, per-core dvfs using on-chip switching regulators, in: 2008 IEEE 14th International Symposium on High Performance Computer Architecture, IEEE, 2008, pp. 123–134.

[40] W. Godycki, C. Torng, I. Bukreyev, A. Apsel, C. Batten, Enabling realistic fine-grain voltage scaling with reconfigurable power distribution networks, in: 2014 47th Annual IEEE/ACM International Symposium on Microarchitecture, IEEE, 2014, pp. 381–393.

[41] R. Ausavarungnirun, K.K.-W. Chang, L. Subramanian, G.H. Loh, O. Mutlu, Staged memory scheduling: achieving high performance and scalability in heterogeneous systems, in: 2012 39th Annual International Symposium on Computer Architecture (ISCA), IEEE, 2012, pp. 416–427.

[42] A. Bakhoda, G.L. Yuan, W.W. Fung, H. Wong, T.M. Aamodt, Analyzing cuda workloads using a detailed gpu simulator, in: 2009 IEEE International Symposium on Performance Analysis of Systems and Software, IEEE, 2009, pp. 163–174.

[43] N. Jiang, D.U. Becker, G. Michelogiannakis, J. Balfour, B. Towles, D.E. Shaw, J. Kim, W.J. Dally, A detailed and flexible cycle-accurate network-on-chip simulator, in: 2013 IEEE International Symposium on Performance Analysis of Systems and Software (ISPASS), IEEE, 2013, pp. 86–96.

[44] M. Sadrosadati, A. Mirhosseini, S. Roozkhosh, H. Bakhishi, H. Sarbazi-Azad, Effective cache bank placement for gpus, in: Design, Automation & Test in Europe Conference & Exhibition (DATE), 2017, IEEE, 2017, pp. 31–36.

[45] D. Abts, N.D. Enright Jerger, J. Kim, D. Gibson, M.H. Lipasti, Achieving predictable performance through better memory controller placement in many-core CMPs, ACM SIGARCH Comput. Archit. News 37 (3) (2009) 451–461.

[46] J. Lee, H. Kim, Tap: A tlp-aware cache management policy for a cpu-gpu heterogeneous architecture, in: IEEE International Symposium on High-Performance Comp Architecture, IEEE, 2012, pp. 1–12.

[47] M.K. Jeong, M. Erez, C. Sudanthi, N. Paver, A qos-aware memory controller for dynamically balancing gpu and cpu bandwidth use in an mpsoc, in: Proceedings of the 49th Annual Design Automation Conference, ACM, 2012, pp. 850–855.

[48] A. Jog, E. Bolotin, Z. Guz, M. Parker, S.W. Keckler, M.T. Kandemir, C.R. Das, Application-aware memory system for fair and efficient execution of concurrent gpgpu applications, in: Proceedings of Workshop on General Purpose Processing Using GPUs, ACM, 2014, p. 1.

[49] A. Jog, O. Kayiran, A.K. Mishra, M.T. Kandemir, O. Mutlu, R. Iyer, C.R. Das, Orchestrated scheduling and prefetching for gpgpus, in: Proceedings of the 40th Annual International Symposium on Computer Architecture, 2013, pp. 332–343.

[50] A. Jog, O. Kayiran, N. Chidambaram Nachiappan, A. K. Mishra, M. T. Kandemir, O. Mutlu, R. Iyer, and C. R. Das, "Owl: cooperative thread array aware scheduling techniques for improving gpgpu performance," in ACM SIGPLAN Not., vol. 48, no. 4. ACM, 2013, pp. 395–406.

[51] O. Kayiran, A. Jog, M.T. Kandemir, C.R. Das, Neither more nor less: optimizing thread-level parallelism for gpgpus, in: Proceedings of the 22nd International Conference on Parallel Architectures and Compilation Techniques, IEEE Press, 2013, pp. 157–166.

[52] T.G. Rogers, M. O'Connor, T.M. Aamodt, Cache-conscious wavefront scheduling, in: Proceedings of the 2012 45th Annual IEEE/ACM International Symposium on Microarchitecture, IEEE Computer Society, 2012, pp. 72–83.

[53] R. Das, O. Mutlu, T. Moscibroda, C.R. Das, Aergia: exploiting packet latency slack in on-chip networks, in: Proceedings of the 37th Annual International Symposium on Computer Architecture, 2010, pp. 106–116.

[54] A. Kumar, L.-S. Peh, P. Kundu, N.K. Jha, Express virtual channels: towards the ideal interconnection fabric, ACM SIGARCH Comput. Archit. News 35 (2) (2007) 150–161.

[55] C.A. Nicopoulos, D. Park, J. Kim, N. Vijaykrishnan, M.S. Yousif, C.R. Das, Vichar: a dynamic virtual channel regulator for network-on-chip routers, in: 2006 39th Annual IEEE/ACM International Symposium on Microarchitecture (MICRO'06), IEEE, 2006, pp. 333–346.

[56] W. Choi, K. Duraisamy, R.G. Kim, J.R. Doppa, P.P. Pande, R. Marculescu, D. Marculescu, Hybrid network-on-chip architectures for accelerating deep learning kernels on heterogeneous manycore platforms, in: Proceedings of the International Conference on Compilers, Architectures and Synthesis for Embedded Systems, 2016, pp. 1–10.

[57] S. Abadal, A. Cabellos-Aparicio, E. Alarcón, J. Torrellas, Wisync: an architecture for fast synchronization through on-chip wireless communication, ACM SIGPLAN Not. 51 (4) (2016) 3–17.

[58] M. Sadrosadati, S.B. Ehsani, H. Falahati, R. Ausavarungnirun, A. Tavakkol, M. Abaee, L. Orosa, Y. Wang, H. Sarbazi-Azad, O. Mutlu, Itap: idletime-aware power management for gpu execution units, ACM Trans. Archit. Code Optim. 16 (1) (2019).

[59] T.N. Miller, X. Pan, R. Thomas, N. Sedaghati, R. Teodorescu, Booster: reactive core acceleration for mitigating the effects of process variation and application imbalance in low-voltage chips, in: 2012 IEEE 18th International Symposium on High Performance Computer Architecture (HPCA), IEEE, 2012, pp. 1–12.

[60] R.G. Dreslinski, B. Zhai, T. Mudge, D. Blaauw, D. Sylvester, An energy efficient parallel architecture using near threshold operation, in: 16th International Conference on Parallel Architecture and Compilation Techniques (PACT 2007), IEEE, 2007, pp. 175–188.

[61] H. Aghilinasab, M. Sadrosadati, M.H. Samavatian, H. Sarbazi-Azad, Reducing power consumption of gpgpus through instruction reordering, in: Proceedings of the 2016 International Symposium on Low Power Electronics and Design, Association for Computing Machinery, 2016, pp. 356–361.

[62] G. Ruhl, S. Dighe, S. Jain, S. Khare, S.R. Vangal, Ia-32 Processor with a wide-voltage-operating range in 32-nm cmos, IEEE Micro 33 (2) (2013) 28–36.

About the authors

Amirhossein Mirhosseini received his PhD in Computer Science and Engineering from the University of Michigan in 2021. His research interests center around datacenters and cloud computing, with a particular focus on microarchitecture enhancements and accelerator design, as well as queueing and scheduling optimizations for latency sensitive microservices. He did his undergraduate studies in Sharif University of Technology, where he participated in a series of research studies on Network-on-Chips and GPU architectures.

Mohammad Sadrosadati received his PhD degree in Computer Engineering from Sharif University of Technology, Tehran, Iran, in 2019 under the supervision of Prof. H. Sarbazi-Azad. He spent one year from April 2017 to April 2018 as an academic guest at ETH Zurich hosted by Prof. O. Mutlu during his PhD program. He is currently a senior researcher in the SAFARI research group at ETH Zurich. His research interests are in the areas of heterogeneous computing, processing-in-memory, memory systems, and interconnection networks. Due to his achievements and impact on improving the energy efficiency of GPUs, he won Khwarizmi Youth Award, one of the most prestigious awards, as the first laureate in 2020, to honor and embolden him to keep taking even bigger steps in his research career.

Behnaz Soltani is currently a PhD student at the Department of Computing, Macquarie University, Sydney, Australia. She received her MSc Degree in Computer Architecture from Sharif University of Technology, Tehran, Iran, in 2017, and her BSc degree in Computer Engineering from the University of Tabriz, Tabriz, Iran, in 2015. Her research interests include computer architecture, interconnection networks, distributed computing, IoT, and machine learning.

Hamid Sarbazi-Azad is currently a professor of computer science and engineering at the Sharif University of Technology, Tehran, Iran. His research interests include high-performance computer/memory architectures, NoCs and SoCs, parallel and distributed systems, social networks, and storage systems, on which he has published over 400 refereed papers. He received Khwarizmi International Award in 2006, TWAS Young Scientist Award in engineering sciences in 2007, Sharif University Distinguished Researcher awards in the years 2004, 2007, 2008, 2010, and 2013, the Iranian Ministry of Communication and Information Technology's award for contribution to IT research and education in 2014, and distinguished book author of the Sharif University of Technology in 2015 and 2020. Dr Sarbazi-Azad is now an associate editor of ACM Computing Surveys, IEEE Computer Architecture Letters, and Elsevier's Computers and Electrical Engineering.

CHAPTER FOUR

Routerless networks-on-chip

Fawaz Alazemi[a], Lizhong Chen[b], and Bella Bose[b]

[a]Kuwait University, Kuwait, Kuwait
[b]Oregon State University, Corvallis, OR, United States

Contents

Advances in Computers, Volume 124
ISSN 0065-2458
https://doi.org/10.1016/bs.adcom.2021.11.002

Abstract

Traditional bus-based interconnects are simple and easy to implement, but the scalability is greatly limited. While router-based networks-on-chip (NoCs) offer superior scalability, they also incur significant power and area overhead due to complex router structures. In this work, we explore a new class of on-chip networks, referred to as *Routerless NoCs*, where routers are completely eliminated. We propose a novel design that utilizes on-chip wiring resources smartly to achieve comparable hop count and scalability as router-based NoCs. Several effective techniques are also proposed that significantly reduce the resource requirement to avoid new network abnormalities in routerless NoC designs. Evaluation results show that, compared with a conventional mesh, the proposed routerless NoC achieves 9.5× reduction in power, 7.2× reduction in area, 2.5× reduction in zero-load packet latency, and 1.7× increase in throughput. Compared with a state-of-the-art low-cost NoC design, the proposed approach achieves 7.7× reduction in power, 3.3× reduction in area, 1.3× reduction in zero-load packet latency, and 1.6× increase in throughput.

1. Introduction

As technologies continue to advance, tens of processing cores on a single-chip multiprocessor (CMP) have already been commercially offered. Intel Xeon Phi Knight Landing [1] is an example of a single CMP that has 72 cores. With hundreds of cores in a CMP around the corner, there is a pressing need to provide efficient networks-on-chip (NoCs) to connect the cores. In particular, recent chips have exhibited the trend to use many but simple cores (especially for special-purpose many-core accelerators), as opposed to a few but large cores, for better power efficiency. Thus, it is imperative to design highly scalable and ultra-low cost NoCs that can match with many simple cores.

Prior to NoCs, buses have been used to provide on-chip interconnects for multi-core chips [2–7]. While many techniques have been proposed to improve traditional buses, it is hard for their scalability to keep up with modern many-core processors. In contrast, NoCs offer a decentralized solution by the use of routers and links. Thanks to the switching capability of routers to provide multiple paths and parallel communications, the throughput of NoCs is significantly higher than that of buses. Unfortunately, routers have been notorious for consuming a substantial percentage of chip's power and area [8, 9]. Moreover, the cost of routers increases rapidly as link width increases. Thus, except for a few ad hoc designs, most on-chip networks do not employ link width higher than 256-bit or 512-bit, even though additional wiring resources may be available. In fact, our study shows that, a 6 × 6 256-bit Mesh only uses 3% of the total available wiring resources (more details in Section 3).

The high overhead of routers motivates researchers to develop *routerless NoCs* that eliminate the costly routers but use wires more efficiently to achieve scalable performance. While the notion of routerless NoC has not been formally mentioned before, prior research has tried to remove routers with sophisticated use of buses and switches, although with varying success. The goal of routerless NoCs is to select a set of smartly placed loops (composed of wires) to connect cores such that the average hop count is comparable to that of conventional router-based NoCs. However, the main roadblocks are the enormous design space of loop selection and the difficulty in avoiding deadlock with little or no use of buffer resources (otherwise, large buffers would defeat the purpose of having routerless NoCs).

In this work, we explore efficient design and implementation to materialize the promising benefits of routerless NoCs. Specifically, we propose a layered progressive method that is able to find a set of loops that meet the requirement of connectivity and the limitation of wiring resources. The method progressively constructs the design of a large routerless network from good designs of smaller networks, and is applicable to any $n \times m$ many-core chips with superior scalability. Moreover, we propose several novel techniques to address the challenges in designing routerless interface to avoid network abnormalities such as deadlock, livelock, and starvation. These techniques result in markedly reduced buffer requirement and injection/ejection hardware overhead. Compared with a conventional router-based Mesh, the proposed routerless design achieves 9.48× reduction in power, 7.2× reduction in area, 2.5× reduction in zero-load packet latency, and 1.73× increase in throughput. Compared with the current state-of-the-art scheme that tries to replace routers with less costly structures (IMR [10]), the proposed scheme achieves 7.75× reduction in power, 3.32× reduction in area, 1.26× reduction in zero-load packet latency, and 1.6× increase in throughput.

2. Background and motivation

2.1 Related work

Prior work on-chip interconnects can be classified into *bus based* and *network based*. The latter can be further categorized as *router-based NoCs* and *routerless*

NoCs. The main difference between bus-based interconnects and routerless NoCs is that bus-based interconnects use buses in a direct, simple, and primitive way, whereas routerless NoCs use a network of buses in a sophisticated way and typically need some sort of switching that earlier bus systems do not need. Each of the three categories is discussed in more detail below.

Bus-based interconnects are centralized communication systems that are straightforward and cheap to implement. While buses work very well for a few cores, the overall performance degrades significantly as more cores are connected to the bus [3, 4]. The two main reasons for such degradation are the length of the bus and its capacitive load. Rings [5–7] can also be considered as variants of bus-based systems where all the cores are attached to a single bus/ring. IBM Cell processor [2] is an improved bus-based system which incorporates a number of bus optimization techniques in a single chip. Despite having a better performance over conventional bus/ring implementations, IBM Cell process still suffers from serious scalability issues [11].

Router-based NoCs are decentralized communication systems. A great deal of research has gone into this (e.g., Refs. [12–19], too many to cite all here). The switching capability of routers provides multiple paths and parallel communications to improve throughput, but the overhead of routers is also quite substantial. Bufferless NoC (e.g., Ref. [20]) is a recent interesting line of work. In this approach, buffer resources in a router are reduced to the minimal possible size (i.e., one flit buffer per input port). Although bufferless NoC is a clever approach to reduce area and power overhead, the router still has other expensive components that are eliminated in the routerless approach (Section 7.5 compares the hardware cost).

Routerless NoCs aim to eliminate the costly routers while having scalable performance. While the notion of routerless NoC has not been formally mentioned before, there are several works that try to remove routers with sophisticated use of buses and switches. However, as discussed below, the hardware overhead in these works is quite high, some requiring comparable buffer resources as conventional routers, thus not truly materializing the benefits of routerless NoCs. zOne approach is presented in Ref. [21], where the NoC is divided into segments. Each segment is a bus, and all the segments are connected by a central bus. Segments and central bus are linked by a switching element. In large NoCs, either the segments or the central bus may suffer from scalability issues due to their bus-based nature. A potential solution is to increase the number of wires in the central bus and the number of cores in a segment. However, for NoCs larger than 8 × 8, it would be challenging to find the best size for the segments and central bus without affecting scalability. Hierarchical rings (HRs) [22] have a similar design approach to Ref. [21]. The NoC is divided into disjoint sets of cores, and each set is connected by a ring. Such rings are called local rings. Additionally, a set of global rings bring together the local rings. Packets switch between local and global rings through a low-cost switching element. Although the design has many nice features, the number of switching element is still not small. For example, for an 8 × 8 NoC, there are 40 switching elements, which is close to the number of routers in the 8 × 8 network. Recently, a multi-ring-based NoC called isolated multiple ring (IMR) is proposed in Ref. [10] and has been shown to be superior than the above Hierarchical rings. To our knowledge, this is the latest and best scheme so far along the line of work on removing routers. While the proposed concept is promising, the specific IMR design has several major issues and the results are far from optimal, as discussed in the next section.

2.2 Need for new routerless NoC designs

2.2.1 Principles and challenges

We use Fig. 1 to explain the basic principles of routerless NoCs. This figure depicts an example of a 16-core chip. The 4 × 4 layout specifies only the positions of the cores, not any topology. A straightforward but naive way to achieve routerless NoC is to use a long loop (e.g., a Hamiltonian cycle) that connects every node on the chip as shown in Fig. 1A. Each node injects packets to the loop and receives packets from the loop through a simple interface (referred to as RL interface hereafter). Apparently, even if a flit on the loop can be forwarded to the next node at the speed of one hop per cycle, this design would still be very slow because of the average $O(n^2)$ hop count, assuming an $n \times n$ many-core chip. Scalability is poor in this case, as conventional topology such as Mesh has an average hop count of $O(n)$.

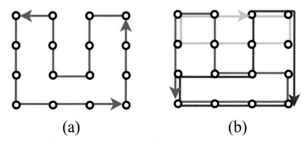

(a)   (b)

Fig. 1 An example of loops in a 4 × 4 grid.

To reduce the hop count, we need to select a better set of loops to connect the nodes, while guaranteeing that every pair of nodes is connected by at least one loop (so that a node can reach another node directly in one loop). Fig. 1B shows an example with the use of three loops, which satisfies the connectivity requirement and reduces the all-pair average hop count by 46% compared with Fig. 1A. Note that, when injecting a packet, a source node chooses a loop that connects to the destination node. Once the packet is injected into a loop, it stays on this loop and travels at the speed of one hop per cycle all the way to the destination node. No changing loops are needed at RL interfaces, thus avoiding the complex switching hardware and per-hop contention that may occur in conventional router-based on-chip networks.

Several key questions can be asked immediately. Is the design in Fig. 1B optimal? Is it possible to select loops that achieve comparable hop count as conventional NoCs such as Mesh? Is there a generalized method that we can use to find the loops for any $n \times n$ network? How can this be done without exceeding the available on-chip wiring resources? Unfortunately, answering these questions is extremely challenging due to the enormous design space. We calculated the number of possible loops for $n \times n$ chips based on the method used in Ref. [23], where a loop can be any unidirectional circular path with the length between 4 and n. Table 1 lists the results up to $n = 8$. As can be seen, the number of possible loops grows extremely rapidly. To make things more challenging, because the task is to find a set of loops, the design space that the routerless NoC approach is looking at is not the number of loops, but the combination of these loops! A large portion of the combinations would be invalid, as not all combinations can provide the connectivity where there is at least one loop between any source and destination pair.

Meanwhile, any selected final set of loops needs to comfortably fit in the available wiring resources on the chip. Specifically, when loops are superpositioned, the number of overlapped loops between any neighboring

Table 1 Number of unidirectional loops in $n \times n$ grid [23].

n	# of loops	n	# of loops
1	0	2	2
3	26	4	426
5	18,698	6	2,444,726
7	974,300,742	8	1,207,683,297,862

node pairs should not exceed a limit. In what follows, we use *overlapping* to refer to the number of overlapped loops between two neighboring nodes (e.g., in Fig. 1B some neighboring nodes have two loops passing through them, while others have only one loop passing), and use *overlapping cap* to refer to the limit of the overlapping. Note that the cap should be much lower than the theoretical wiring resources on chip due to various practical considerations (analyzed in Section 3). As an example, if the overlapping cap is 1, then Fig. 1A has to be the final set. If the overlapping cap increases to 2, it provides more opportunity for improvement, e.g., the better solution in Fig. 1B. The overlapping cap is a hard limit and should not be violated. However, as long as this cap is met, it is actually beneficial to approach this cap for as many neighboring node pairs as possible. Doing this indicates more wires are being utilized to connect nodes and reduce hop count.

2.2.2 Major issues in current state of the art
There are several major issues that must be addressed in order to achieve effective routerless NoCs. We use IMR [10] as an example to highlight these issues. IMR is a state-of-the-art design that follows the above principle to deploy a set of rings such that each ring joins a subset of cores. While IMR has been shown to outperform other schemes with or without the use of routers, the fundamental issues in IMR prevent it from realizing the true potential of routerless NoCs. This calls for substantial research on this topic to develop more efficient routerless designs and implementations.

1. *Large overlapping.* For example, IMR uses a large number of super-positioned rings (equivalent to the above-defined overlapping cap of 16) without analyzing the actual availability of wiring resources on-chip.

2. *Extremely slow search.* A genetic algorithm is used in IMR to search the design space. This general-purpose search algorithm is very slow (taking several hours to generate results for 16×16, and is not able to produce good results in a reasonable time for larger networks). Moreover, the design generated by the algorithm is far from optimal with high hop counts, as evaluated in Section 6. Thus, efforts are much needed to utilize clever heuristics to speed up the process.

3. *High buffer requirement.* Currently, the network interface of IMR needs one packet-sized buffer per ring to avoid deadlock. Given that up to 16 rings can pass through an IMR interface, the total number of buffers at each interface is very close to a conventional router.

The above issues are addressed in the next three sections. Section 3 analyzes the main contributing factors that determine the wiring availability in practice, and estimates reasonable overlapping caps using a contemporary many-core processor. Section 4 proposes a layered progressive approach to select a set of loops, which is able to generate highly scalable routerless NoC designs in less than a second (up to 128 × 128). Section 5 presents our implementation of routerless interface. This includes a technique that requires only one flit-sized buffer per loop (as opposed to one packet-sized buffer per loop). This technique alone can save buffer area by multiple times.

3. Analysis on wiring resources
3.1 Metal layers

As technology scales to smaller dimensions, it provides a higher level of integration. With this trend, each technology comes with an increasing number of routing metal layers to meet the growing demand for higher integration. For example, Intel Xeon Phi (Knights Landing) [1] and KiloCore [24] are fabricated in the process technology with 11 and 13 metal layers, respectively. Each metal layer has a pitch size which defines the minimum wire width and the space between two adjacent wires. The physical difference between metal layers results in various electrical characteristics. This allows designers to meet their design constraints such as delay on the critical nets by switching between different layers. Typically, lower metal layers have narrower width and are used for local interconnects (e.g., within a circuit block); higher metal layers have wider width and are used for global interconnects (e.g., power supply, clock); and middle metal layers are used for semi-global interconnects (e.g., connecting neighboring cores). Table 2 lists several key physical parameters of Xeon Phi including the middle layers that can be used for on-chip networks.

3.2 Wiring in NoC

To estimate the actual wiring resources that can be used for routing, several important issues should be considered when placing wires on the metal layers.

Routing strategy: In general, two approaches can be considered for routing interconnects over cores in NoCs. In the first approach, dedicated routing channels are used to route wires in NoCs. This method of routing was widely used in earlier technology nodes where only three metal layers were typically provided [28], and it has around 20% area overhead. In the second

Table 2 Wiring resources in a many-core processor chip.

Many core processor	Xeon Phi, Knights landing	
Number of cores	72	
NoC size	6×6	
Die area	(31.9mm × 21.4mm) 683 mm^2 [25]	
Technology	FinFET 14nm	
Interconnect	13 Metal layers	
Inter-core metal layers	Metal layer	Pitch [26,27]
	M4	80 nm
	M5	104 nm

approach, wires are routed over the cores at different metal layers [29]. In the modern technology nodes with 6–13 metal layers, this approach of routing over logic becomes more common for higher integration. This can be done in two ways: (1) several metal layers are dedicated for routing wires and (2) a fraction of each metal layer is used to route the wires. The first way is preferable given that many metal layers are available in advanced technology nodes [28, 29].

Repeater. Wires have parasitic resistance and capacitance which increase with the length of wires. To meet a specific target frequency, a long wire needs to be split into several segments, and repeaters (inverters) are inserted between the segments, as shown in Fig. 2. The size of repeaters should be considered in estimating the available wiring resources. For a long wire in the NoC, the size of each repeater (h times of an inverter with minimum size) is usually not small, but the number of repeaters (k) needed is small [30]. In fact, it has been shown that increasing K has negligible improvement in reducing the delay [30]. For a 2GHz operating frequency, using only one repeater with the size of 40 times W/L of the minimum sized inverter can support a wire length of 2 mm [29], which is longer than the typical distance between two cores in a many–core processor [31].

Coping with cross-talk: Cross-talk noises can occur either between the wires on the same metal layer or between the wires on different metal layers, both of which may affect the number of wires that can be placed. The impact of cross-talk noises on voltage can be calculated by Eq. (1) as the voltage changes on a floated victim wire [32].

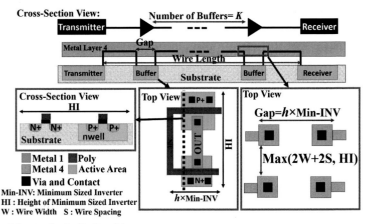

Fig. 2 A long wire in NoCs with repeaters.

$$\Delta V_{victim} = \frac{C_{adj}}{C_{victim} + C_{adj}} \times \Delta V_{aggressor} \qquad (1)$$

where ΔV_{victim} is the voltage variation on the victim wire, $\Delta V_{aggressor}$ is the voltage variation on the aggressor, C_{victim} is the total capacitance (including load capacitance) of the victim wire, and C_{adj} is the coupling capacitance between the aggressor and the victim. It can be observed from Eq. (1) that the impact of cross-talk on the victim wire depends on the ratio of C_{adj} to C_{victim}. Hence, the cross-talk on the same layer has much larger impact on the power, performance, and functionality of the NoC because the adjacent wires which run in parallel on the same metal layer has larger coupling capacitance (C_{adj}) [32]. There are two major techniques to mitigate cross-talk noises, shielding and spacing. In the shielding approach, crosstalk noises are largely avoided between two adjacent wires by inserting another wire (which is usually connected to the ground or supply voltage) between them. In the spacing approach, adjacent wires are separated by a certain distance that would keep the coupling noise below a level tolerable by the target process and application. Compared with spacing, shielding is much more effective as it can almost remove crosstalk noises [33]. However, shielding also incurs more area overhead as the distance used in the spacing approach is usually smaller than that of inserting a wire.

3.3 Usable wires for NoCs

To gain more insight on how many wiring resources are usable for on-chip networks under current manufacturing technologies, we estimated the number of usable wires by taking into account the above factors.

The estimation is based on using two metal layers to route wires over the cores. The area overhead of the repeater insertion including the via contacts and the area occupation of the repeaters are considered based on the layout design rules of each metal layer. We used the conservative way of shielding to reduce crosstalk noises (and the inserted wires are not counted toward usable wires), although spacing may likely offer more usable wires. In addition, in practice, 20% to 30% of each dedicated metal layer for routing wires over the cores is used for I/O signals, power, and ground connections [29]. This overhead is also accounted for. The maximum values of h and K are used for worst-case estimation. As such, the above method gives a very conservative estimation of the usable wires. Assuming that there is a chip with similar physical configuration as described in Table 2, the two metal layers M4 and M5 under 14 nm technology can provide 101,520 wires in the cross-section. This translates into 793 unidirectional links of 128-bit, or 396 unidirectional links of 256-bit, or 198 unidirectional links of 512-bit in the cross-section. In contrast, a 6×6 mesh only uses 12 unidirectional 256-bit links in the bisection, which is about 3% of the usable wires. It is important to note that the conventional router-based NoCs do not use very wide links for good reasons. For instance, router complexity (e.g., the number of crosspoints in switches, the size of buffers) increases rapidly as the link width increases. Also, although wider links provide higher throughput, it is difficult to capitalize on wider links for lower latency. The reduction in serialization latency by using wider links quickly becomes insignificant as link width approaches the packet size. This motivates the need for designing routerless NoCs where wiring resources can be used more efficiently.

　　The above estimation of the number of usable wires helps to decide the overlapping cap mentioned previously. To avoid taxing too much on the usable wiring resources and to have a scalable design, we propose to use an overlapping cap of n for $n \times n$ chips. In the above 6×6 case, this translates into 4.5% of the usable wires for 128-bit loop width, or 9.1% for 256-bit loop width. This parameterized overlapping cap helps to provide the number of loops that is proportional to chip size, so the quality of the routerless designs can be consistent for larger chips.

4. Designing routerless NoCs
4.1 Basic idea

Our proposed routerless NoC design is based on what we call *layered progressive* approach. The basic idea is to select the loop set in a progressive way where the design of a large routerless network is built on top of the design of smaller networks. Each time the network size increments, the newly selected loops are conceptually bundled as a layer that is reused in the next network size.

Specifically, let M_k be the final set of selected loops for $k \times k$ grid ($2 \leq k \leq n$) that meets the connectivity, overlapping, and low hop count requirements. We construct M_{k+2} by combining M_k with a new set (i.e., layer) of smartly placed loops. The new layer utilizes new wiring resources that are available when expanding from $k \times k$ to $(k + 2) \times (k + 2)$. The resulting M_{k+2} can also meet all the requirements and deliver superior performance. For example, as shown in Fig. 3, the grid is logically split into multiple layers with increasing sizes. Let L_k be the set of loops selected for Layer k. First, suppose that we already find a good set of loops for 2×2 grid that connects all the nodes with a low hop count and does not exceed an overlapping of 2 between any neighboring nodes. That set of loops is M_2, which is also L_1 as this is the base case. Then we find another set of loops L_2, together with M_2, which can form a good set of loops for 4×4 grid (i.e., $M_4 = L_2 \cup M_2$). The resulting M_4 can connect all the nodes with a low hop count and does not exceed an overlapping of 4 between any neighboring nodes and so on so forth, until reaching the targeted $n \times n$ grid. In general, we have $M_n = L_{\lfloor n/2 \rfloor} \cup M_{n-2} = L_{\lfloor n/2 \rfloor} \cup L_{\lfloor n/2 \rfloor - 2} \cup M_{n-4} = \ldots = L_{\lfloor n/2 \rfloor} \cup L_{\lfloor n/2 \rfloor - 2} \cup L_{\lfloor n/2 \rfloor - 4} \cup \ldots \cup L_1$.

Apparently, the key step in the above progressive process is how to select the set of loops in Layer k, which enables the progression to the next sized grid with low hop count and overlapping. In the next sections, we walk through several examples to illustrate how it is done to progress from 2×2 grid to 8×8 grid.

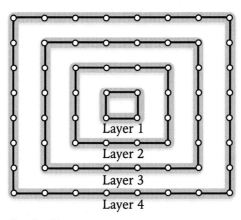

Fig. 3 Layers of an 8 × 8 grid.

4.2 Examples

4.2.1 2 × 2 Grid

This is the base case with one layer. There are exactly two possible loops, one in each direction, in a 2 × 2 grid. Both of them are included in $M_2 = L_1$, as shown in Fig. 4. The resulting M_2 satisfies the requirement that every source and destination pair is connected by at least one loop. The maximum number of loops overlapping between any neighboring nodes is 2, which meets the overlapping cap. This set of loops achieves a very low all-pair average hop count of 1.333, which is as good as the Mesh.

4.2.2 4 × 4 Grid

M_4 consists of loops from two layers. Based on our layered progressive approach, L_1 is from M_2. We select eight loops to form L_2, as illustrated in Fig. 5. The eight loops fall into four groups (from this network size and forward, each new layer is constructed using four groups with the similar heuristics as discussed below). The first group, A_4 (the subscript indicates the size of the grid), has only one anticlockwise loop. It provides connectivity among the 12 new nodes when expanding from Layer 1 to Layer 2. The loops in the second group, B_4, have the first column as the common edge of the loops, but the opposite edge of the loops moves gradually toward the right (this is more evident in group B_6 in Fig. 6). Similarly, the third group, C_4, uses the last column as the common edge of the loops and gradually moves the opposite edge toward the left. It can be verified that groups B_4 and C_4 provide connectivity between the 12 new nodes in Layer 2 and the 4 nodes in Layer 1. Since the connectivity among the 4 inner nodes has already been provided by L_1, the connectivity requirement of 4 × 4 grid is met by having L_1, A_4, B_4, and C_4. The fourth group, D_4, offers additional "shortcuts" in the horizontal dimension.

A very nice feature of the selected M_4 is that the wiring resources are efficiently utilized, as the overlapping between many neighboring node pairs is close to the overlapping cap of 4. For example, for the first (or the last) column, each group of loops has exactly one loops passing through that column, totaling an overlapping of 4, which is the same as the cap. Thus, no overlapping "ration" is underutilized. For the second column (or the third) column, groups A_4 and D_4 have no loop passing through, and groups B_4 and C_4 have two loops passing through in total. However, note that the final

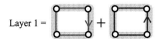

Fig. 4 Loops in L_1, and $M_2 = L_1$.

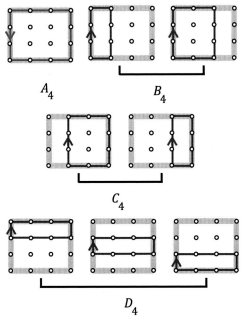

Fig. 5 Loops in L_2. $M_4 = L_2 \cup L_1$.

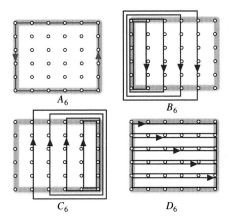

Fig. 6 Loops in L_3. $M_6 = L_3 \cup L_2 \cup L_1$.

M_4 also includes L_1 which has two loops passing through the second (or the third) column. Hence, the total overlapping of the middle columns is also 4, exactly the same as the cap. Simple counting can show that the overlapping on the horizontal dimension is also 4 for each row. Owing to this efficient use of wiring resource "ration," the all-pair average hop count is 3.93 for the selected set of loops in M_4. The final set is $M_4 = L_2 \cup M_2 = L_2 \cup L_1$.

4.2.3 6 × 6 Grid

M_6 consists of loops from three layers. L_1 and L_2 are from M_4, and L_3 is formed in a similar fashion as 4×4 grid from four groups, as illustrated in Fig. 6. Again, connectivity is provided by M_4 and groups A to C. Together with group D, the number of overlapping on each column and row is 6, thus fully utilizing the allocated wiring resources.

Additionally, for the purpose of reducing hop count and balancing horizontal and vertical wiring utilization, when we combine M_4 and L_3 to form M_6, every loop in M_4 is reversed and then rotated for 90° clockwise.[a] If this slightly changed M_4 is denoted as M'_4, the final set can be expressed as $M_6 = L_3 \cup M'_4 = L_3 \cup (L_2 \cup L_1)'$, with an all-pair average hop count of 6.07.

4.2.4 8 × 8 Grid

Similar to earlier examples, L_4 consists of loops as shown in Fig. 7. The final set M_8 is $M_8 = L_4 \cup M'_6 = L_4 \cup (L_3 \cup (L_2 \cup L_1)')'$ with an all-pair average hop count of 8.32.

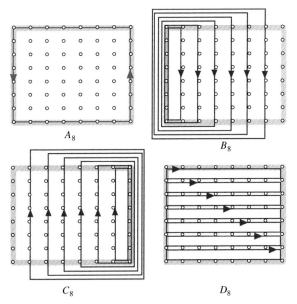

Fig. 7 Loops in L_4. $M_8 = L_4 \cup L_3 \cup L_2 \cup L_1$.

---

[a] In 4×4 grid, reversal and rotation of M_2 is not necessary because M_2 and M'_2 have the same effect on L_1.

4.3 Formal procedure

For an $n \times n$ grid, the loops for a routerless NoC design can be recursively found by the procedure shown in Algorithm 1. The procedure is recursive and denoted as RLrec. The procedure begins by generating loops for the outer layer, say layer i, and then it recursively generates loops for layer $i - 1$ and so on until the base case is reached or the layer has a single node or empty. Procedure $G(r_1, r_2, c_1, c_2, d)$ is a simple function that generates a rectangular shape loop with corners (r_1, c_1), (r_1, c_2), (r_2, c_1), and (r_2, c_2) and direction d. When processing each layer in this algorithm, procedure G is called repeatedly to generate four groups of loops. Additionally, the

---

ALGORITHM 1 RLrec.

Input : N_L, N_H; the low and high numbers

1 **begin**

2   **if** $N_L = N_H$ **then**

3   **return** $\{\}$

4   Let $M = \{\}$

5   **if** $N_H - N_L = 1$ **then**

6     $M = M \cup G(N_L, N_H, N_L, N_H, \text{clockwise})$

7     $M = M \cup G(N_L, N_H, N_L, N_H, \text{anticlockwise})$

8   **return** M

9   $M = M \cup G(N_L, N_H, N_L, N_H, \text{anticlockwise})$ // Group A

10   **for** $i = N_L + 1 \rightarrow N_H - 1$ **do**

11     $M = M \cup G(N_L, N_H, N_L, i, \text{clockwise})$ // Group B

12     $M = M \cup G(N_L, N_H, i, N_H, \text{clockwise})$ // Group C

13   **for** $i = L \rightarrow H - 1$ **do**

14     $M = M \cup G(i, i+1, N_L, N_H, \text{clockwise})$ // Group D

15   $M' = \text{RLrec}(N_L + 1, N_H - 1)$

16   Reverse and rotate for $90°$ every loop in M'

17   **return** $M \cup M'$

generated loops rotate 90 degrees and reverse directions after processing each layer to balance wiring utilization and reduce hop count, respectively. The final loops generated by the RLrec algorithm have an overlapping of at most n.

While it would be ideal if an analytical expression can be derived to calculate the average hop count for this heuristic approach, this seems to be very challenging at the moment. However, it is possible to calculate the average hop count numerically. This result is presented in the evaluation, which shows that our proposed design is highly scalable.

5. Implementation details

After addressing the key issue of finding a good set of loops, the next important task is to efficiently implement the routerless NoC design in hardware. Because of the routerless nature, no complex switching or virtual channel (VC) structure is needed at each hop, so the hardware between nodes and loops has a small area footprint in general. However, due to various potential network abnormalities such as deadlock, livelock, and starvation, a certain number of resources are required to guarantee correctness. If not addressed appropriately, this may cause substantial overhead that is comparable to router-based NoCs. In this section, we propose a few effective techniques to minimize those overhead.

In a routerless NoC, each node uses an interface (RL interface) to interact with one or multiple loops that pass through this node. Fig. 8 shows the main components of a RL interface. While details are explained in the following sections, the essential function of the interface includes injecting packets into a matching loop based on connectivity and availability, forwarding packets to the next hop on the same loop, and ejecting packets at the destination node. Notice that packets cannot switch loops once injected. All the loops have the same width (e.g., 128-bit wires).

5.1 Injection process
5.1.1 Extension buffer technique
A loop is basically a bundle of wires connected with flip-flops at each hop (Fig. 8). At clock cycle i, a flit arriving at the flip-flop of loop l must be consumed immediately by either being ejected at this node or forwarded to the next hop on loop l through output l. If no flit arrives at loop l (thus not using output l), the RL interface can inject a new flit on loop l through output l. However, it is possible that an injecting packet consists of multiple flits and

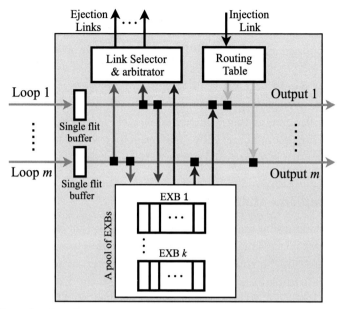

Fig. 8 Routerless interface components.

requires several cycles to finish the injection, during which other flits on loop l may arrive at this RL interface. Therefore, addition buffer resources are needed to hold the incoming flits temporarily.

If routerless NoC uses the scheme proposed in prior ring-based work (e.g., IMR [10]), a full packet-sized buffer per loop at each hop would be needed to ensure correctness, which is very inefficient. As illustrated in Fig. 9, a long packet B with multiple flits is waiting for injection (there is no issue if it is a short single-flit packet). At clock cycle i, the injection is allowed because packet B sees that no other flit in Interface Y is competing with B for the output to Interface Z. From cycle $i + 1$ to $i + 3$, the flits of B are injected sequentially. However, while packet B is being injected during these cycles, another long packet A may arrive at Interface Y. Because RL interfaces do not employ flow control to stop the upstream node, Interface Y needs to provide a packet-sized buffer to temporarily store the entire packet A. A serious inefficiency lies in the fact that, if there are m loops passing through a RL interface, the interface needs to have m packet-sized buffers, one for each loop.

To address this inefficiency, we notice that an interface injects packets one at a time, so not all the loops are affected simultaneously. Based on this observation, we propose the extension buffer technique to share the

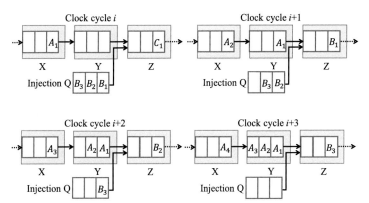

Fig. 9 Injecting a long packet requires a packet-sized buffer per loop at each hop in prior implementation (*X*, *Y*, and *Z* are interfaces).

packet-sized buffer among loops. As shown in Fig. 8, each loop has only a flit-sized buffer, but the interface has a pool of extension buffers (EXBs). The size of each EXB is the size of a long packet, so when a loop is "extended" with an EXB, it would be large enough to store a long packet. Minimally, only one EXB is needed in the pool, but having multiple EXBs may have slight performance improvement. This is because another injection might occur while the previous EXB is not entirely released (drained) due to a previous injection (e.g., clock cycle $i + 3$ in Fig. 9). However, as shown later in the evaluation, the performance difference is negligible. As a result, our proposed technique of using one shared EXB can essentially achieve the same objective of ensuring correctness as IMR but reduces the buffer requirement by m times. This is equivalent to an $8\times$ saving in buffer resources in 8×8 networks and $16\times$ saving in 16×16 networks.

5.1.2 Injection process

The injection process with the use of EXBs is straightforward. To inject a packet p of n_f flits, the first step is to look up a small routing table to see which loop can reach p's destination. The routing table is precomputed since all the loops are predetermined. The packet p then waits for the loops to become available (i.e., having sufficient buffer space). Assume l is a loop that has the shortest distance to the destination among all the available loops. When the injection starts, the interface holds the output port of l for n_f cycles to inject p, and assigns a free *EXB* to l if $n_f > 1$ and l is not already connected to another

EXB. During those n_f cycles, any incoming flit through the input port of *l* is enqueued in the extension buffer. The EXB is released later when its buffer slots are drained.

5.2 Ejection process

The ejection process starts as soon as the head flit of a packet *p* reaches the RL interface of its destination node. The interface ejects *p*, one flit per cycle. Once *p* is ejected, the interface will wait for another packet to eject. There is, however, a potential issue with the ejection process. While unlikely, a RL interface with *m* loops may receive up to *m* head flits simultaneously in a given cycle that are all destined to this node. Because any incoming packets need to be consumed immediately and the packets are already at the destination, the interface needs to have *m* ejection links in order to eject all the packets in that cycle. As each eject link has the same width as the loop (i.e., 128-bit), this incurs substantial hardware overhead.

To reduce this overhead, we utilize the fact that the actual probability of having *k* packets ($1 < k \leq m$) arriving at the same destination in the same cycle is low, and this probability decreases drastically as *k* increases. Based on this observation, we propose to optimize for the common case where only *e* ejection links are provided ($e \ll m$). If more than *e* packets arrive at the same cycle, ($k - e$) packets are forwarded to the next hop. Those deflected packets will continue on their respective loops and will circle back to the destination later. As shown in the evaluation, having two ejection links can reduce the percentage of circling packets to be below 1% on average (1.6% max) across the benchmarks. This demonstrates that this is a viable and effective technique to reduce overhead.

5.3 Avoiding network abnormalities

As network abnormalities are theoretically possible but practically unlikely scenarios, our design philosophy is to place very relaxed conditions to trigger the handling procedures so as to minimize performance impact while guaranteeing correctness.

5.3.1 Livelock avoidance

A livelock may occur if a packet circles indefinitely and never gets a chance to eject. We address this issue by having a byte-long circling counter at each head flit with an initial value of zero. Every time a packet reaches its destination interface and is forced to be deflected, the counter is incremented by 1. If the circling counter of a packet p reaches 254 but none of the ejection link is available, the interface marks one of its ejection links as reserved and then deflects p for the last time. The marked ejection link will not eject any more packets after finishing the current one, until p circles back to the ejection link (by then the marked ejection link will be available; otherwise there is a possible protocol-level deadlock, discussed shortly). Once p is ejected, the interface will unmark the ejection link for it to function normally. Due to the extremely low circling percentage (maximum three times of circling for any packet in our simulations), this livelock avoidance scheme has minimal performance impact.

5.3.2 Deadlock avoidance

With no protocol-level dependence at injection/ejection endpoints, routing-induced deadlock is not possible in routerless NoCs as packets arriving at each hop are either ejected or forwarded immediately. Hence, a packet can always reach its destination interface without being blocked by other packets. The above livelock avoidance ensures that the packet can be ejected within a limited number of circlings.

With more than one dependent packet types, the marked ejection link in the above livelock avoidance scheme may not be able to eject the current packet (say a request packet) in the ejection queue, because the associated cache controller cannot accept new packets from the ejection queue (i.e., input of the controller). This may happen when the controller itself is waiting for packets (say a reply packet) in the injection queue (i.e., output of the controller) to be injected into the network. A potential protocol-level deadlock may occur if that reply packet cannot be injected, such as the loop is full of request packets that are waiting to be ejected.

To avoid such protocol-level deadlock, the conventional approach is to have a separate physical or virtual network for each dependent packet type. While similar approach can be used for routerless NoCs, here we propose a less resource demanding solution which is made possible by the circling property of loops. This solution only needs an extra reserved EXB, as well as a separate injection and ejection queue for each dependent packet type. The separate injection/ejection queues can come from duplicating original queues or from splitting the original queues to multiple queues. In either case, the loops and wiring resources are *not* duplicated, which is important to keep the cost low. Following the above livelock avoidance scheme, when a packet p on loop l completes the final circling (counter value of 255) and finds that the marked ejection link is still not available, p is temporarily buffered in the reserved EXB instead of forwarding to output l. Meanwhile, we allow the head packet q in the injection queue of the terminating packet type (e.g., a reply packet in the request-reply example) to inject into loop l through output l. Once q is injected, the cache controller is able to put another reply packet in its output (i.e., the injection queue) which, in turn, allows the controller to accept a new request from its input (i.e., the ejection queue). This creates space in the ejection queue to accept packet p that is previously stored in the reserved EXB. Once p moves to the ejection queue, the EXB is freed. Essentially, the reserved EXB acts as a temporary exchanging space, while the separate injection/ejection queues avoid blocking of different packet types at the endpoints.

5.3.3 Starvation avoidance

The last corner case we address is starvation. With the previous livelock and deadlock handling, if a packet is consumed at its destination RL interface, the interface can use the free output to inject a new packet. However, it is possible that a particular interface X is not the destination of any packets and there is always a flit passing through X every single cycle. This never occurred in any of our experiments as it is practically impossible that a cache bank is not accessed by any other cores. However, it is theoretically possible and, when occurred, prevents X from injecting new packets. We propose the following technique to avoid starvation for the completeness of the routerless NoC design. If X cannot inject a packet after a certain number of clock cycles (a very long period, e.g., hundreds of thousand cycles or long enough to have negligible impact on performance), X piggybacks the next passing head flit f with the ID of X. When f is ejected at its destination interface Y, instead of injecting a new packet, Y injects a single-flit no-payload dummy packet that is destined to X. When the dummy packet arrives at X, X can now inject a new packet by using the free slot created by the dummy packet. This breaks the starvation configuration.

5.4 Interface hardware implementation

Fig. 8 depicts the main components of a RL interface. We have explained the extension buffers (EXBs), single-flit buffers, routing table, and multiple ejection links in the previous sections. The arbitrator receives flits from input buffers and selects up to e input loops for ejection based on the oldest first policy. The arbitrator contains a small register that holds the arbitration results. The link status selector is a simple state machine associated with the loops. It monitors the input loops and arbitration results, and changes the state of the loops (e.g., ejection, stall in extension buffers, etc.) in the state machine. There are several other minor logic blocks that are not shown in Fig. 8 for better clarity. Note that the RL interface does not use the information of neighboring nodes, which differs from most conventional router-based NoCs that need credits or on/off signals for handshaking.

To ensure the correctness of the proposed interface hardware, we implement the design in RTL Verilog that includes all the detailed components. The Verilog implementation is verified in ModelSim, synthesized in Synopsys Design Compiler, and placed and routed using Cadence Encounter tool. We use the latest 15nm process NanoGate FreePDK 15 Cell Library [34] for more accurate evaluation. As a key result, the RL interface is able to operate at up to 4.3 GHz frequency while keeping the packet forwarding process in one clock cycle. This is fast enough to match up with most commercial many-core processors. Injecting packets may take an additional cycle for table look-up. In the main evaluation below, both the interfaces and cores are operating at 2 GHz.

6. Evaluation methodology

We evaluate the proposed routerless NoC (RL) extensively against Mesh, EVC, and IMR in BookSim [35]. For synthetic traffic workloads, we use uniform, transpose, bit reverse, and hotspot (with 8 hotspots nodes). BookSim is warmed up for 10, 000 clock cycles and then collects performance statistics for another 100, 000 cycles at various injection rates. The injection rate starts at 0.005 flit/node/cycle and is incremented by 0.005 flit/node/cycle until the throughput is reached. Moreover, we integrate BookSim with Synfull [36] for performance study of PARSEC [37] and SPLASH-2 [38] benchmarks. Power and area studies are based on Verilog postsynthesis simulations, as described in Section 5.4.

In the synthetic study, each router in Mesh is configured with relatively small buffer resources, having 2 VCs per link and 3 flits per VC. The link width is set to 256-bit. Also, the router is optimized with lookahead routing and speculative switch allocation to reduce pipeline stages to 2 cycles per router and 1 cycle per link. EVC has the same configuration as Mesh except for one extra VC that is required to enable express channels. For IMR, the ring set is generated by the evolutionary approach described in [10]. To allow a fair comparison with RL, the maximum number of overlapping cap, for both RL and IMR, is set to n for $n \times n$ NoC. We also follow the original paper to faithfully implement IMR's network interface.

Each input link in an IMR's interface is attached with a buffer of 5 flits and the link width is set to 128-bit (the same as the original paper). In RL, loops are generated by RLrec algorithm and accordingly the routing table for each node is calculated. Each interface is configured with two ejection links and each input link has a flit-size buffer. Also, an EXB of 5 flits is implemented in each interface. The link width is 128-bit (the same as IMR). In all the designs, packets are categorized into data and control packets where each control packet has 8 bytes and each data packet has 72 bytes. Accordingly, data packets in Mesh, EVC, IMR, and RL are of 3, 3, 5, and 5 flits, respectively, and the control packets are of a single flit.

For benchmark performance study, we also add 2D Mesh with various configurations as well as a 3D Cube design into the comparison. RL has the same configuration as the synthetic study. For 2D Mesh, we use nine configurations, each having the configuration M(x, y) where $x \in \{1, 2, 3\}$ is the router delay and $y \in \{1, 2, 3\}$ is the buffer size, i.e., routers with 1-cycle, 2-cycle, and 3-cycle delay, and with 1-flit, 2-flit, and 3-flit buffer size. 3D Cube is configured with 2 VCs per link, 3 flits per VC, and 2 cycleS per hop latency.

7. Results and analysis

7.1 Ejection links and extension buffers

The proposed RL scheme is flexible to use any number of ejection links and EXBs. On the ejection side, the advantages of having more ejection links are higher chance for packet ejection and lower chance for packet circling in a loop. However, adding more ejection links complicates the design of the interface and leads to additional power and area overhead in the interface and the receiving node. On the injection side, EXBs have a direct effect on the injection latency of long packets. Recall that a loop must be already attached with an EXB or a free EXB is available to be able to inject a long packet. Similar to ejection links, having more EXBs can lower injection latency but incur larger area and power overhead.

We studied the throughput of RL with different configurations of ejection links and EXBs on various synthetic traffic patterns. The NoC size for this study is 8 × 8. The results are shown in Fig. 10. In the figure, each configuration is denoted by (x, y) where x is the number of ejection links and y is the number of EXBs. The basic and best in terms of area and power overhead is (1, 1) configuration, but it has the worst performance. By adding up to three EXBs with a single ejection link, the throughput is only slightly

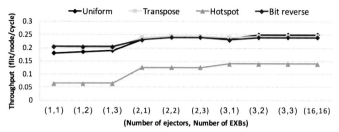

Fig. 10 Throughput of routerless NoC under different number of ejection links and extension buffers (EXBs).

changed (less than 5%). This indicates that the number of EXBs is not very critical to performance, and it is possible to use only one EXB for injecting long packets while saving buffer space.

For (2, 1) configuration, it doubles the chance for packet ejection when compared to (1, x) configurations. The throughput is notably improved by an average of 38% for all the patterns when compared to (1, 1) configurations. For instance, hotspot traffic pattern has 0.125 throughput in (2, 1) configuration but only 0.065 in (1, 1), a 92.5% improvement. However, on top of (2, 1) configuration, adding up-to three EXBs (i.e., (2, 3)) improves throughput only by 5% on average.

Given all the results, we choose the (2, 1) configuration as the best trade-off point, and use it for the remainder of this section. We also plot the (16, 16) configuration which is the ideal case (no blocking in injection or ejection may happen). As can be seen, (2, 1) is very close to the ideal case. Section 7.3 provides a detailed study for the number of times packet circling in loops for the (2, 1) configuration.

7.2 Synthetic workloads

Fig. 11 plots the performance results of four synthetic traffic patterns for an 8 × 8 NoC. RL has the lowest zero-load packet latency in all four traffic patterns. For example, in uniform random, the zero-load packet latency is 21.2, 14.9, 10.5, and 8.3 cycles for Mesh, EVC, IMR, and RL, respectively. When averaged over the four patterns, RL has an improvement of 1.59×, 1.43×, and 1.25× over Mesh, EVC, and IMR, respectively. RL achieves this due to low per hop latency (one cycle) and low hop count.

In terms of throughput, the proposed RL also has advantage over other schemes. For example, the throughput for hotspot is 0.08, 0.05, 0.06, and 0.125 (per flit/node/cycle) for Mesh, EVC, IMR, and RL, respectively.

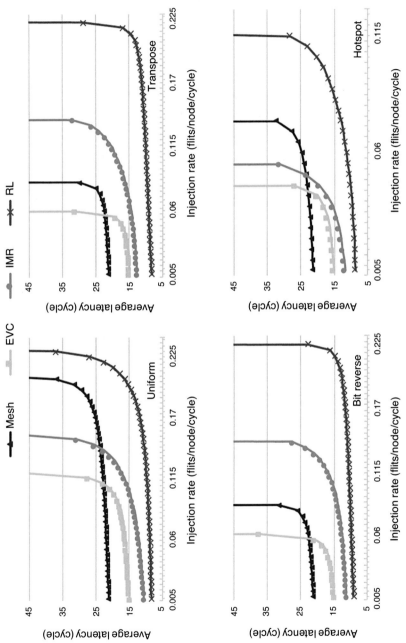

Fig. 11 Performance comparison for synthetic traffic patterns.

In fact, RL has the highest throughput for all the traffic patterns. When averaged over the four patterns, RL improves throughput by 1.73x, 2.70x, and 1.61x over Mesh, EVC, and IMR, respectively. This is mainly owing to the better utilization of wiring resources in RL. Note that EVC has a lower throughput than Mesh as EVC is essentially a scheme that trades off throughput for lower latency at low traffic load.

7.3 PARSEC and SPLASH-2 workloads

We utilize Synfull and Booksim to study the performance of RL, 2D Mesh with different configurations, EVC, IMR, and a 3D Cube under 16 PARSEC and SPLASH-2 benchmarks. The NoC sizes under evaluation are 4 × 4, 8 × 8, and 16 × 16 for RL, 2D Mesh, EVC, and IMR, and 4 × 4 × 4 for 3D cube. Fig. 12 shows the results.

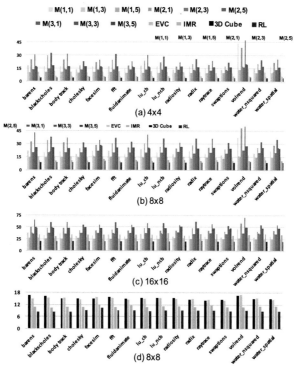

Fig. 12 PARSEC and SPLASH-2 benchmark performance results (y-axis represents average pack latency in cycles). RL is compared with different Mesh configurations, EVC, and IMR in (A), (B), and (C). In (D), RL is also compared with a 3D Cube.

In Fig. 12A–C, RL is compared against 2D Mesh, EVC, and IMR. From the figures, the best configuration for Mesh is M(1,5) (i.e., per hop latency of 1 and buffer size of 5) and the worst is M(3,1). Lowering per hop latency in Mesh helps to improve overall latency, and reducing buffer sizes may cause packets to wait longer for credits and available buffers. The average packet latency of RL in 4 × 4, 8 × 8, and 16 × 16 are 4.3, 8.9, and 20.1 cycles, respectively. This translates into an average latency reduction of RL over M(1,5) by 57.8%, 38.4%, and 22.2% in 4 × 4, 8 × 8, and 16 × 16, respectively. The IMR rings in 16×16 are very long and seriously affect its latency. RL reduces the average latency by 23.3% over EVC and 41.2% over IMR.

In Fig. 12D, the performance of 3D cube is clearly better than all the Mesh configurations in Fig. 12B mainly due to lower hop count and larger bisection bandwidth. Despite this, RL still offers better performance than 3D cube. The average latency of RL is 8.9 cycles, which is 41% lower than the 15.2 cycles of 3D cube.

7.4 Power

Fig. 13 compares the power consumption of Mesh (i.e., M(2,3)), EVC, IMR, and RL for different benchmarks, normalized to the Mesh. All the power consumption shown in this figure are reported after P&R in NanGate FreePDK 15 Cell Library [34] by Cadence Encounter. The activity factors for the power measurement are obtained from Booksim, and the power consumption includes that of all the wires.

The average dynamic power consumption for RL is only 0.26mW, and for Mesh, EVC, and IMR the average is 2.88 mW, 4.27 mW, and 2.91 mW, respectively. Because RL has no crossbar, it requires only 9%, 6.1%, and 8.9% of the dynamic power consumed by Mesh, EVC, and IMR, respectively. Meanwhile, static power is mostly consumed by buffers. Unlike Mesh, EVC, and IMR, RL has a much lower buffer requirement. As a result, RL consumes very low static power of 0.18 mW on average, while Mesh, EVC, and IMR consume 1.39 mW, 1.64 mW, and 0.58mW, respectively. Adding dynamic and static power together, on average, RL reduces the total NoC power consumption by 9.48×, 13.1×, and 7.75× over Mesh, EVC, and IMR, respectively.

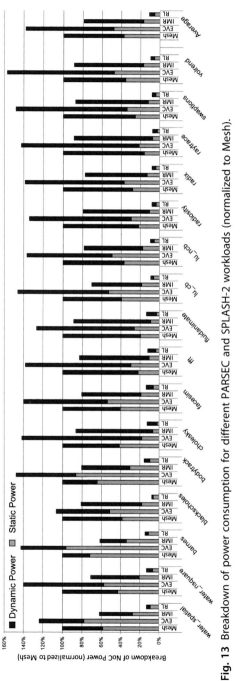

Fig. 13 Breakdown of power consumption for different PARSEC and SPLASH-2 workloads (normalized to Mesh).

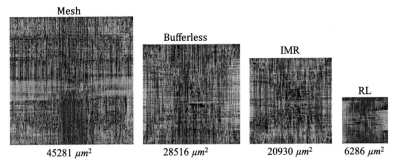

Fig. 14 Area comparison under 15 nm technology.

7.5 Area

Fig. 14 compares the router or interface area of the different schemes we are studying. The results are obtained from Cadence Encounter after P&R.[b] We also add a bufferless design to the comparison. The largest area is $60731\mu m^2$ for EVC (not shown in the figure) followed by 45281 μm^2, 28516 μm^2, 20930 μm^2, and 6286 μm^2 for Mesh, Bufferless, IMR, and RL, respectively. The EXB and ejection link sharing techniques as well as the simplicity of the RL interface are the main contributors for the significant reduction of area overhead. Overall, RL has an area saving of 89.6%, 86.1%, 77.9%, and 69.9% compared with EVC, Mesh, Bufferless,[c] and IMR, respectively.

The wiring area is not included as wires are spread throughout the metal layers and cannot be compared directly. We do acknowledge that IMR and RL use more wiring resources than other designs. RL uses a small percentage of middle metal layers for wires and, as a result, more repeaters are needed. The total area for all the link repeaters is $0.127mm^2$ which is 4.3% of the mesh router area. However, as middle layers are above the logic area, RL is unlikely to increase the chip size.

8. Discussion

8.1 Scalability and regularity

Figs. 11 and 12 already showed the advantage of RL in terms of latency and throughput for large networks. Fig. 15 further compares the average hop count (zero-load hop count) of RL, IMR, and optimal Mesh. As can be

---

[b] Our CAD tools limit P&R for processing cores.
[c] In addition to area reduction, RL also has 2.8× higher throughput (under UR) and 64.3% lower latency than bufferless NoC.

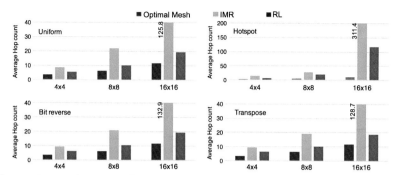

Fig. 15 Average hop count for synthetic workloads.

seen, IMR has very high average hop count because of its lengthy rings. In contrast, the average hop count of RL is only slightly higher than optimal Mesh. Note that RL achieves this low hop count without having the switch capability of conventional routers.

Routerless NoC is not as irregular as it appears in the figures. In our actual design and evaluation, all the RL interfaces use the same design (some ports are left unused if no loops are connected), so the main irregularity is the way that links form loops. One way to quantify the degree of link irregularity is how many different possible lengths of links, which is $n - 1$ for $n \times n$ NoC. This degree is similar to that of Flattened Butterfly [18] and MECS [19].

8.2 Average overlapping

We discussed before that as long as the overlapping cap is met, it is beneficial to approach this cap for as many neighboring node pairs as possible to increase resource utilization and improve performance. Table 3 presents this statistics for RL and IMR. It can be seen that the average overlapping between adjacent nodes in RL is at least 20% more than that of IMR. Also, the longest loop in RL is always shorter than the longest ring in IMR, and the difference increases as the NoC gets bigger. Shorter loops reduce average hop count and offer a lower latency. For example, in 16×16 the longest loop in RL is of 60 nodes, while in IMR it is of 240 nodes.

8.3 Impact on latency distribution

The extension buffer technique and the reduced ejection link technique save buffer resources at the risk of increasing packet latency. Fig. 16 shows distribution of average packet latency, averaged over different benchmarks.

Table 3 Average overlapping and loops/rings in RL/IMR.

Network	Overlap cap	Avg overlap (%) of links	Max loops/ rings in node	Avg loops/ rings (%) in node	Longest loop/ring
RL 4 × 4	4	3.33 (83.3%)	6	5 (62%)	12
IMR 4 × 4	4	2.33 (58.3%)	4	3.5 (43%)	14
RL 8 × 8	8	6 (75%)	14	10.5 (65%)	28
IMR 8 × 8	8	4.71 (58.9%)	10	8.2 (54%)	48
RL 16 × 16	16	11.33 (70.8%)	30	21.2 (66%)	60
IMR 16 × 16	16	8.13 (50.8%)	18	15.2 (47%)	240

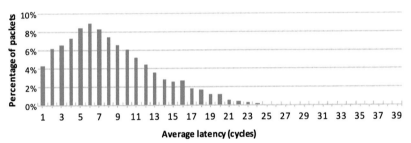

Fig. 16 Latency distribution of benchmarks for RL 8 × 8 NoC.

The RL interface is configured the same as previous sections with one EXB and two ejectors. The takeaway message from the figure is that the two techniques have minimal impact on latency under tight resource allocation. For example, the average packet latency is only 8.3 cycles for RL, and only 0.71% of the packets have latency larger than 20 cycles, with the largest being 39 cycles. The tail in the latency distribution is thin and short.

8.4 RL for $n \times m$ chip

The RL design can be easily extended to any $n \times m$ network sizes. The RL interface design and functionalities remain unchanged. The RLrec algorithm needs to be modified slightly. With rectangular shapes instead of squares, N_L and N_H are not sufficient to denote the four corners of a layer. Two more variables are needed to specify the corners of a layer correctly, for instance, $N_L r$ and $N_H r$ for low and high rows, and $N_L c$ and $N_H c$ for low and high columns. Once a layer is correctly specified, the four groups of loops can be generated in similar fashion. The rotation step is skipped as this is not possible for rectangular networks, but the reversing direction step remains. The overlapping calculation needs to reflect the orientation of the rectangular loops as well.

9. Conclusion

Current and future many-core processors demand highly efficient on-chip networks to connect hundreds or even thousands of processing cores. In this work, we analyze on-chip wiring resources in detail and propose a novel routerless NoC design to remove the costly routers in conventional NoCs while still achieving scalable performance. We also propose an efficient interface hardware implementation and evaluate the proposed scheme extensively. Simulation results show that the proposed routerless NoC design offers significant advantage in latency, throughput, power, and area, compared with other designs. These results demonstrate the viability and potential benefits of the routerless approach and also call for future works that continue to improve various aspects of routerless NoCs such as performance, reliability, and power efficiency.

Acknowledgments

We appreciate the authors of IMR [10] for sharing the source code of generating IMR. We also thank Arash Azizimazreah and Timothy Pinkston for their help and valuable feedback to the work. This research was supported, in part, by the National Science Foundation (NSF) grants #2006571 and ##1750047.

References

[1] I. Cutress, SuperComputing 15: Intel's Knights landing Xeon Phi silicon on display, 2015. november.

[2] S. Williams, J. Shalf, L. Oliker, S. Kamil, P. Husbands, K. Yelick, The potential of the cell processor for scientific computing, in: Computing Frontiers, ACM, 2006.

[3] P. Gratz, C. Kim, R. McDonald, S.W. Keckler, D. Burger, Implementation and evaluation of on-chip network architectures, in: International Conference on Computer Design, IEEE, 2006.

[4] D. Wentzlaff, P. Griffin, H. Hoffmann, L. Bao, B. Edwards, C. Ramey, M. Mattina, C.-C. Miao, J.F. Brown III, A. Agarwal, On-chip interconnection architecture of the tile processor, IEEE Micro 27 (2007) 15–31.

[5] G.S. Delp, D.J. Farber, R.G. Minnich, J.M. Smith, M.C. Tam, Memory as a network abstraction, IEEE Network 5 (4) (1991) 34–41.

[6] L. Barroso, M. Dubois, Cache coherence on a slotted ring, in: ICPP, 1991.

[7] L.A. Barroso, M. Dubois, The performance of cache-coherent ring-based multiprocessors, in: ISCA, 1993.

[8] Y. Hoskote, S. Vangal, A. Singh, N. Borkar, S. Borkar, A 5-GHz mesh interconnect for a teraflops processor, IEEE Micro 27 (2007) 51–61.

[9] J. Howard, S. Dighe, Y. Hoskote, S. Vangal, et al., A 48-core IA-32 processor in 45 nm CMOS using on-die message-passing and DVFS for performance and power scaling, IEEE J. Solid-State Circuits 46 (2011) 173–183.

[10] S. Liu, T. Chen, L. Li, X. Feng, Z. Xu, H. Chen, F. Chong, Y. Chen, IMR: high-performance low-cost multi-ring NoCs, IEEE Trans. Parallel Distrib. Syst. 27 (6) (2016) 1700–1712.

[11] T.W. Ainsworth, T.M. Pinkston, On characterizing performance of the cell broadband engine element interconnect bus, in: International Symposium on Networks-on-Chip (NOCS), 2007.

[12] W.J. Dally, B. Towles, Route packets, not wires: on-chip interconnection networks, in: DAC, 2001.

[13] C.A. Nicopoulos, D. Park, J. Kim, N. Vijaykrishnan, M.S. Yousif, C.R. Das, ViChaR: a dynamic virtual channel regulator for Network-on-Chip routers, in: MICRO, 2006.

[14] A.K. Kodi, A. Sarathy, A. Louri, iDEAL: inter-router dual-function energy and area-efficient links for Network-on-Chip (NoC) architectures, in: ISCA, 2008.

[15] L. Chen, T.M. Pinkston, NoRD: node-router decoupling for effective power-gating of on-chip routers, in: MICRO, 2012.

[16] N.E. Jerger, L.S. Peh, M. Lipasti, Virtual circuit tree multicasting: a case for on-chip hardware multicast support, in: ISCA, 2008.

[17] M.K. Papamichael, J.C. Hoe, The CONNECT Network-on-Chip generator, Computer 48 (12) (2015) 72–79.

[18] J. Kim, W.J. Dally, D. Abts, Flattened butterfly: a cost-efficient topology for high-radix networks, in: ISCA, 2007.

[19] B. Grot, J. Hestness, S.W. Keckler, O. Mutlu, Express cube topologies for on-chip interconnects, in: HPCA, 2009.

[20] C. Fallin, C. Craik, O. Mutlu, CHIPPER: a low-complexity bufferless deflection router, in: HPCA, 2011.

[21] A.N. Udipi, N. Muralimanohar, R. Balasubramonian, Towards scalable, energy-efficient, bus-based on-chip networks, in: HPCA, 2010.

[22] C. Fallin, X. Yu, G. Nazario, O. Mutlu, A high-performance hierarchical ring on-chip interconnect with low-cost routers, tech. rep, 2011.

[23] D. E. Knuth, The Art of Computer Programming, Volume 4A, Section 7.1.4.

[24] C.-H. Jan, U. Bhattacharya, R. Brain, S.-J. Choi, G. Curello, G. Gupta, W. Hafez, M. Jang, M. Kang, K. Komeyli, et al., A 22nm SoC platform technology featuring 3-D tri-gate and high-k/metal gate, optimized for ultra low power, high performance and high density SoC applications, in: Electron Devices Meeting (IEDM), IEEE, 2012.

[25] S. Natarajan, M. Agostinelli, S. Akbar, M. Bost, A. Bowonder, V. Chikarmane, S. Chouksey, A. Dasgupta, K. Fischer, Q. Fu, et al., A 14nm logic technology featuring 2 nd-generation FinFET, air-gapped interconnects, self-aligned double patterning and a 0.0588 μm 2 SRAM cell size, in: IEEE International Electron Devices Meeting, 2014.

[26] http://ark.intel.com/products/95830/intel-xeon-phi-processor-7290-16gb-1_50-ghz-72-core/.

[27] B. Bohnenstiehl, A. Stillmaker, J. Pimentel, T. Andreas, B. Liu, A. Tran, E. Adeagbo, B. Baas, A 5.8 pJ/Op 115 billion ops/sec, to 1.78 trillion ops/sec 32nm 1000-processor array, in: Symposium on VLSI Circuits, 2016.

[28] L.-T. Wang, Y.-W. Chang, K.-T.T. Cheng, Electronic Design Automation: Synthesis, Verification, and Test, Morgan Kaufmann Publishers Inc., 2009.

[29] D. Pamunuwa, J. Oberg, L.R. Zheng, M. Millberg, A. Jantsch, H. Tenhunen, Layout, performance and power trade-offs in mesh-based Network-on-Chip architectures, in: International Conference on Very Large Scale Integration, 2003.

[30] J. Liu, L.R. Zheng, D. Pamunuwa, H. Tenhunen, A global wire planning scheme for network-on-chip, in: International Symposium on Circuits and Systems (ISCAS), 2003.

[31] S. Vangal, J. Howard, G. Ruhl, S. Dighe, H. Wilson, J. Tschanz, D. Finan, P. Iyer, A. Singh, T. Jacob, et al., An 80-tile 1.28 TFLOPS network-on-chip in 65nm CMOS, in: ISSCC, 2007.

[32] D. Harris, N. Weste, CMOS VLSI Design: A Circuits and Systems Perspective, Pearson/Addison-Wesley, 2005.

[33] R. Arunachalam, E. Acar, S.R. Nassif, Optimal shielding/spacing metrics for low power design, in: IEEE Annual Symposium on VLSI, 2003.

[34] NanGate, Inc, NanGate freePDK15 open cell library, URL http://www.nangate.com.

[35] N. Jiang, J. Balfour, D.U. Becker, B. Towles, W.J. Dally, G. Michelogiannakis, J. Kim, A detailed and flexible cycle-accurate network-on-chip simulator, in: ISPASS, IEEE, 2013.

[36] M. Badr, N.E. Jerger, SynFull: synthetic traffic models capturing cache coherent behaviour, in: ISCA, 2014.

[37] B. Christian, Benchmarking modern multiprocessors, Ph.D. thesis, Princeton University 2011, (January).

[38] S.C. Woo, M. Ohara, E. Torrie, J.P. Singh, A. Gupta, The SPLASH-2 Programs: Characterization and Methodological Considerations, in: ISCA, 1995.

About the authors

Fawaz Alazemi received the B.S. degree from Kuwait University and the M.S. degree from Oregon State University. In 2019, he received the Ph.D. degree from Oregon State University. He started as an assistant professor of Computer Science at Kuwait University. His research focuses on Network-on-chips, parallel systems, and interconnection networks. He has served as a reviewer for several IEEE journals. He is a member of the ACM, IEEE.

Lizhong Chen is an Associate Professor in the School of Electrical Engineering and Computer Science at Oregon State University and Director of the System Technology and Architecture Research (STAR) Lab. He received his Ph.D. in Computer Engineering and M.S. in Electrical Engineering from the University of Southern California in 2014 and 2011, respectively, and B.S. in Electrical Engineering from Zhejiang University in 2009. His research interests are in the broad area of computer architecture, machine learning accelerators, interconnection networks, GPUs, and emerging IoT technologies. Dr. Chen is the recipient of National Science Foundation (NSF) CAREER Award, several Best Paper Awards/Nominations at major architecture conferences, and Chu Kochen Award (the highest honor from Zhejiang University). He served as an associate editor of IEEE Transactions on Computers, program committee member in top computer architecture conferences, reviewer for IEEE and ACM journals, and panelist of multiple NSF panels related to computer systems architecture. Dr. Chen is a Senior Member of IEEE and ACM, and an inducted of the IEEE HPCA Hall of Fame.

Bella Bose (S'78–M'80–SM'94–F'95) received the B. E. degree in Electrical Engineering from Madras University, India in 1973; the M. E. degree in Electrical Engineering from the Indian Institute of Science, Bangalore, in 1975; and the M. S. and Ph. D. degrees in Computer Science and Engineering from Southern Methodist University, Dallas, Texas, in 1979 and 1980, respectively. Since 1980, he has been with Oregon State University where he is a Professor in the School of Electrical Engineering and Computer Science. His current research interests include error control codes, fault tolerant computing, parallel processing, and computer networks. He is a fellow of both the ACM and the IEEE.

CHAPTER FIVE

Routing algorithm design for power- and temperature-aware NoCs

Kun-Chih (Jimmy) Chen[a] and Masoumeh Ebrahimi[b]
[a]National Sun Yat-sen University, Kaohsiung, Taiwan
[b]Royal Institute of Technology, Stockholm, Sweden

Contents

Abstract

The Network-on-Chip (NoC) interconnection is a popular way to build up contemporary large-scale multi-processor System-on-Chip (MPSoC) systems. However, due to the high integration density with high operation frequency, the larger power density leads to serious temperature problems. The thermal issue limits the performance and results in higher leakage power and lower system reliability. The thermal and power issues become worsen in the modern 3D stacking NoC structure and become the primary design challenge. In this chapter, we first investigate the correlation between power and temperature in NoC systems and introduce a thermal model for such systems. With this thermal model, we introduce novel routing design methodologies for power- and temperature-aware NoCs by using Game theory and reinforcement learning.

Advances in Computers, Volume 124
ISSN 0065-2458
https://doi.org/10.1016/bs.adcom.2021.11.012

1. Introduction

In traditional System-on-Chip (SoC) designs, *bus* was a standard structure to interconnect different Intellectual Property (IP) units. Nowadays, it is possible to integrate hundreds of processing elements (PE) and billions of gates in a single chip, thanks to the dramatic advances in semiconductor technologies. Communication among a large number of PEs requires scalable, reliable, and high-performance interconnection networks while meeting the power limitations on a chip. Network-on-Chip (NoC) has been introduced as a viable solution to meet these requirements and address the communication demands of advanced SoC designs. The main characteristics of NoCs are scalability, parallelism, and inherent reliability. A NoC architecture is composed of a network of routers, each connected to at least a processing element or a memory block. Fig. 1A and B shows a 2D and 3D NoC, respectively, based on a mesh topology where each PE is connected to a router via a network interface (NI). Data processed by each PE is first packetized in NI and then transferred to another PE through NoC. Depending on the NoC architecture, packets are temporarily stored in the input/output buffers of each router before transmitting to the next router.

By scaling up the NoC size, the average communication latency of packets also increases, leading to higher power consumption and lower performance. A proper routing algorithm design could alleviate the power and latency problem to an extent. However, scaling over a 2D surface imposes significant chip area as well. Considering the aforementioned challenges, the trend is moving toward 3D ICs where the benefits of NoC interconnects and 3D ICs can be combined. In 3D ICs, active silicon layers are vertically stacked, leading to shorter interconnect paths among the cores, and thus

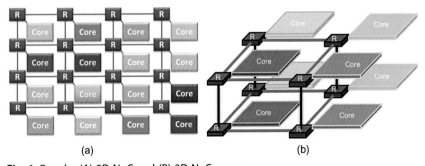

(a)   (b)

Fig. 1 Regular (A) 2D NoC and (B) 3D NoC.

higher performance and lower power consumption. 3D ICs also enable different logic and memory technologies to be integrated into a single chip [1]. Different interconnect technologies, such as micro-bump, wire bonding, and through-silicon via (TSV), can be utilized to facilitate the vertical connections [2]. Among different technologies, TSVs have received much attention due to offering higher density and communication bandwidth [3]. However, 3D NoC design comes with its own challenges. One challenge is the large area occupied by TSVs due to the need for a pad for bonding [4]. Temperature is another issue in 3D designs as it is trapped in middle layers and cannot be easily dissipated. Different architectures and techniques are proposed to address these challenges, such as placing the processing cores closer to the heat sink or taking advantage of flexible routing strategies.

In both 2D and 3D designs, routing algorithms play an important role in reducing communication latency, power consumption, and temperature. Routing algorithms define how packets should be routed in the network. Some of these algorithms are very restrictive, while others are flexible and adaptive. The choice of routing algorithms defines their contributions in reducing power consumption and temperature and improving performance.

This chapter's organization is as follows: In Section 2, we introduce the correlation between several physical phenomena, such as thermal, traffic load, and power. We then propose a proper thermal model for NoC systems. In Section 3, we show a novel routing design methodology by employing the Game theory to properly distribute packets among downstream NoC nodes. Finally, we show the roadmap to design a flexible routing algorithm that could be dynamically adjusted depending on the network condition in order to manage power and temperature.

2. Fundamental of thermal- and traffic-aware NoC

2.1 Correlation between thermal energy and traffic in a NoC system

Heat is a physical phenomenon, which is defined as the transfer of thermal energy across a well-defined boundary around a thermodynamic system. The heat transfer can be classified into conduction, convection, and radiation. The behavior and characteristics of these methods have been described by various physical laws. Regarding the real systems, they often exhibit a complicated combination of different heat transfer methods.

An important concept in thermodynamics is the thermodynamic system. A thermodynamics system is defined as that portion of the universe's mass

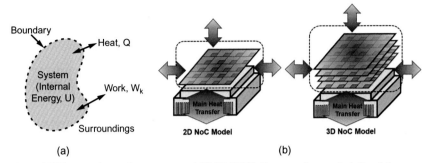

Fig. 2 (A) Thermodynamics system and (B) 2D/3D NoC system is a well-defined thermodynamics system. ©2018 IEEE. Reprinted with permission from K. Chen, Game-based thermal-delay-aware adaptive routing (GTDAR) for temperature-aware 3D Network-on-Chip systems, IEEE Trans. Parallel Distrib. Syst. 29 (9) (2018) 2018–2032, https:/doi.org/10.1109/TPDS.2018.2812164.

chosen for a thermodynamics analysis. The system is separated from the surroundings by a boundary, which encloses the whole characteristics of the given system, as shown in Fig. 2A. Therefore, the thermodynamic system has a fixed mass, and all matter is either in the system or in the surroundings. Exchanges of work or heat between the system and the surroundings take place across this boundary. Anything that passes across the boundary that affects a change in the internal energy needs to be accounted for in the energy balance equation. In thermodynamics, the objective is to determine energy (or heat) transfers and the change in the system's state. This chapter focuses on the thermal behavior in the 2D/3D NoC system, as shown in Fig. 2B. In practice, the boundary of a NoC system is defined as the package. Besides, there is no mass transfer between the system and the surroundings by crossing the boundary of the NoC system (i.e., the package). Hence, the NoC system is mainly solid and fits the definition of a thermodynamics system.

The correlation analysis between the thermal and traffic in a NoC system can be done macroscopically with the constant-volume constant-pressure case of the work process. According to the first law of thermodynamics, the thermal energy is the portion of the internal energy, U, that is responsible for a system temperature. The sources of thermal energy are packet switching at each NoC router, packet delivery between routers, and data processing at each processing element. The thermal energy is usually seen to be the context of the ideal gas, which is well approximated by a monatomic gas at low pressure. The ideal gas is a gas of particles, which are seen as the points of object with perfect spherical symmetry. In other words, each ideal gas particle will interact only by elastic collisions and fill a volume. Therefore, their mean free path between collisions is much larger than their diameter. The mechanical kinetic energy of a single particle is:

$$E_k = \frac{1}{2}mv^2, \tag{1}$$

where m is the particle's mass and v is its velocity. The thermal energy of the gas sample consisting of N atoms is given by the sum of these energies. In an ideal situation (i.e., no energy losses to the container or the environment), the internal energy, U, of an entire thermodynamics system can be described as:

$$U = \frac{1}{2}Nm\bar{v} = \frac{3}{2}Nk_bT = U_{thermal}, \tag{2}$$

where the $\bar{v}$ indicates the average velocity of N atoms. The total thermal energy, $U_{thermal}$, of the sample is proportional to the macroscopic temperature by a constant factor accounting for the three translational degrees of freedom of each particle and the Boltzmann constant k_b, which converts units between the microscopic model and the macroscopic temperature T. This formalism is the basic assumption that directly yields the ideal gas law. Obviously, Eq. (2) shows that for the ideal gas, the internal energy consists of only its thermal energy is:

$$U = U_{thermal} \tag{3}$$

Therefore, the traffic behavior in the NoC system directly affects the thermal behavior in that system.

2.2 Fourier's law and heat conduction

Various mathematical methods have been developed to approximate the results of heat transfer in systems. According to the first law of thermodynamics, a closed system takes the following form on a rate basis:

$$Q = W_k + \frac{dU}{dt}, \tag{4}$$

where Q represents the positive heat transfer rate toward the system and W_k denotes the positive work transfer rate away from the system. Q and W_k may be expressed in joules per second (J/s) or watts (W). For a thermal-aware NoC system, Q is from an electric power used to switch circuits, and then

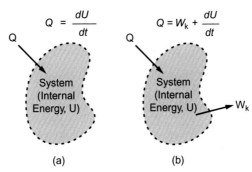

(a)   (b)

Fig. 3 The first law of thermodynamic for (A) the system without heat conduction and (B) the system with heat conduction.

it becomes heat. W_k is the heat conducted through the heat sink and package interfaces toward surroundings. $\frac{dU}{dt}$ denotes the rate of change of internal thermal energy, U, with time, t. For a system without work transfer, the W_k becomes zero, and the total internal energy U equals the Q, as shown in Fig. 3A. Hence, Eq. (3) is a special case as there is no heat conduction. In the general case of thermal-aware NoC system, the main heat transfer is through the heat sink to the surroundings, and the others go through the less thermal-conductive package interfaces, as shown in Fig. 2B. The interaction between heat transfer and system's energy is sketched schematically in Fig. 3B. In this section, we will focus on the W_k (also called heat conduction).

Heat conduction is the mechanism of heat flow in which energy is transported from the region of high temperature to the region of low temperature, as required by the second law of thermodynamics. When an object is at a different temperature from another body or its surroundings, heat flows so that the body and the surroundings reach the same temperature, at which point they are in thermal equilibrium.

The law of heat conduction is also called Fourier's law. Fourier's law states that the rate of heat flow by conduction is proportional to the negative gradient of the temperature and to the area. For many simple applications, we can apply the one-dimensional heat conduction property to estimate the thermal behavior in the system, as shown in Fig. 4. Therefore, according to Fourier's law, the heat flow by conduction in a given one-dimensional direction (i.e., the x direction) is proportional to:

• the gradient of temperature in that direction, dT/dx,
• the area normal to the direction of heat flow, A.

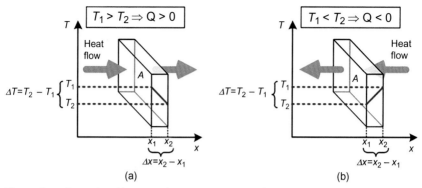

Fig. 4 One-dimensional heat conduction: (A) the heat flow in the positive direction and (B) the heat flow in the negative direction.

Then, the heat flow in the x direction can be formulated as:

$$Q = -kA\frac{dT}{dx} = -kA\frac{\Delta T}{\Delta x}, \tag{5}$$

which is defined as the rate of heat flow, Q, in the positive x direction through a surface of area A. Please note that $\frac{\Delta T}{\Delta x}$ is the gradient of temperature in the x direction, and k is the thermal conductivity which is a property of the material.

The minus sign in Eq. (5) depends on the direction of the heat flow, as illustrated in Fig. 4. If the temperature decreases in the positive x direction, dT/dx is negative and Q becomes a positive quantity (Fig. 4A). Otherwise, if the temperature increases in the positive x direction, dT/dx is positive and Q becomes negative and the heat flow is in the negative x direction, as shown in Fig. 4B. As mentioned in Ref. [5], the primary heat conduction path from silicon to heat spreader and heat sink accounts for about 90% of heat transfer with forced-air cooling. Therefore, in this work, we apply the property of one-dimensional heat conduction to analyze the heat flow from the upper layer die of a NoC system to surrounding through the heat sink (i.e., the arrow of main heat transfer of Fig. 2B).

2.3 Electrical analogy

The most usual approach to analyze the one-dimensional heat-transfer phenomena is to map the target thermodynamics system into the corresponding thermal circuit. It means that we can use "voltage" to model the temperature difference and use the "current" to describe the heat flow. There exists a well-known duality between heat transfer and electrical phenomena. As

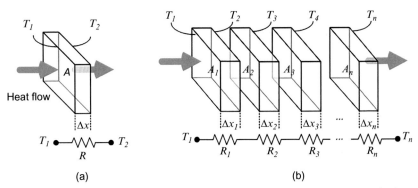

Fig. 5 Temperature distribution and equivalent thermal resistance concept for heat flow through (A) a slab and (B) multilayer partition.

shown in Fig. 5A, the temperature difference is resulted from the heat flow pass against a resistance, which fits the well-known Ohm's law, and the rate of heat flow can be described as:

$$Q = \frac{Overall\ temperature\ difference}{Thermal\ resistance} = \frac{\Delta T}{R}. \tag{6}$$

According to Eq. (5), the thermal resistance is:

$$R = \frac{\Delta x}{kA}. \tag{7}$$

Regarding the 3D stacking structure, such as 3D NoC, the thermal resistance is additive when several conducting layers lie between the hot and cool regions. In a multilayer partition with n series slabs, as shown in Fig. 5B, the total conductance is related to the conductance of its layers by:

$$R = R_1 + R_2 + R_3 + \cdots + R_n. \tag{8}$$

With Eqs. (7) and (8), Eq. (5) can be derived as:

$$
\begin{aligned}
Q &= -\frac{\Delta T}{\dfrac{\Delta x_1}{A_1 K_1} + \dfrac{\Delta x_2}{A_2 K_2} + \dfrac{\Delta x_3}{A_3 K_3} + \cdots + \dfrac{\Delta x_n}{A_n K_n}} \\
&= -\frac{T_n - T_1}{\dfrac{\Delta x_1}{A_1 K_1} + \dfrac{\Delta x_2}{A_2 K_2} + \dfrac{\Delta x_3}{A_3 K_3} + \cdots + \dfrac{\Delta x_n}{A_n K_n}}
\end{aligned} \tag{9}
$$

Therefore, the rules for combining resistance (in series and in parallel) are the same for both heat flow and electric current.

Thermal capacitance C is also necessary for modeling transient thermal behavior, which is to capture the time delay before the current temperature reaching the steady-state temperature. Lumped values of thermal R and C can be computed to represent the heat flow among regions of a chip. Besides, the thermal R and C lead to exponential rise and fall time, which can be characterized by the thermal RC time constants (it is analogous to the electrical RC time constants). The rationale behind this duality is that current flow and heat flow are both described by the same differential equations for a potential difference.

2.4 The proposed thermal model

In the thermal-design community, these equivalent circuits are called thermal models and dynamic thermal models if they include thermal capacitors. This duality provides a convenient basis for an architecture-level thermal model. For example, Fig. 6A shows the structure of the conventional 2D NoC system. As mentioned before, the heat flow within a package can

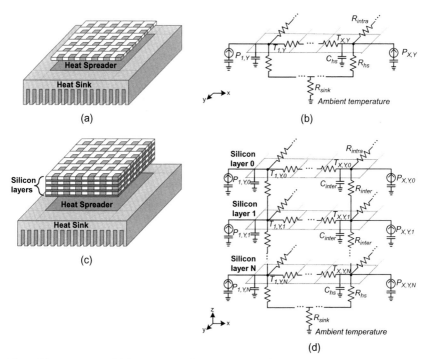

Fig. 6 The structure of (A) 2D NoC and (C) 3D NoC, and the corresponding (B) thermal model of a 2D NoC system and (D) thermal model of a 3D NoC system.

be modeled by an electrical resistor-capacitor (RC) circuit. Therefore, we can build a Thermal RC model with a heat sink, as shown in Fig. 6B. At the location of (x,y) on the NoC system, the corresponding temperature and power consumption are represented by $T_{x,\,y}$ and $P_{x,\,y}$, respectively. R_{intra} represents the thermal resistance in the horizontal silicon layer. Regarding the heat sink, we can use R_{hs} and C_{hs} to represent the thermal resistance and thermal capacitance of the involved heat sink, respectively. Because the heat sink is usually bounded on one side of the NoC package, we use the R_{sink} as the convective resistance of the heat sink to model the heat transfer between the heat sink and ambiance. Regarding the 3D NoC, there are multiple stacked silicon layers, as shown in Fig. 6C. By following the analogous electrical resistor-capacitor (RC) network, Fig. 6D is the corresponding Thermal RC model of a 3D NoC system with a heat sink. Because of the additional vertical dimension in 3D NoC system, the $T_{x,\,y,\,z}$ and $P_{x,\,y,\,z}$ mean the corresponding temperature and power consumption of the node at the location of (x,y,z), respectively. Moreover, the additional R_{inter} and C_{inter} represent the thermal resistance and thermal capacitance between vertical silicon layers, respectively.

The behavior of the transient temperature of each object follows Fourier's Law [6]. According to a NoC node, the temperature behavior in a time unit, depending on the sensing period of the involved thermal sensor, can be modeled as [6]:

$$\frac{dT(t)}{dt} = \frac{P(t)}{C} - \frac{T(t)}{RC}, \tag{10}$$

where $T(t)$ and $P(t)$ are the temperature and total power consumption of the NoC node at time t; R and C are the effective thermal resistance and thermal capacitance toward the ambiance, respectively. To leverage the model analysis, Eq. (10) can be rewritten as:

$$\frac{dT(t)}{dt} = a \cdot P(t) - b \cdot T(t), \tag{11}$$

where a (i.e., it is equal to $\frac{1}{C}$) and b (i.e., it is equal to $\frac{1}{RC}$) are the physic constants and depend on the material of the given technology. Afterward, we set the initial temperature at t_0 to T_0 (i.e., $T(t_0) = T_0$). By using the boundary condition, the linear differential Eq. (10) can be solved as:

$$T(t) = \int_{\tau=t_0}^{\tau=t} aP(\tau)e^{-b(t-\tau)}d\tau + T_0 \cdot e^{-b(t-\tau)}. \tag{12}$$

We assume the involved Dynamic Thermal Management (DTM) is performed every $(t - t_0)$, which is called as the DTM period. In a practical way, each NoC node usually retains its working activity in a DTM period. Hence, each node consumes P, the total power consumption, during the DTM period. In other words, $P(t)$ in Eq. (12) equals P in the time period $(t - t_0)$. Because the temperature of every object eventually approaches the steady-state temperature (i.e., $T(t) = T_{ss}$, when $t = \infty$), Eq. (12) can be rewritten as:

$$T(t) = T_{ss} - (T_{ss} - T_0) \cdot e^{-bt}. \tag{13}$$

3. Thermal-aware proactive routing algorithms in 3D NoCs

For the 2D NoC structure, the system temperature can be balanced by balancing the traffic distribution because of the identical cooling efficiency everywhere on the NoC system. However, the varying cooling efficiency in the vertical direction of 3D NoC reveals the unexpected shortage of the traditional power-aware routing design mechanisms, originally designed for 2D systems. In this section, we will introduce the issues of traffic- and thermal-aware routing algorithm design for 3D NoC system and provide some techniques to solve these design challenges. In Section 4, we will further discuss different power-aware routing algorithms to balance the traffic load to reduce the transmission power during the packet delivery.

3.1 Design issues of traffic- and thermal-aware routing

To make the 3D NoC system thermal safety, it is a popular way to involve the dynamic thermal managements (DTMs). In this direction, there are many kinds of DTMs proposed in recent years. In general, DTMs can be classified into reactive DTM (RDTM) and proactive DTM (PDTM) [7]. The RDTM is triggered to control the system temperature, while the operation temperature of the NoC system achieves the alarming level [8]. On the other hand, PDTM could early control the temperature of the potential thermal-emergent computing units based on the information of temperature prediction [9], such as Eq. (13).

According to the temperature control strategies involved in DTMs, they can be further classified into temporal DTMs and spatial DTMs [9, 10]. The temporal DTM approaches aim to control the computing speed of those thermal-emergent NoC nodes [8, 11]. The advantage of the temporal DTM is that the temperature can be controlled within short cooling time. However, the dramatic performance degradation is usually happened in the 3D NoC system with temporal DTM mechanism because of the slower processing speed of the thermal-emergent computing units. On the other hand, the spatial DTM approaches employ the power migration to regulate the system temperature of the 3D NoC system [10, 12, 13]. Through the specific routing algorithms, the spatial DTM aims to detour the packets away from the thermal-emergent NoC nodes. In this way, it is not necessary to downgrade the processing speed of those thermal-emergent NoC nodes. As a result, it mitigates the impact on the performance during the temperature regulation period. Nevertheless, due to the heterogeneous thermal conductance among interlayer and intralayer of the 3D NoC systems, the thermal conduction of each 3D NoC layer is different, leading to longer cooling time as compared with the temporal DTM approaches.

Based on the aforementioned analysis, temporal and spatial DTM approaches both have some negative influence on the system performance because of the scenario of heterogeneous temperature and traffic behavior in 3D NoC. Consequently, traffic and temperature behavior information should be considered simultaneously during the NoC operation to combine the advantages of the temporal and spatial DTM approaches. To achieve this goals, many thermal- and traffic-aware packet routing techniques were proposed in recent years [14–17]. Generally speaking, these routing techniques deliver the packets away from those thermal-controlled NoC nodes based on a certain real-time information on the NoC system. Among them, [14, 15] deliver the packets by referring to the traffic information. Although the traffic distribution becomes more balanced, the temperature distribution will become unbalanced in 3D silicon structure, which may worsen the heat problem, as shown in Fig. 7A, because of the heterogeneous thermal conduction environment in 3D NoC system. On the contrary, some methods [16, 17] only consider the temperature information to deliver the packets. Although the temperature distribution of the system is balanced, it is easy to cause heavy traffic congestion problem, as shown in Fig. 7B. The reason is that the changing rates of the temperature and traffic are different (i.e., temperature changing rate is much slower than the traffic changing rate).

Obviously, it is not efficient to manage the traffic load and temperature distribution by only considering traffic information or temperature information. Table 1 shows different situations, considering various combinations of temperature and traffic information. The traditional traffic- and thermal-aware adaptive routing approaches assume that the heavy traffic load infers to the high-temperature situation, while a low-temperature situation is

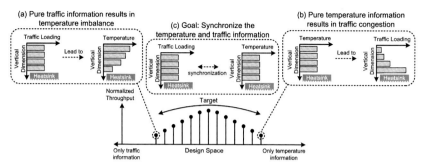

Fig. 7 Design space of traffic- and thermal-aware routing algorithms. ©2018 IEEE. *Reprinted with permission from K. Chen, Game-based thermal-delay-aware adaptive routing (GTDAR) for temperature-aware 3D Network-on-Chip systems, IEEE Trans. Parallel Distrib. Syst. 29 (9) (2018) 2018–2032, https://doi.org/10.1109/TPDS.2018.2812164.*

Table 1 Full case of the relationship between temperature and traffic load information.

		Temperature information	
		High	**Low**
Traffic Load	Heavy	Case 1: Consecutively heavy traffic load	**Case 2: Traffic block**
	Light	**Case 3: Highly historical traffic load**	Case 4: Small traffic load

Source: ©2018 IEEE. Reprinted with permission from K. Chen, Game-based thermal-delay-aware adaptive routing (GTDAR) for temperature-aware 3D Network-on-Chip systems, IEEE Trans. Parallel Distrib. Syst. 29 (9) (2018) 2018–2032, https://doi.org/10.1109/TPDS.2018.2812164.

caused by light traffic load (i.e., Case 1 and Case 4 in Table 1). However, temperature is a phenomenon of a long-term traffic situation. In other words, the changing rate of temperature situation and traffic load is much different (i.e., the temperature changing rate is slow and the traffic changing rate is fast). Therefore, the conventional approaches do not consider the situation of Case 2 and Case 3 in Table 1, and these two cases are the design issues of the current traffic- and thermal-aware routing approaches. Furthermore, the design goal of this field is to synchronize the temperature and traffic information during the thermal control period, as shown in Fig. 7C, which helps to encounter Case 2 and Case 3 in Table 1.

3.2 Routing algorithm design with proactive thermal management in 3D NoC

3.2.1 Introduction of thermal-delay metric

To synchronize the changing rate of temperature and traffic information, authors in Ref. [18] propose to worsen the traffic load around the hot region,

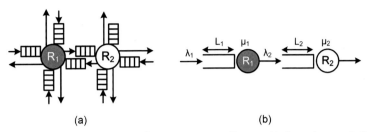

(a)   (b)

Fig. 8 (A) The interconnection between two adjacent NoC nodes and (B) its corresponding queueing system.

which helps to transfer the Case 3 to the Case 1 in Table 1. In this way, the hot NoC nodes will be blocked eventually due to the much heavy traffic load (i.e., Case 2 in Table 1). Therefore, the temperature can be degraded due to blocking switching activity. To prove the situation of Case 2, the Queueing theory is applied to well analyze the interaction between two adjacent NoC nodes in Ref. [18]. We use Fig. 8A as an example to demonstrate this queueing analysis. In this example, the NoC nodes R_1 and R_2 are adjacent and R_1 is a hot node. Fig. 8B shows the effective cascaded queueing system of Fig. 8A, and we assume the arrival rate of the two nodes are λ_1 and λ_2; the service rate of the two nodes are μ_1 and μ_2; the total input buffer length of R_1 is L_1; and the length of the R_2's input buffer toward the R_1 is L_2. Regarding the R_2 subqueueing system, the full probability of the input buffer of R_2, $P_B^{R_2}$, is

$$P_B^{R_2} = \frac{1 - \rho_{R_2}}{1 - \rho_{R_2}^{L_2+1}} \times \rho_{R_2}^{L_2}, \tag{14}$$

where

$$\rho_{R_2}^{L_2} = \frac{\lambda_2}{\mu_2}. \tag{15}$$

To simplify the problem, we adopt the M/M/1/K queueing model and assume the service rate (μ_2) and arrival rate (λ_2) of R_2 are both Poisson distribution. Consequently, the ρ_{R_2} is less than or equal to 1 (i.e., $\rho_{R_2} \leq 1$) for a stable system. According to the involved cascaded M/M/1/K queueing model, the service rate of the upstream node depends on the full probability of the input buffer in the downstream node. Hence, the effective service rate of R_1 can be approximated by:

$$E(\mu_1) = (1 - P_B^{R_2})\mu_1. \tag{16}$$

Obviously, R_1 will be blocked as long as the input buffer of R_2 toward the R_1 is full. Besides, the $E(\mu_1)$ in Eq. (16) is less than μ_1 (i.e., the service rate of R1 is reduced) because $(1 - P_B^{R_2})$ is less than one. With Fourier's Law in Eq. (10), the lower service rate of a NoC node leads to lower power consumption (i.e., smaller $P(t)$ in Eq. (10), which further degrades the temperature of the NoC node. Consequently, Case 2 of Table 1 is proven. In other words, it is a workable way to downgrade the system temperature by creating a situation of a traffic block.

As shown in Eq. (16), the blocking probability of switching activity in R_1 depends on the full probability of the input buffer of R_2. With larger $P_B^{R_2}$, the effective service rate of the NoC node will be degraded, thereby the temperature will be degraded. Regarding the given example in Fig. 8, by following Eqs. (14) and (15), we can observe that it is necessary to reduce the output buffer length (L_2) of the hot R_1 node. Therefore, in addition to the switching activity (i.e., service rate), this is the other way to adjust the input buffer length of the hot NoC node to control the temperature, as shown in Fig. 9. In this way, we can transfer Case 3 in Table 1 to Case 2 in Table 1, which increases the efficiency of the involved DTM mechanism.

To provide a guideline of buffer length adjustment, we apply the Poisson Arrival See Time Average (PASTA) property in Queueing theory. We define the average waiting time, W_{avg}, as the total processing time for the total amount of data in the current input buffer, if the current packet is the last one for this queueing system. In the example of Fig. 8B, the congestion situation of the hot R_1 node's input buffers depends on the average waiting time for the coming packet. Obviously, the longer L_1 leads to longer W_{avg} based on the Little's formula in Queueing theory:

$$W_{avg} = \frac{L_1}{E(\mu_1)}. \tag{17}$$

Based on the previous queueing analysis, Corollary 1 can be conducted. With Corollary 1, the thermal problem will be transferred to the traffic problem, which leverages the traffic-aware routing to solve the temperature problem.

Corollary 1. *For a hot node, to synchronize the changing rate of temperature and traffic information, it is necessary to decrease the output buffer length and increase the input buffer length of this hot node.*

By using the aforementioned buffer length adjustment, we can transfer the information of the current temperature to the real-time traffic load

Baseline 3D NoC Router

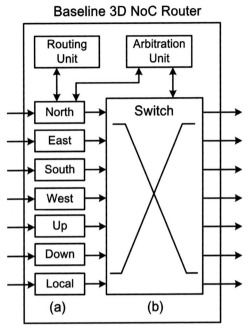

Fig. 9 To worsen the traffic load in a hot 3D NoC router, we can (A) congest the input buffers and (B) reduce the switching activity. *©2018 IEEE. Reprinted with permission from K. Chen, Game-based thermal-delay-aware adaptive routing (GTDAR) for temperature-aware 3D Network-on-Chip systems, IEEE Trans. Parallel Distrib. Syst. 29 (9) (2018) 2018–2032, https://doi.org/10.1109/TPDS.2018.2812164.*

information (i.e., hotter node leads to larger W_{avg} in Eq. (17). To further consider the information of the future temperature as the reference information for the packet delivery, a novel metric, called Thermal Delay (TD), can be introduced by:

$$TD = \frac{1}{W_{avg}} \times T(t + \Delta t), \qquad (18)$$

where the W_{avg} is the average waiting time and the $T(t + \Delta t)$ is the predictive temperature at time $(t + \Delta t)$. With the temperature prediction model, proposed by Chen et al. [9], the $T(t + \Delta t)$ can be described by:

$$T(t + \Delta t) = T(t) + (T(t) - T(t - \Delta t)) \cdot e^{-b\Delta t}. \qquad (19)$$

Because the average waiting time is positive proportional to the total input buffer length L (i.e., Eq. (17)), Eq. (18) can be reformulated to

$$TD = \frac{1}{L} \times T(t + \Delta t). \tag{20}$$

With this TD metric in Eq. (20), the real-time information of the temperature stress and performance status in each NoC node can be addressed. The first term of Eq. (20) is used to calibrate the influence of the temperature stress to the NoC node. For example, the temperature stress of a hot node with long input buffer length may be degraded by using Corollary 1 because of the blocked traffic situation in this node (i.e., the Case 2 in Table 1).

3.2.2 Analysis of thermal-delay information

By viewing the discipline of the thermal-delay metric in Eq. (20), the input buffer length (L) of each NoC node should be adjusted based on the current temperature stress. Based on Corollary 1, it has the higher priority to adjust the length of the input and output buffers for the NoC node with higher TD value. To mitigate the performance impact, bringing from the buffer length adjustment, Chen [18] proposed a router architecture, which supports the dynamic buffer length arrangement. Fig. 10 illustrates an example. Obviously, the length of the hot node's input buffers will be increased to make an environment of traffic congestion, which is used to synchronize the information of traffic and temperature.

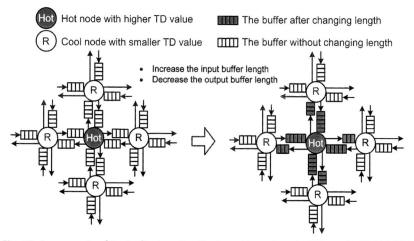

Fig. 10 An example of the buffer length adjustment based on the TD metric. ©2018 IEEE. Reprinted with permission from K. Chen, Game-based thermal-delay-aware adaptive routing (GTDAR) for temperature-aware 3D Network-on-Chip systems, IEEE Trans. Parallel Distrib. Syst. 29 (9) (2018) 2018–2032, https://doi.org/10.1109/TPDS.2018.2812164.

With the different length of each NoC node's input buffer, the situation of heavy traffic congestion will be usually happened at the hot nodes. Consequently, it is necessary to analyze the Congestion Level (*CL*) of the downstream NoC nodes before the packet delivery. As mentioned in Ref. [19], the congestion level of an NoC node can be classified into Buffer Transfer Delay (*BTD*) and Output Contention Delay (*OCD*). The *BTD* of the input buffer on the *i* side represents the waiting time before switching the input packet, which can be described as:

$$BTD(i) = BL_{DR}(i) \times \mu_{DR}, \tag{21}$$

where the $BL_{DR}(i)$ is the numbers of currently used input buffer slots on the *i* side of the downstream NoC node and the μ_{DR} means the service rate of the downstream NoC node, which is influenced by the involved DTM mechanism. By following the Corollary 1, the input buffer length of each NoC node may be adjusted after comparing the temperature information with each other adjacent NoC nodes. Consequently, the *BTD* will be increased as long as the NoC node's input buffer is full or the switching activity is low. Therefore, the *BTD* information of the downstream NoC node can leverage the efficient routing path selection in the involved congestion-aware adaptive routing algorithm. It means that the packets will be delivered to the noncongested region, which is also the cooler region.

In addition to the *BTD*, the *OCD* is the other metric, which may affect the effective service rate for the input packet in the downstream NoC node. Because of the contention issue, some packets may be blocked in the input buffer until they obtain the grant from the switch allocator. Hence, the *OCD* of the input buffer on the *i* side of the downstream NoC node relies on contention probability between each input buffer of the same router, which can be expressed as:

$$OCD(i) = \sum_{\substack{o,\, j \in Ch_{dir} \\ j \neq i,\, o \neq i}}^{N_{ch}} c_{jo} BTD(j), \tag{22}$$

where the Ch_{dir} is the direction of the input buffer. Besides, the N_{ch} is defined as the number of competing input buffers. The packets in the competing input buffers have the identical output direction. Furthermore, the $BTD(j)$ is the expected buffer transfer delay of the other competing input buffers on the *j* side. The c_{jo} is the contention probability between the packets, toward the identical output direction *o*, in the current input buffer on the i side and the other competing input buffers on the j side. Because the packet arrival rate is usually a Poisson distribution and the round-robin policy is usually adopted for packet switching, the c_{jo} can be approximated by:

$$c_{jo} \approx \frac{1}{N_{ch}}. \tag{23}$$

With the BTD and OCD, the Congestion Level (CL) of the downstream NoC node can be estimated. The CL is defined as the traffic status of the path from the input buffer on the i side to the output direction o, and it can be expressed as:

$$CL_{io} = r_{io} \times OCD(i), \tag{24}$$

where the r_{io} is the selection ratio of the routing path from the input direction i to the output direction o. For the 3D NoC structure, we only need to consider at most three candidates of routing directions toward the destination by using the minimal adaptive routing method. In other words, it is necessary to check two lateral routing paths selection and one vertical routing path selection, as shown in Fig. 11. Here, we assume the packet arrival rate is Poisson distribution, thereby the r_{io} is approximated to 1/3 in steady state of the system. By considering Eqs. (22), (23), and (24) together, the CL can be computed by:

$$CL_{io} \approx \frac{1}{3} \times \frac{1}{N_{ch}} \times \sum_{\substack{i \in Ch_{dir} \\ j \neq i}}^{N_{ch}} BTD(j), \tag{25}$$

where j is the competing input buffer direction and the N_{ch} is the number of competing input buffer directions.

3.2.3 Game-based routing algorithm design
With the CL information of each downstream NoC node, it is necessary to determine the ratio to distribute the packets to each candidate of output directions in the downstream NoC nodes. In Ref. [18], Chen proposed a novel way to apply the Nash Equilibrium property in Game Theory and introduce a novel routing path selection method. We adopt Fig. 11 as an example to analyze the possible routing direction between the routers in Layer N and Layer ($N+1$) of 3D NoC. By using the minimal adaptive routing method in the 3D NoC system, we need to consider not only the lateral routing direction but also the vertical routing direction (i.e., Up and Down).

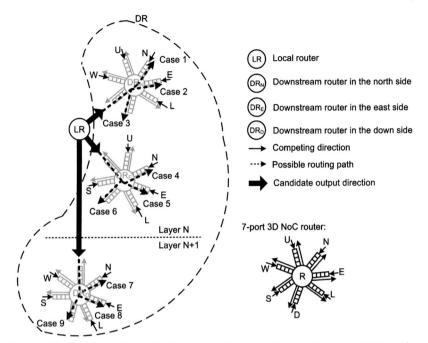

Fig. 11 The game model of the interaction between three adjacent 3D NoC nodes. ©2018 IEEE. Reprinted with permission from K. Chen, Game-based thermal-delay-aware adaptive routing (GTDAR) for temperature-aware 3D Network-on-Chip systems, IEEE Trans. Parallel Distrib. Syst. 29 (9) (2018) 2018–2032, https://doi.org/10.1109/TPDS.2018. 2812164.

As mentioned in Ref. [8], the up-then-lateral turns should be prohibited in the vertical turn model for the deadlock freedom. Regarding the example in Fig. 11, we only consider the at most three candidates of routing directions toward the destination (i.e., North, East, and Down). To simplify the involved game model, we aggregate all the downstream NoC nodes in the north, east, and down side (i.e., D_{RN}, D_{RE}, and D_{RD}) into the DR group. In this way, we can model the routing problem as a 2-player game model and the players' name are LR and DR. Because this routing problem is a 2-player static game with mixed strategies (i.e., each rational player has a probability to decide an action), the Nash Equilibrium property, defined in Definition 1, is a proper way to find the proportion of each decision selection.

Definition 1. A pair of strategies $(\sigma_{LR}^{*}, \sigma_{DR}^{*})$ is a Nash Equilibrium for each player (i.e., LR and DR) such that

$$\pi_{LR}\left(\sigma_{LR}^{*}, \sigma_{DR}^{*}\right) \geq \pi_{LR}\left(\sigma_{LR}, \sigma_{DR}^{*}\right) \quad \forall \sigma_{LR} \in \Sigma_{LR} \tag{26}$$

and

$$\pi_{DR}(\sigma_{LR}^*, \sigma_{DR}^*) \geq \pi_{DR}(\sigma_{LR}^*, \sigma_{DR}) \quad \forall \sigma_{DR} \in \Sigma_{DR}, \tag{27}$$

where the π_i is the payoffs to player i under the mixed strategy σ_j of each player. Besides, the set of all possible mixed strategies for player k is denoted by Σ_k. Note that the i, j, and k can be LR or DR.

To clarify the game in Fig. 11, Table 2 is the strategy form, showing the payoff of each player under each mixed strategy. The Case 1 in Table 2 indicates a scenario that a packet will be delivered to North side by LR and be delivered to North side by DR sequentially. Regarding other cases in Table 2 (i.e., Case 2 to Case 9), the packet delivery path can be traced by the same way. Note that the payoff of player LR is zero because the LR is the packet launcher to the player DR and thereby without any payoff. On the other hand, the DR is a coalition player including D_{RN}, D_{RE}, and D_{RD}, and the payoff of DR depends on the different packet delivery path (i.e., D_1 to D_9).

As mentioned before, the Nash Equilibrium property is applied to find the ratio of routing path selection toward each candidate of output directions in the downstream NoC nodes. By using Definition 1, the LR should deliver the packets to the North, East, and Down side with a proportion of $P1$, $P2$, and $(1 - P_1 - P_2)$, respectively, and a set of identical equations can be derived as:

$$\begin{aligned} P_1 \times D_1 + P_2 \times D_4 + (1 - P_1 - P_2) \times D_7 = \\ P_1 \times D_2 + P_2 \times D_5 + (1 - P_1 - P_2) \times D_8 =, \\ P_1 \times D_3 + P_2 \times D_6 + (1 - P_1 - P_2) \times D_9 \end{aligned} \tag{28}$$

Table 2 The corresponding strategy form of the game model in Fig. 11.©2018 IEEE.

		DR		
		North	East	Down
LR	*North*	Case 1: 0, D_1	Case 2: 0, D_2	Case 3: 0, D_3
	East	Case 4: 0, D_4	Case 5: 0, D_5	Case 6: 0, D_6
	Down	Case 7: 0, D_7	Case 8: 0, D_8	Case 9: 0, D_9

Source: Reprinted with permission from K. Chen, Game-based thermal-delay-aware adaptive routing (GTDAR) for temperature-aware 3D Network-on-Chip systems, IEEE Trans. Parallel Distrib. Syst. 29 (9) (2018) 2018–2032, https://doi.org/10.1109/TPDS.2018.2812164.

and the P_1 and P_2 can be derived as:

$$
\begin{cases}
P_1 = \dfrac{\mu\xi - \beta\nu}{\alpha\mu - \beta\gamma} \\[2mm]
P_2 = \dfrac{\gamma\xi - \alpha\nu}{\beta\gamma - \alpha\mu}
\end{cases}
\tag{29}
$$

where

$$
\begin{cases}
\alpha = D_1 - D_2 - D_7 + D_8 \\
\beta = D_4 - D_5 - D_7 + D_8 \\
\gamma = D_2 - D_3 - D_8 + D_9 \\
\mu = D_5 - D_6 - D_8 + D_9 \\
\xi = D_8 - D_7 \\
\nu = D_9 - D_8
\end{cases}
\tag{30}
$$

Based on the Game theory, the mixed strategy exists if and only if P_1 and P_2 are both within 0 and 1. Nevertheless, because all the D_1 to D_9 belong to real number, the range of P_1 and P_2 is from $-\infty$ to ∞. Consequently, if the probability of this strategy selection is negative, we can know that the current strategy is dominated by other strategies (i.e., the probability is zero). On the contrary, the current strategy is seen as a dominant strategy as long as the probability of this strategy selection is bigger than 1 (i.e., the probability to select the current strategy is 1 because the probability is between 0 and 1).

Obviously, it is necessary to compute the payoff parameters (i.e., D_1 to D_9) in Eq. (30) before finding the ratio of each routing path direction. We make the Case 1 of Table 2 as an example to demonstrate the calculation of the payoff parameters. In the Case 1 of Table 2, a specific packet delivering path is from the local node LR to the south of the downstream NoC node D_{RN} and to the north of the downstream NoC node D_{RN} afterward. Hence, the input buffer of the D_{RN} in the south side is not a competing input buffer. In addition, because the up-then-lateral turn is prohibited for the deadlock-free packet routing, the input buffer of the D_{RN} in the south side is not a competing input buffer. Furthermore, due to the involved minimal adaptive routing mechanism, the packets are not allowed to be returned to the original direction. Therefore, we eliminate the competing input buffers in the D_{RN}'s target output direction in the north side. Based on the above analysis, the competing buffers in this case are the buffers in the East, West, Local, and Up sides. Therefore, we can compute the payoff parameter D_1 by:

$$D_1 = -CL_{SN} = \frac{1}{3} \times \frac{1}{4} \times (BLD_{DR_N}(E) + BLD_{DR_N}(W) \\ + BLD_{DR_N}(L) + BLD_{DR_N}(U)) \quad . \quad (31)$$

By using the similar way, the other payoff parameters (i.e., D_2 to D_9) in Eq. (30) can be computed. By substituting all payoff parameters into Eqs. (29) and (30), the ration of each packet delivering direction to the downstream NoC nodes can be obtained. Therefore, each packet delivery can be distributed to the downstream NoC nodes properly with $O(1)$ computational complexity.

To see the advantage of the game-based routing algorithm for the thermal-aware 3D NoC system, Chen [18] employs a traffic-thermal co-simulator [20] to simulate the traffic and temperature behavior on an $8 \times 8 \times 4$ 3D NoC. Figs. 12–14 show the analysis of the traffic and

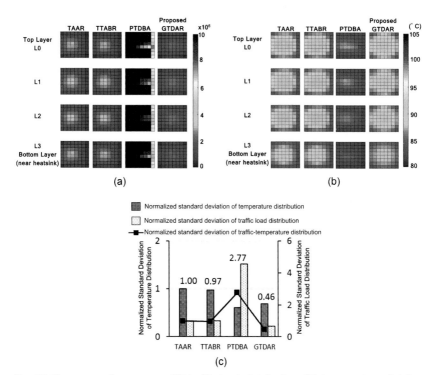

Fig. 12 The comparison among (A) traffic load distribution, (B) temperature distribution, and (C) the corresponding standard deviation under Uniform Random traffic pattern. ©2018 IEEE. Reprinted with permission from K. Chen, Game-based thermal-delay-aware adaptive routing (GTDAR) for temperature-aware 3D Network-on-Chip systems, IEEE Trans. Parallel Distrib. Syst. 29 (9) (2018) 2018–2032, https://doi.org/10.1109/TPDS.2018.2812164.

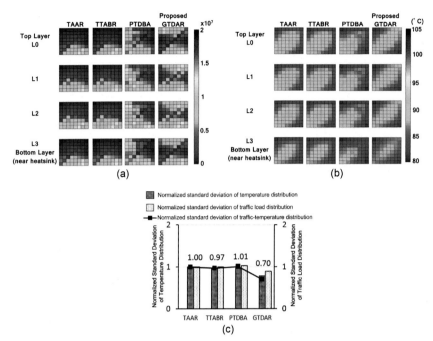

Fig. 13 The comparison of (A) traffic load distribution, (B) temperature distribution, and (C) the corresponding standard deviation under Transpose-1 traffic pattern. *©2018 IEEE. Reprinted with permission from K. Chen, Game-based thermal-delay-aware adaptive routing (GTDAR) for temperature-aware 3D Network-on-Chip systems, IEEE Trans. Parallel Distrib. Syst. 29 (9) (2018) 2018–2032, https://doi.org/10.1109/TPDS.2018.2812164.*

temperature distribution under three synthetic traffic patterns. Obviously, compared with other related works [14, 15, 21], the traffic and temperature distribution become more balanced by considering the thermal–delay metric for the packet delivery.

4. Automated power and thermal management in 2D and 3D NoCs

4.1 Regular and irregular 2D and 3D interconnection networks

Many-core systems can be classified into homogeneous or heterogeneous systems. In the homogeneous many-core systems, cores are identical, demanding a regular NoC to connect the cores. However, in the heterogeneous many-core systems, cores might have different shapes and sizes, forming an irregular network. An example of irregular 2D and 3D systems is shown in Fig. 15A and B, respectively. In the homogeneous many-core

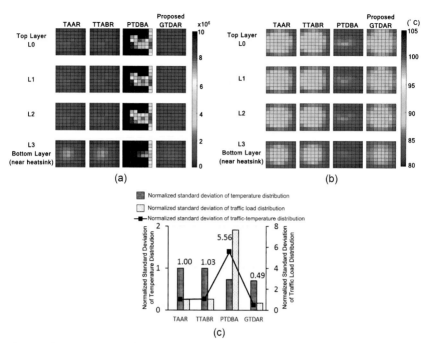

Fig. 14 The comparison of (A) traffic load distribution, (B) temperature distribution, and (C) the corresponding standard deviation under Hotspot traffic pattern. ©2018 IEEE. Reprinted with permission from K. Chen, Game-based thermal-delay-aware adaptive routing (GTDAR) for temperature-aware 3D Network-on-Chip systems, IEEE Trans. Parallel Distrib. Syst. 29 (9) (2018) 2018–2032, https://doi.org/10.1109/TPDS.2018.2812164.

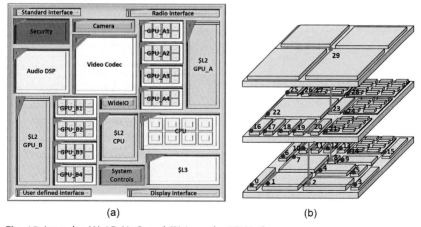

Fig. 15 Irregular (A) 2D NoC, and (B) Irregular 3D NoC.

systems, faults in links or routers may cause the interconnection network to change from a regular to an irregular shape. This shows the importance of algorithms that could manage the design parameters such as latency, power, and temperature automatically.

Combining 3D IC designs and Networks-on-Chip schemes provides an efficient solution to the interconnect problem of future Systems-on-Chip. In 3D NoCs, TSVs come with significant cost and reliability issues, moving 3D-NoCs toward heterogeneous designs. Such irregular networks introduce new challenges such as complexity in designing high-performance routing algorithms. Thereby, it is even more challenging to propose methods to reduce power consumption and temperature. On the other hand, congestion frequently occurs in NoCs when the resources occupied by the packets in the network exceed the network capacity [22]. Congestion also leads to increased hotspot regions, transmission delay, and power consumption. Putting all together, it is not feasible to design a heuristic algorithm for any given regular/irregular network in order to manage design parameters. To address this issue, an automated management technique is demanded which could dynamically adjust and distribute traffic in the network to able to control different design parameters. This section shows the steps to design a method to dynamically manage latency, power, and temperature in 2D/3D regular/irregular NoCs.

4.2 Power and temperature management techniques

In NoCs, power and temperature could be managed in several ways, such as: (1) setting different power modes for each router so that the overall power consumption of the network can be managed. For example, clock-gating modes can be used to disable the ports that are idle for some period of time; (2) controlling the voltage-frequency islands so that power can be distributed over the network without exceeding the power budget; (3) employing congestion-aware routing avoiding hot-spot formation in parts of networks with high traffic; 4) In 3D NOCS, TSVs can be configured to pass packets in double data rate (DDR). This allows a low-power vertical transmission contributing to the overall reduction of network power consumption.

The focus of the power and temperature management in NoCs is commonly concentrated on congestion-aware routing algorithms in order to distribute the traffic load over the available resources in the platform. In this direction, there are several works presented for 2D NoCs [23, 24]. On the other hand, attempts are much limited in 3D NoCs due to the complexity of designing deadlock-free routing algorithms in the wormhole switching network. In Ref. [25], the effort is made toward power optimization for 3D regular NoCs, while in Ref. [26], the performance is optimized with temperature consideration. In this work, for optimizing performance, the routing algorithm propagates the power toward the bottom layer and adjusts the amount of traffic to prevent contention among packets. However, the presented approach is static, with a limited ability to balance the traffic load. Although in Ref. [26], the power consumption has not been directly discussed, it can be deduced that lower traffic congestion and a better thermal distribution lead to lower power consumption. Most of the prior work tries to address the latency, power, and thermal issues in regular networks, while research in irregular networks is much scarce.

4.3 Flexibility in routing

To manage design parameters by the means of routing algorithms, such algorithms should be able to offer the maximum flexibility to route packets through any of the possible minimal and nonminimal routes while avoiding deadlock and livelock. Many adaptive routing algorithms have been presented in literature [27, 28], but most of them suggest limited flexibility and are not able to use the full potential of network resources such as virtual channels for routing packets. Various factors are affecting the NoC routing design, such as the increasing number of cores, the high expected level of performance, topology choice, and the irregularity involved in connecting the cores. Considering various factors, it is very challenging to reach an optimal routing algorithm for each network design. EbDa [29, 30] is a novel approach showing a roadmap to develop maximally adaptive routing algorithms for any given NoCs (regular/irregular, 2D/3D) and any number of physical/virtual channels.

To design a fully flexible routing algorithm, EbDa suggests taking a few steps by knowing the number of virtual channels in the network, as shown in Fig. 16A. In the first step, as shown in Fig. 16B, channels are divided into disjoint groups. In each group, channels should be selected from different dimensions, with the exception that only one dimension may represent channels from both directions. Several virtual channels from the same direction can also exist within a group. In this example, all channels are placed into two disjoint groups, one covering only X1- channel and the second group covering all the remaining channels. In the second step, all possible 90-degree turns and U-turns are extracted from each formed group as illustrated in Fig. 16C and D,

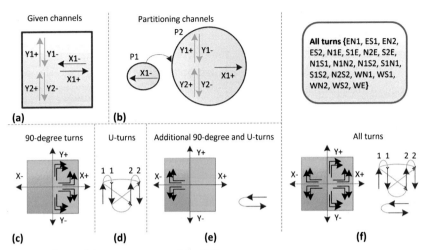

Fig. 16 All possible turns suggested by EbDa.

respectively. In the third step, groups are ordered, and all possible turns can be extracted by taking a channel from any group in a higher-order to a channel in any lower-order group. Doing this, some additional 90-degree and U-turns can be extracted which are listed in Fig. 16E. By taking these steps, all possible turns for a deadlock-free maximally adaptive routing can be extracted, which are listed in Fig. 16F.

Using EbDa, we can take advantage of all available routes to distribute packets and balance the traffic load whenever power consumption or temperature exceeds a predefined threshold limit. However, longer routes cannot be carelessly chosen for avoiding congested areas and hotspots for many reasons [31] such as: (1) longer routes may lead to increased power consumption and temperature; (2) the location of hotspots might change dynamically; and (3) performance may decrease due to taking longer routes by packets. Therefore, the algorithms must be reconfigured with regard to the new hotspot location and would be able to handle a transition phase when the hotspot location changes. To overcome this shortcoming, some algorithms allow dropping packets in the transition phase, which is not an appropriate approach for networks-on-chip. Thereby, an advanced method is needed to decide on the proper routing path, and such algorithms can be based on learning methods.

4.4 Automated power and temperature management in NoCs

Several learning-based algorithms and methods have been proposed in the context of interconnection networks [24, 32–34]. In [32], a reinforcement learning (RL) approach is utilized to improve energy efficiency in NoCs by automatically learning an optimal control policy. In Bi-LCQ [33], the Q-routing method is applied on a cluster-based NoCs. A different attempt than routing is presented in Ref. [34], named SVR-NoC. This approach tries to predict the traffic flow latency and the channel average waiting time by applying a support vector regression method. In Ref. [35], the Q-leaning-based method is applied to manage power in 3D irregular NoCs based on deflection routing. HARAQ [24] combines the Q-routing and nonminimal routing to achieve a high-performance design in the wormhole switching network.

In a Q-learning based method, as suggested by HARAQ, each router has a table with an n number of rows, where each row corresponds to one router, region, or cluster in the network. The table has m columns, each referring to one output port from which a packet can be delivered from a router. For each destination target, the output channel with the lowest value is selected to deliver a packet to the next router. After sending the packet to the next router, an acknowledgment packet is sent back to the previous router, containing local and global information. The local information includes the status of the router concerning the considered metric (e.g., congestion, power, and temperature), while the global information returns the smallest value in the corresponding entry of the table from that router to the packet's destination. In this way, each router will have updated information about the network status, and this information improves as more packets are routed in the network.

In general, by having a flexible routing, such as the one suggested by EbDa [29, 30] and by taking advantage of the Q-learning method similar to HARAQ [24, 35], latency, power, or temperature can be automatically adjusted by using optimal or near-optimal routes between each pair of source and destination nodes. This includes both minimal and nonminimal paths, which are required to balance the traffic over the network. To apply the learning method in NoC, we need to define the cost functions, targeting latency, power, or temperature. The method can be extended to 2D/3D regular/irregular NoC.

The power consumption of individual routers may dynamically change because of the uneven network traffic distribution. When system resources in one part of the network are allocated to an application, the traffic in that part of the network becomes active as packets are routed in different directions. To manage such traffic, the corresponding routers regulating the packet flow must dissipate proportional dynamic power based on the switching activity.

To be able to measure the power dissipation accurately, a power meter should be designed in the router microarchitecture. The power meter reads the rate of link-level packet flow and translates it to the proportional power in watt. The translation depends on the technology used to synthesize the routers. It should be noted that the power consumption of a 3D router is significantly larger than a 2D router (e.g., 35.7 vs 2.6 mW in our simulation under TSMC 65 nm technology). This increase is mainly because of adding buffers in up and down directions and extra logic to the crossbar unit (i.e., increases from $5 \times 5 = 25$ in a 2D NoC to $7 \times 7 = 49$ in a 3D NoC).

The switching activity occurs when packets are flowing through the router. The flow rate affects the overall switching activity. In zero-flow, when there is no packet passing through any port of the network, the dynamic power consumption is zero, and only the leakage power exists. In the maximum packet flow (max-flow), packets enter and exit each port in every cycle. In max-flow, all ports are active, contributing to the maximum dynamic power consumption. In general, to estimate the total power consumption in a router, the dynamic power consumption of active ports is summed up to the leakage power.

While reducing switching activity results in lower power consumption, as was mentioned in Section 3, the temperature is a phenomenon of a long-term traffic situation. To automatically manage power and temperature, both factors have to be considered at the same time. Selecting shorter routing paths helps to reduce switching activities and thus the power consumption, while longer routes with lower temperature can be used, helping to alleviate temperature in hotter areas of the network. Thereby, a proper cost function should be defined, taking these factors into account. As a result, to reduce power consumption, shortest paths are selected while to control temperature, Eq. (13) in Section 2 can be used as a cost metric. Integrating these metrics as the cost function into the Q-learning methods, the routing algorithm can adapt itself at run time. In this way, the network is guaranteed to reach its optimal state with regard to the given metrics shortly after a change occurs in the network. The change might be due to a new application entering the platform, faults in a link, router, or TSV, or a change in the traffic pattern.

5. Conclusion

NoC is a proper interconnection network to address the communication complexity among hundreds to thousands of cores in a many-core systems-on-chip. However, NoC comes with some challenges, such as larger power density and the increased temperature on the chip, resulting in lower system performance and reduced reliability. In this chapter, we first investigated the correlation between power and temperature to come up with a new thermal model. Then, we introduced a power- and temperature-aware routing algorithm based on the Game theory. Finally, we described an approach to automatically manage power and temperature in the network by designing a maximally routing algorithm combined with reinforcement learning.

References

[1] A. Rahman, R. Reif, System-level performance evaluation of three-dimensional integrated circuits, IEEE Trans. Very Large Scale Integr. VLSI Syst. 8 (6) (2000) 671–678, https://doi.org/10.1109/92.902261.

[2] A.W. Topol, D.C.L. Tulipe, L. Shi, D.J. Frank, K. Bernstein, S.E. Steen, A. Kumar, G.U. Singco, A.M. Young, K.W. Guarini, M. Ieong, Three-dimensional integrated circuits, IBM J. Res. Dev. 50 (4.5) (2006) 491–506, https://doi.org/10.1147/rd. 504.0491.

[3] A.Y. Weldezion, M. Grange, D. Pamunuwa, Z. Lu, A. Jantsch, R. Weerasekera, H. Tenhunen, Scalability of Network-on-Chip communication architecture for 3-D meshes, in: 2009 Third ACM/IEEE International Symposium on Networks-on-Chip, 2009, pp. 114–123, https://doi.org/10.1109/NOCS.2009.5071459.

[4] G. Katti, M. Stucchi, K. De Meyer, W. Dehaene, Electrical modeling and characterization of through silicon via for three-dimensional ICs, IEEE Trans. Electron Devices 57 (1) (2010) 256–262, https://doi.org/10.1109/TED.2009.2034508.

[5] W. Huang, M. Allen-Ware, J.B. Carter, M.R. Stan, K. Skadron, E. Cheng, Temperature-aware architecture: lessons and opportunities, IEEE Micro 31 (3) (2011) 82–86, https://doi.org/10.1109/MM.2011.60.

[6] S. Wang, R. Bettati, Reactive speed control in temperature-constrained real-time systems, in: Real-Time Systems, 39, Springer, 2008, pp. 73–95.

[7] K.J. Chen, C. Chao, A.A. Wu, Thermal-aware 3D Network-on-Chip (3D NoC) designs: routing algorithms and thermal managements, IEEE Circuits Syst. Mag. 15 (4) (2015) 45–69, https://doi.org/10.1109/MCAS.2015.2484139.

[8] C.-H. Chao, K.-C. Chen, T.-C. Yin, S.-Y. Lin, A.-Y.A. Wu, Transport-layer-assisted routing for runtime thermal management of 3D NoC systems, ACM Trans. Embed. Comput. Syst. 13 (1) (2013) 1–22, https://doi.org/10.1145/2512468.

[9] K. Chen, E. Chang, H. Li, AWu, RC-based temperature prediction scheme for proactive dynamic thermal management in throttle-based 3D NoCs, IEEE Trans. Parallel Distrib. Syst. 26 (1) (2015) 206–218, https://doi.org/10.1109/TPDS.2014. 2308206.

[10] Inchoon Yeo, C.C. Liu, E. JKim, Predictive dynamic thermal management for multi-core systems, in: 2008 45th ACM/IEEE Design Automation Conference (DAC), 2008, pp. 734–739.

[11] Shang Li, L. Peh, A. Kumar, N.K. Jha, Thermal modeling, characterization and management of on-chip networks, in: 37th International Symposium on Microarchitecture (MICRO-37'04), 2004, pp. 67–78, https://doi.org/10.1109/MICRO.2004.35.

[12] A.K. Coskun, T.S. Rosing, K.C. Gross, Proactive temperature balancing for low cost thermal management in MPSoCs, in: 2008 IEEE/ACM International Conference on Computer-Aided Design, 2008, pp. 250–257, https://doi.org/10.1109/ICCAD. 2008.4681582.

[13] A.-M. Rahmani, K.R. Vaddina, K. Latif, P. Liljeberg, J. Plosila, H. Tenhunen, Design and management of high-performance, reliable and thermal-aware 3D networks-on-chip, IET Circuits Devices Syst. 6 (2012). 308–321(13).

[14] K. Chen, S. Lin, H. Hung, A.A. Wu, Topology-aware adaptive routing for non-stationary irregular mesh in throttled 3D NoC systems, IEEE Trans. Parallel Distrib. Syst. 24 (10) (2013) 2109–2120, https://doi.org/10.1109/TPDS.2012.291.

[15] K. Chen, C. Kuo, H. Hung, A.A. Wu, Traffic- and thermal-aware adaptive beltway routing for three dimensional Network-on-Chip systems, in: 2013 IEEE International Symposium on Circuits and Systems (ISCAS), 2013, pp. 1660–1663, https://doi.org/10.1109/ISCAS.2013.6572182.

[16] F. Liu, H. Gu, Y. Yang, DTBR: a dynamic thermal-balance routing algorithm for Network-on-Chip, Comput. Electr. Eng. 38 (2) (2012) 270–281, https://doi.org/10.1016/j.compeleceng.2011.12.006.

[17] C. Kuo, K. Chen, E. Chang, A. Wu, Proactive thermal-budget-based beltway routing algorithm for thermal-aware 3D NoC systems, in: 2013 International Symposium on System on Chip (SoC), 2013, pp. 1–4, https://doi.org/10.1109/ISSoC.2013.6675281.

[18] K. Chen, Game-based thermal-delay-aware adaptive routing (GTDAR) for temperature-aware 3D Network-on-Chip systems, IEEE Trans. Parallel Distrib. Syst. 29 (9) (2018) 2018–2032, https://doi.org/10.1109/TPDS.2018.2812164.

[19] E. Chang, H. Hsin, S. Lin, A. Wu, Path-congestion-aware adaptive routing with a contention prediction scheme for Network-on-Chip systems, IEEE Trans. Comput. Aided Des. Integr. Circuits Syst. 33 (1) (2014) 113–126, https://doi.org/10.1109/TCAD.2013.2282262.

[20] K.-Y. Jheng, C.-H. Chao, H.-Y. Wang, A.-Y. Wu, Traffic-thermal mutual-coupling co-simulation platform for three-dimensional Network-on-Chip, in: Proceedings of 2010 International Symposium on VLSI Design, Automation and Test, 2010, pp. 135–138, https://doi.org/10.1109/VDAT.2010.5496709.

[21] Y. Lee, H. Hsin, K. Chen, E. Chang, A.A. Wu, Thermal-aware dynamic buffer allocation for proactive routing algorithm on 3D Network-on-Chip systems, in: Technical Papers of 2014 International Symposium on VLSI Design, Automation and Test, 2014, pp. 1–4, https://doi.org/10.1109/VLSI-DAT.2014.6834908.

[22] M. Ebrahimi, M. Daneshtalab, F. Farahnakian, J. Plosila, P. Liljeberg, M. Palesi, H. Tenhunen, HARAQ: congestion-aware learning Model for highly adaptive routing algorithm in on-chip networks, in: 2012 IEEE/ACM Sixth International Symposium on Networks-on-Chip, 2012, pp. 19–26, https://doi.org/10.1109/NOCS.2012.10.

[23] G. Ascia, V. Catania, M. Palesi, D. Patti, Implementation and analysis of a new selection strategy for adaptive routing in Networks-on-Chip, IEEE Trans. Comput. 57 (6) (2008) 809–820, https://doi.org/10.1109/TC.2008.38.

[24] M. Ebrahimi, M. Daneshtalab, F. Farahnakian, J. Plosila, P. Liljeberg, M. Palesi, H. Tenhunen, HARAQ: congestion-aware learning model for highly adaptive routing algorithm in on-chip networks, in: 2012 IEEE/ACM Sixth International Symposium on Networks-on-Chip, 2012, pp. 19–26, https://doi.org/10.1109/NOCS.2012.10.

[25] A.Y. WeldeZion, M. Ebrahimi, M. Daneshtalab, H. Tenhunen, Automated power and latency management in heterogeneous 3D NoCs, in: Proceedings of the Eighth International Workshop on Network on Chip Architectures, Association for Computing Machinery, New York, NY, USA, 2015, pp. 33–38, ISBN: 9781450339636, https://doi.org/10.1145/2835512.2835517.

[26] C. Chao, K. Jheng, H. Wang, J. Wu, A. Wu, Traffic- and thermal-aware run-time thermal management scheme for 3D NoC systems, in: 2010 Fourth ACM/IEEE International Symposium on Networks-on-Chip, 2010, pp. 223–230, https://doi.org/10.1109/NOCS.2010.32.

[27] M. Ebrahimi, M. Daneshtalab, P. Liljeberg, J. Plosila, H. Tenhunen, LEAR–a low-weight and highly adaptive routing method for distributing congestions in on-chip

networks, in: 2012 20th Euromicro International Conference on Parallel, Distributed and Network-based Processing, 2012, pp. 520–524, https://doi.org/10.1109/PDP.2012.52.

[28] P. Majumder, S. Kim, J. Huang, K.H. Yum, E.J. Kim, Remote control: a simple deadlock avoidance scheme for modular systems-on-chip, IEEE Trans. Comput. 70 (11) (2020) 1, https://doi.org/10.1109/TC.2020.3029682. 1.

[29] M. Ebrahimi, M. Daneshtalab, EbDa: a new theory on design and verification of deadlock-free interconnection networks, in: 2017 ACM/IEEE 44th Annual International Symposium on Computer Architecture (ISCA), 2017, pp. 703–715, https://doi.org/10.1145/3079856.3080253.

[30] M. Ebrahimi, M. Daneshtalab, A general methodology on designing acyclic channel dependency graphs in interconnection networks, IEEE Micro 38 (3) (2018) 79–85, https://doi.org/10.1109/MM.2018.032271064.

[31] M. Ebrahimi, M. Daneshtalab, P. Liljeberg, H. Tenhunen, Fault-tolerant method with distributed monitoring and management technique for 3D stacked meshes, in: The 17th CSI International Symposium on Computer Architecture Digital Systems (CADS 2013), 2013, pp. 93–98, https://doi.org/10.1109/CADS.2013.6714243.

[32] H. Zheng, A. Louri, An energy-efficient Network-on-Chip design using reinforcement learning, in: Proceedings of the 56th Annual Design Automation Conference 2019, Association for Computing Machinery, New York, NY, USA, 2019, ISBN: 9781450367257, https://doi.org/10.1145/3316781.3317768.

[33] F. Farahnakian, M. Ebrahimi, M. Daneshtalab, P. Liljeberg, J. Plosila, Bi-LCQ: a low-weight clustering-based Q-learning approach for NoCs, Microprocess. Microsyst. 38 (1) (2014) 64–75, https://doi.org/10.1016/j.micpro.2013.11.008.

[34] Z. Qian, D. Juan, P. Bogdan, C. Tsui, D. Marculescu, R. Marculescu, SVR-NoC: a performance analysis tool for Network-on-Chips using learning-based support vector regression model, in: 2013 Design, Automation Test in Europe Conference Exhibition (DATE), 2013, pp. 354–357, https://doi.org/10.7873/DATE.2013.083.

[35] A.Y. WeldeZion, M. Ebrahimi, M. Daneshtalab, H. Tenhunen, Automated power and latency management in heterogeneous 3D NoCs, in: Proceedings of the Eighth International Workshop on Network on Chip Architectures, Association for Computing Machinery, New York, NY, USA, 2015, pp. 33–38, ISBN: 9781450339636, https://doi.org/10.1145/2835512.2835517.

About the authors

Kun-Chih (Jimmy) Chen (IEEE S'10-M'14-SM'21) is currently an Associate Professor of Computer Science and Engineering Department of National Sun Yat-sen University. His research interests include Multiprocessor SoC (MPSoC) design, Neural network learning algorithm design, Reliable system design, and VLSI/CAD design. Dr. Chen received the Best Paper Award of 2014 International Symposium on VLSI Design, Automation and Test (VLSI-DAT'14), Best Paper Award of 2016

International Joint Conference on Convergence (IJCC'16), PhD Dissertation Award of IEEE Taipei Section, TCUS Young Scholar Innovation Distinction Award, Best Student Paper Award of IEEE ISCAS, IEEE Tainan Section Best Young Professional Member Award, Exploration Research Award of Pan Wen Yuan Foundation, Taiwan IC Design Society Outstanding Young Scholar Award, and Chinese Institute of Electrical Engineering (CIEE) Outstanding Youth Electrical Engineer Award. Dr. Chen serves as the referee of many IEEE journals and conferences, including TC, TPDS, TCAS-I, ISCAS, ICASSP, VLSI-DAT, etc. Dr. Chen is invited to be the Guest Editor of Journal of Systems Architecture (JSA), Nano Communication Network (NanoComNet), and IEEE Journal on Emerging and Selected Topics in Circuits and Systems (JETCAS). Besides, he serves as a Technical Program Committee (TPC) chair of the International Workshop on Network on Chip Architectures (NoCArc) in 2018 and the General Chair of the same workshop in 2019. He is an IEEE senior member, ACM member, APSIPA member, life member of Chinese Institute of Electrical Engineering (CIEE) and Taiwan IC Design Society (TICD).

Masoumeh (Azin) Ebrahimi (Senior Member, IEEE) received the Ph.D. degree (with Honors) from the University of Turku, Finland, in 2013, and the joint M.B.A. degree from the University of Turku and the EIT-ICT School in 2015. She is currently an Associate Professor with KTH Royal Institute of Technology, Sweden, and an Adjunct Professor with the University of Turku, Finland. She has led several national and international projects, such as EU-MarieCurie-Vinnova, Academy of Finland, SSF, STINT, and Vetenskapsrådet (VR). Her scientific work contains more than 100 publications, and her main areas of interests include interconnection networks, near-memory processing, and neural network accelerators. She is a member of HiPEAC. She actively acts as a Guest Editor, an Organizer, and the Program Chair in different venues and conferences.

CHAPTER SIX

Approximate communication for energy-efficient network-on-chip

Ling Wang[a] and Xiaohang Wang[b]
[a]School of Telecommunication Engineering, Xidian University, Xi'an, The People's Republic of China
[b]School of Software Engineering, South China University of Technology, Guangzhou, The People's Republic of China

Contents

Abstract

Approximate Computing is being touted as a viable solution for high-performance computation by relaxing the accuracy constraints of applications. This trend has been accentuated by emerging data intensive applications in domains like image/video processing, machine learning, and big data analytics that allow inaccurate outputs within an acceptable variance. With the increasing communication demand as well

Advances in Computers, Volume 124
ISSN 0065-2458
https://doi.org/10.1016/bs.adcom.2021.09.004

as the optimization bottleneck of NoC performance and energy consumption, approximate communication, which leverage relaxed accuracy for energy-efficiency Networks-on-Chip (NoC), have become the accepted method for connecting a large number of on-chip components. We, respectively, proposed approximate designs for traffic regulation, bufferless NoC, and multiplane NoC. These designs improve network performance and reduce power consumption by reducing network load, optimizing data transmission, and optimizing network architecture design. The approximate communication designs show a huge improvement in energy-efficient NoCs while maintaining low application error.

1. Introduction

Approximation by trading off output accuracy for benefits in performance and energy efficiency has gained a high degree of recognition as a solution for satisfying energy-efficiency hardware design [1]. Approximate designs rely on the ability of applications to tolerate computation on noisy/erroneous data or imprecision in the computation results. There are large number applications of machine learning, searching, scientific computing, and multimedia that are inherently tolerant approximation [2]. Since inexactness is acceptable, these applications allow a presence of approximate data in storing, computing, or transmitting. These applications, which exhibit some level of error tolerance, motivate the approximate hardware designs to achieve high performance and energy efficiency.

Now approximate computing, as an emerging performance-efficient paradigm, has been widely used in computer architecture design, such as approximate memory system [3, 4], value approximation in CPU-based [5, 6] and GPU-based system [7, 8], relaxes synchronization [9], resilience-aware circuit clocking scheme [10], and so on. Compute-based approximation techniques use inexact compute units [11–19] or neural network models [20, 21] for code acceleration. Memory-based techniques exploit data similarity across memory hierarchies to achieve larger capacity, energy efficiency, or lifetime optimization . A significant portion of research on hardware approximation techniques has focused on either the computation units for accelerated inaccurate execution, or the storage hierarchy (cache/DRAM-based) for low-overhead (area/power) memory.

Approximate communication techniques also deserve attention. With increasing on-chip core counts, network-on-chip (NoC) has emerged as the most competent method for on-chip communication in large-scale parallel systems. It connects varied on-chip components, such as cores, caches, and memory controllers. And it allows the communication necessary for exchanging data of parallel threads and ensuring data coherence. However, NoCs consume a significant amount of power in modern chip multiprocessors (CMP) [22]. Energy efficiency has been a primary concern in NoC designs [23]. Reducing the NoC power while increasing performance is essential for scaling up to larger chip multiprocessor systems. Relaxing accuracy in exchange for performance improvement and energy saving, approximate techniques show their bright future on the research of energy-efficiency designs.

This chapter, with a focus on the approximate communication design for energy-efficient NoC, mainly conducts exploratory research in the following three aspects:

First, a dynamic traffic regulation scheme is proposed for approximate communication of NoC. Network congestion is one of the main factors that affect transmission delay, and different traffic flows have different impacts on network congestion. This method designs an approximation–based traffic regulation structure in the network interface, which reduces the amount of injected data through data approximation, and can regulate the injection rate of each node. In addition, it designs a dynamic traffic regulation algorithm to dynamically adjust the injection rate of each node according to the impact of traffic flow on network congestion. Thus, it improves the NoC performance. Based on the PARSEC benchmark experiments, the results show that this method can reduce the average transmission delay by 30.9% on average, reduce application execution time by 15.8%, and achieve dynamic power saving by 24.4% within 10% quality loss.

Second, a performance optimization method for bufferless NoC based on approximate communication is proposed. By removing the buffers, the bufferless NoC reduces power consumption and area overhead but also leads to an increase in transmission delay and a decrease in network throughput. Through the performance analysis of the bufferles NoC, in the retransmission-based bufferless NoC, packet retransmission is a key factor affecting the NoC performance. In order to improve the performance of the bufferless NoC, this method designs a new bufferless NoC architecture, which reduces packet retransmission through lossy transmission and improves the NoC performance. Moreover, it also proposes a packet approximate codec design to approximate the missing data. Thus, rhis method improves the performance of bufferless NoC with extremely low quality loss. Based on the PARSEC benchmark experiments, the results show that compared with the existing bufferless NoC, this design reduces the retransmission by 83.6%, reduces the transmission delay by 46.7%, increases the network throughput by 92%, and achieves application acceleration 1.2 times, while maintaining low application error.

Third, an NoC energy optimization method based on multiplane network design is proposed. The NoC performance optimization usually leads to an increase in area overhead and affects the energy consumption of NoC. In order to reduce the energy consumption of NoC, this method designs a two-plane network structure which includes a lossy subnetwork and a lossless subnetwork. Based on lossy transmission, the lossy subnetwork realizes a lightweight, low-delay, bufferless architecture design. In addition, based on the multiplane transmission design, this method speeds up part of the data transfer and achieves transmission quality control. Thus, this method improves NoC performance while reducing NoC area overhead and power consumption. Based on the PARSEC benchmark experiments, the results show that compared with the single-plane NoC, this method reduces the transmission delay by 41.9%, and saves 48.6% of the NoC area overhead and 25.7% of the NoC power consumption under the same throughput.

The rest of the paper is organized as follows. Section 2 details the related work. In Section 3, we present the approximation-based dynamic traffic regulation design. Section 4 explains the approximate bufferless NoC implementation. Section 5 presents the design of approximate multiplane NoC.

2. Related work

Recent studies have been conducted regarding approximate computing in NoC architecture design for applications that allow inaccurate outputs [24–29]. These articles explore the performance improvement or energy efficiency of approximate computing for reducing communication bottlenecks by two techniques: communication reduction and dynamic power management. The APPROX-NoC [25], DAPPER [24], and DEC-NoC [26] belong to the former. APPROX-NoC reduces injected flits by approximate compression, which improves the compression ratio based on a value approximation matching technique [25]. DAPPER coalesces approximable data waiting for transmission to reduce the number of injected packets in the NoC of the GPGPU architecture [24]. DEC-NoC relaxes transmission accuracy in order to reduce retransmissions due to communication errors [26]. These approximation designs focus primarily on the size of the workload and do not improve packet transmissions in the NoC. Ahmed et al. [27] and Ascia et al. [28, 29] use dynamic power management to transfer approximable data at lower power in order to trade the result quality for energy savings. In AxNoC [27], a dual-voltage look-ahead scheme is proposed for the power management in approximate NoC design, which isolates a critical path for providing headers and critical flits with perfect transmission under high voltage and the remaining flits with bit flips by decreasing the supply voltage. In Ref. [28], Ascia et al. selectively reduce the voltage swing of a link based on the forgiving nature of the approximable communication flow for a trade-off between quality degradation and energy saving. In Ref. [29], Ascia et al. introduce a dynamic power management into approximate wireless NoC design, in which both wireless and wired communications are controlled based on their expected reliability levels. Approximable communications will be applied with low power and transferred with low expected reliability. These approximate NoCs achieve a high energy benefit with dynamic power management, but exert little effort in network contention. Channel contending during transmission greatly reduces network performance, and different traffic flows have different impacts on network congestion. We proposed the approximation mechanisms in traffic regulation and router architecture to reduce transmission conflicts and improve performance. Betzel et al. [30] and Reza et al. [31] summarize the approximate techniques in NoCs and present a bright future of approximate communication in energy-efficient and high-performance NoC design.

3. Approximation-based dynamic traffic regulation

Different traffic flows have different impacts on network congestion. For example, Fig. 1 shows the network transmission status of a certain time. Packets from nodes 1, 2, 6 contribute to the network congestion, while packets from node 5 don't. Therefore, controlling the packets injected from nodes 1, 2, 6 can lead to better congestion improvement for the network. However, the transmission state will be very complicated in NoC.

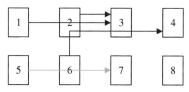

Fig. 1 Different contributions of the flows on network congestion.

Each router is likely to communicate with others. Its complexity will grow exponentially with the network size. Moreover, the transmission state is changing over time.

Furthermore, any viable approximation technique needs to provide strict safety guarantees. It requires security strategy to determine the set of candidate loads for safe approximation. In addition, the loss of packets data adversely affects application's result quantity. And each packet has different effect on the output quality. How to control the drop rate of traffic flow to meet an acceptable quality requirement further complicates the problem.

To address the above challenges, we propose an approximation-based dynamic traffic regulation (ABDTR). This approach dynamically regulates the drop rate of each injected packet according to the network congestion state and its contribution to the congestion. Packets that have a higher impact on network congestion lost more data. And we choose the appropriate drop-rate range to satisfy the output quality.

3.1 Architecture design For ABDTR

3.1.1 Overview

NoC architectures exploit packets transmission through many threads execution in CMP to mitigate the penalties of long transmission latency. The bandwidth is the main bottleneck for NoC performance. High injection rate (IR) will lead to the exceeding of the bandwidth constrains, which will bring network congestion and long transmission latency. ABDTR exploits traffic regulation method, per-router's IR control through dropping a fraction of safe-to-approximate data of packets, which trades transmission integrity for better performance and energy saving.

The key idea of ABDTR is to approximate values instead of their transmission in the network. The ABDTR does not change the memory operations and IP cores operations. It controls the per-router's IR and selectively drops some data of packet before being injected into network. And after the packet transmission, it approximates the missing value for restoring the complete packet. This traffic regulation is based on the congestion information. These packets with high influence in network congestion will be controlled with high drop rate. The drop rate of each router is also controlled to guarantee output quality and approximation safety.

The ABDTR consists of three parts: controller, throttler, and approximator. They are designed between routers and cores. Fig. 2 shows an overview of the ABDTR architecture. In the figure, IP cores refer to

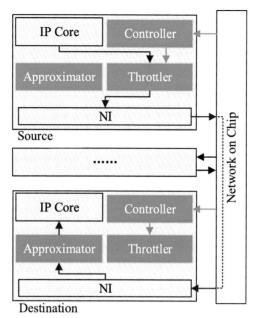

Fig. 2 ABDTR architecture overview.

processing elements in general, which injects packets into the network in a controlled manner. The controller is a key part of traffic regulation, which takes possible NoC congestion information as inputs and makes decisions on regulation parameter. As shown in Fig. 2, the red line represents a traffic flow with traffic regulation and the blue line is regulation signal. In source node, the throttler filters out part of injection data according to the regulation parameter from the controller. The throttler is like a faucet. The former controls traffic flow and the latter controls water flow. The approximator estimates the missing values and inserts them into the transmitted data in destination node, which ensures the integrity of the transmission. In short, the ABDTR works like a lossy compression and decompression. The throttler does compression, the approximator does decompression, and the controller controls the compression ratio and output quality.

The regulation parameter between controller and throttler is defined as *drop rate*. Its value varies with different traffic flows and changes with the congestion information in real time. The drop rate setting in each router is a handle to control the trade-off between performance/energy gains and quality loss. A higher drop rate enables the throttler to filter out more values, thereby reducing the volume of data movement across the network. We explore the drop rate of each router as an architectural handle. In ABDTR, the drop rate can be dynamically regulated to improve NoC performance and save energy as much as possible with an acceptable quality degradation.

3.1.2 Controller design

In ABDTR, a per-traffic control strategy is designed. Each traffic flow will be controlled based on NoC congestion information and its contribution to the congestion. Therefore, the controller needs to collect NoC congestion information and figure out what is the drop rate of each traffic flow. The controller has three main functions: (1) sensing Network congestion state, (2) computing traffic's contribution to NoC congestions, and (3) regulating the drop rate of traffics. We will elaborate these three functions in controller design.

3.1.2.1 Sensing network congestion

The goal of our traffic regulation strategy is to control congestion and bound network latency. Congestion is a resource sharing problem. Links and buffers are the shared resources in packets switched networks. We use buffer fillings rather than link utilization because we think that this is the most direct congestion measure. A link is used for transmission. Even if there is no congestion, the link still has data transmission. However, once congestion occurs, a router's input buffers will store large amounts of data. So the amount of used buffer slots is monitored as a congestion measure. The main idea is that each router monitors its input buffers and whenever the number of packets in one of these buffers is increased to a specific threshold value, the router becomes a congestion point. This method of discriminating congestion has been widely used in many dynamic routing designs [32–34].

3.1.2.2 Computing traffic's contribution to NoC congestions

As discussed in Section 1, traffic flows have different contributions to NoC congestions. For each congestion point, we need to find out which routers' high IRs are causing the congestion. Then these routers with impact on the congestion will be recorded and required to reduce their IRs. The packets stored in the congestion point are the most direct cause of the congestion. So the source routers of these congested packets are considered to have too high injection rates which lead to the congestion. Controller reads the congestion point buffers and gets the congested packets' source router ID. Then, it transports a message to these source routers to increase their drop rate by using connections in the NoC. In order to have a reliable system, congestion messages must have no effect on these connections. Therefore, an independent channel is used to transport control messages. We design an emergency channel which can only store and transmit control messages. The emergency channel is a separate virtual channel and has the highest priority in operation. When there is a packet in the emergency channel, it can be transmitted first. Moreover, the control packets are fixed to two flit size in order to further accelerate the transmission of the control messages. In the system operation, the control messages include the congestion warnings and system-level instructions.

3.1.2.3 Means of control
In a traffic flow regulation, we only control its injection in source node. And drop rate is the control object due to their impact on NoC IR. We design the drop rate as a discrete regulation parameter. At each time, each router needs to choose a fixed option to meet quality requirement. The mean of drop rate control will be elaborated in Section 3.1.3 and the regulation strategy will be detailed in Section 3.

3.1.3 Throttler design
Data are transmitted in NoC in packets. A package typically contains a plurality of data. We design the throttler at a sampling mode. Injection data are omitted at the same interval and then we packet the rest, which truncates packets and throttles the injection rate down. Fig. 3 shows this process where

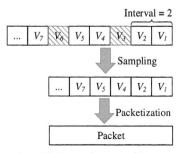

Fig. 3 The process of throttler with interval value of 2.

interval equals 2. The values with shadow are sampled out and dropped without packetization. This throttler has the following features:

(1) The sample interval is closed related to the drop rate of throttler, the larger the interval, the less the drop rate. Their relationship is formulated as,

$$D = \frac{1}{\gamma + 1} \tag{1}$$

where D means drop rate and γ is interval value.

(2) The interval is at least 1, dropping a data per interval data, so the drop rate cannot exceed 50%.

(3) The total data width of omitting data needs to exceed the width of a flit for reducing the length of packets.

(4) Drop rate of throttler is discrete since the interval is an integer.

(5) Drop rate is a key factor of node injection rate and output integrity.

3.1.4 Approximator design

Approximator is another main challenge for ABDTR. It works for predicting the missing value with low-overhead. When a packet reaches the destination node, approximator needs to quickly generate values to recover the packet based on the received data and the interval value. The approximator works like a value predictor which generates new values based on the value pattern. But the value predictor generally needs strict correctness verification. The approximator can accept inexactness values since the error-tolerant applications. Fig. 4 shows the process of approximator where interval value is two. The approximator makes predictions (the shaded part) based on the received data and inserts a prediction value after each two values received. Thus, the integral packet data will be restored.

The complexity and accuracy are the two main challenges of approximator. The complexity determines the hardware overhead and the accuracy

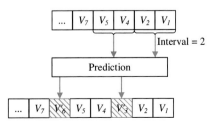

Fig. 4 The process of approximator with interval value of 2.

influences the result quality. So the microarchitectural mechanism of approximator needs to provide a good trade-off between accuracy and complexity. There are several traditional value predictors, such as last value [35], stride[36], FCM [37], and several modern exact value predictors (e.g., VTAGE [38], BeBoP[39], EOLE[40]). We choose the linear interpolation because of its low complexity and reasonable accuracy. The linear interpolation requires only one addition to perform the approximation. The values in a packet are generally fetching from successive memory block. There is some significant value similarity in packet (e.g., adjacent pixels in an image). The dropped value is the most relevant to its previous one and next one. So the average of the two values is used as the predicted value. If the dropped value is the last one, the predicted value is its previous one.

3.2 NoC dynamic traffic regulation strategy

Our design principle for ABDTR is to maximize the opportunities for performance and energy-efficiency improvements, while limiting the adverse effects of approximation on output quality. To achieve this goal, the drop rate in each router is dynamically regulated at runtime with the change of network congestion state. The ABDTR needs to drop more injection data of source routers with higher influence in congestion. In order to avoid the approximation causing catastrophic failures, programmer annotates the transmission data that are safe to approximate. The congestion-critical and safe-to-approximate injection packets are the candidates for regulating with ABDTR. Afterward, we regulate the drop rate of each router for maximizing performance while maintaining high output quality.

3.2.1 Approximation safety guarantees

Not all transmission data can be safely approximated. For example, data that affect critical data segments, array indices, pointer addresses, or control flow conditionals are usually not safe to approximate [7]. So the first step is to ensure that data with safety violations are excluded from ABDTR. In other words, ABDTR can be only applied to the packet data that have a slight impact on the quality degradations and can be approximated without catastrophic failures of applications.

Prior works in approximation [41–43] show that language construction with programmer annotations is necessary to identify which instructions are safe to approximate. Furthermore, some studies have provided language constructs and compiler support for annotating safe-to-approximate operations, such as FlexJava [44] and EnerJ [45]. Load value approximation [5] elaborates annotating guidelines and safe-to-approximate regions of data of applications from PARSEC 3.0 [46]. ABDTR leverage these prior works by programmer annotations to determine the set of candidate data for safe approximation. We make use of ISA extensions to enable the compiler to mark the safely approximable loads.

3.2.2 Drop-rate regulation

The drop rates of routers are knobs that enables ABDTR to trade-off between quality loss and performance-energy improvement. The next step is to pick the best drop rate of each router for maximizing the benefits of ABDTR while satisfying the target output quality requirement. In general, the drop-rate regulation is a NP-hard problem. For example, in a NoC system with n routers, if each router can be regulated with m available drop rates, the total number of regulation combinations is as large as m^n. Moreover, the drop-rate regulation needs to respond quickly to the network congestion state changing and provide formal output quality guarantees.

The drop rate of a traffic flow needs to change with its impact on network congestion. If the impact is strong, the drop rate needs to be high, and vice versa. Whether the traffic has an impact on congestion can be known from the controller messages, but without a specific value. We propose a lightweight heuristic method to dynamically regulate the drop rate of traffic flow based on controller messages and its previous drop rate selection. Each router starts with a drop rate of 0 (no dropping data). Then if the router i has influence on NoC congestion (a controller message), ABDTR in router i will start working with minimum drop rate. At the next time, if the influence persists, router i will increase its drop rate since its impact on congestion is considered larger. Otherwise, the drop rate needs to be decreased. The regulation of drop rate is achieved through the selection of interval value of throttler, as shown in Eqs. (2) and (3). Eq. (2) calculates the drop rate when it needs to be increased while Eq. (3) is for decreasing the drop rate. In Eqs. (2) and (3), the value changes at two time points, respectively, is to decrease and increase packet size at each regulation. Drop rate is ABDTR's knob for navigating quality trade-offs. It needs to be limited to a range (0 to α) in order to ensure the output quality. This regulation brings easy microarchitectural design. To increase drop rate, the interval value only needs to be added one and moved a bit to the left. On the contrary, the interval value being moved only a bit to the right will realize the drop rate reduction. And in order to control the drop rate within the value of α, the interval value should be bigger than the minimum value, which is based on Eq. (1).

$$D_i^t = \frac{1}{\gamma_i^t + 1} = \frac{1}{\lceil \gamma_i^{t-1}/2 \rceil + 1}; \quad 0 \le (D_i^t) \le \alpha \tag{2}$$

$$D_i^t = \frac{1}{\gamma_i^t + 1} = \frac{1}{(\gamma_i^{t-1} \times 2) + 1}; \quad 0 \le (D_i^t) \le \alpha \tag{3}$$

where D_i^t means drop rate of router i at time t; γ_i^t is interval value of router i at time t; α is the maximum drop rate.

This dynamic drop-rate regulation process is simple and rapid. Moreover, the drop-rate regulation needs to satisfy the target output quality requirement. We provide a statistical method to determine the α that satisfies the target output quality. The maximum drop rate is 50% where the interval takes 1. So the α takes a value from 0% to 50%. The α starts with 50%. If the output error is greater than target value, the compiler decreases the α by a small step. The step is usually the reciprocal of the data packet size. This process continues until the drop-rate regulation satisfies the target quality. Ultimately, the parameters of drop-rate regulation are empirically set in our evaluation.

Altogether, the two steps provide a compilation workflow that focuses ABDTR on the safe-to-approximate packets with the highest potential in terms of performance and energy savings while satisfying the target output quality requirement.

3.3 Methodology

3.3.1 Experimental setup

3.3.1.1 Full-system simulation

We evaluate the proposed mechanism using an timing-detailed full-system many-core simulator with mesh NoC microarchitecture [47]. The simulated architecture is a shared memory structure, where a core with a private L1 cache and a shared L2 cache bank forms a tile. NoC interconnects tiles, traffic regulation modules (ABDTR), network-interfaces, and on-chip routers. Routers are organized in a popular mesh topology with wormhole switching on which the XY routing algorithm is implemented to achieve simplicity and deadlock free. Multithreaded applications can run on the simulator with threads running on different cores. Communications between threads are through the NoC. Memory transactions, including read and write, and control instructions will generate messages in NoC. On the target system, a directory-based MESI protocol is implemented to support cache coherency. Once a cache miss occurs in private L1, memory request is

Table 1 Configuration used in the simulation.

Number of processors	64 (MIPS ISA 32 compatible)
Fetch/Decode/Commit size	4/4/4
ROB size	64
L1 D cache (private)	16 KB, two-way, 32B line, two cycles, two ports, dual tags
L1 I cache (private)	32 KB, two-way, 64B line, two cycles
L2 cache (shared) MESI protocol	64 KB slice/node, 64B line, six cycles, two ports
Main memory size	2 GB, latency 200 cycles
On-chip network parameters	
NoC flit size	32-bit
Data packet size	32 flits
Meta packet size	3 flit
NoC latency	router two cycles, link one cycle
NoC VC number	3
NoC buffer	5 × 16 flits
Routing algorithm	XY routing

forwarded to the local shared L2 or via the NoC to remote shared L2/DRAM. Table 1 lists the simulator parameters.

3.3.1.2 Energy modeling

To measure the energy benefits of ABDTR, the DSENT, Cacti, and McPAT libraries are integrated into the simulator for computing the power consumption of interconnection, caches, and processors, respectively. The power and area of the ABDTR circuit are obtained from RTL level synthesis using a 45-nm CMOS technology. We then fed the ABDTR circuit power consumption data into the simulator.

3.3.2 Experimental applications

3.3.2.1 Applications

Table 2 lists the benchmarks used for performance evaluation, and they are selected from PARSEC. Columns 1 and 2 of Table 2 summarize these

Table 2 Experimental applications.

Name	Domain	Quality metric	Profiling set
Blackscholes	Financial Analysis	Percentage of results with error greater than 1%	65,536 financial options with multiple parameters
Canneal	Machine Learning	Relative error of the final routing	Synthetic netlist with 400,000 points
Ferret	Image Search	Percentage of images not being searched out in precise results	256 Images
Fluidanimate	Animation	Percentage of particles whose final location with error greater than 0.2%	305,809 data of a collection of Newtonian particles
Streamcluster	Data Mining	Percentage of points that not being clustered as precise results	8192 16-dimensional points
Vips	Image Processing	Average relative error of final output RGB image's pixel values	1662 × 5500-Pixel Color Vips Image

applications and their domains. The applications can tolerate small errors[5] and represent a wide range of domains including financial analysis, machine learning, image search, image processing, animation, and data mining [46]. Each application is configured with 64 threads and runs on 64 cores. We do not perform any optimizations in the source code in favor of ABDTR.

3.3.2.2 Quality metrics

Column 3 of Table 2 lists each application's quality metric. Quality metric of each application determines the application's output-quality loss as it undergoes approximation. Using application-specific quality metrics is commensurate with previous works on approximation [5]. To measure quality loss, we compare the output from the ABDTR-enabled execution to the output with no approximation. For *Blackscholes* which output a list of prices. We calculate the relative error of each approximated price to its precise price, and then use the percentage of the error greater than 1% as the quality metric. The output error for *canneal* is defined as the difference between the final routing cost for the approximated and exact execution. *Ferret* and *Streamcluster* output a list of images and points. We measure the quality loss using the percentage of outputs that do not match the precise results. For *Fluidanimate*, the particles with location error greater than

0.2% are considered to be mismatch, and we use the mismatch rate as its quality metric. Finally, average relative error of each pixel point of output image is measured for *Vips*.

3.4 Experimental results

3.4.1 Effect of ABDTR

Fig. 5 illustrates the effect of ABDTR on the average network delay reduction (Fig. 5A) and application quality degradation (Fig. 5B) with different drop rate control ranges (α). As α increases, so do the NoC performance benefits. However, the benefits come with some cost in output quality.

As shown in Fig. 5A, ABDTR achieves significant reduction in average network delay. For most of benchmarks, the average network delay have been more than 30% reduction with 50% α value. Fig. 5B shows the influence of drop-rate regulation on quality degradation. We can observe that quality degradation depends consistently on the characteristics of benchmarks. *ferret* is very sensitive to the change of α. Its quality degradation increases sharply as the α increases. In other applications, average network delay benefits increase sharply while the quality degradation increases gradually. Even though *ferret* shows 34.6% average network reduction with a 50% α value, corresponding quality loss of 26% is not acceptable. So ABDTR can work well with error-torrent applications.

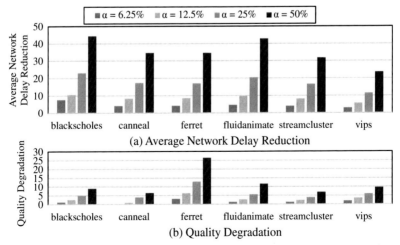

Fig. 5 Effect of ABDTR on (A) network average delay reduction, (B) quality degradation.

3.4.2 Performance and energy trade-offs with quality

Fig. 6A shows the reduction in NoC average delay with ABDTR for maximum 1%, 3%, 5%, and 10% quality degradation. We have explored this trade-off by setting different drop-rate regulation ranges as elaborated above. The baseline is the default system with no-regulation. As shown in Fig. 7A, ABDTR improves NoC average delay by 30.9% on average, with 10% quality loss. The maximum average delay reduction is over 40% (44.4%) for *blackscholes*. As the acceptable quality loss increases, ABDTR brings more benefits of the average network delay reduction. Even at a modest 3% acceptable quality loss, ABDTR reduces average network delay by more than 15% in some applications. The ABDTR consistently reduces the NoC traffic delay for all the benchmarks. Due to different traffic characteristics, the improvement is different case by case. The results show the effectiveness of our proposed mechanism in traffic regulation.

Fig. 6B and C illustrates the speedup and the energy saving of NoC on different benchmarks, respectively. ABDTR yields, on average, and 15.8% speedup and 24.4% energy saving with 10% quality loss. The energy saving is

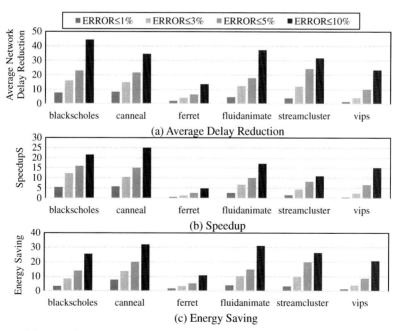

Fig. 6 (A) Network average delay reduction, (B) full-system performance improvement, and (C) NoC energy saving reduction for at most 1%, 3%, 5%, and 10% quality degradation.

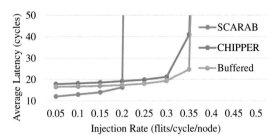

Fig. 7 Average latency for single-flit uniform random traffic on bufferless and buffered NoCs.

high as 32% for *canneal* and 31.1% for *fluidanimate* with 10% quality loss. The maximum speedup is over 25% for *canneal*. Moreover, With 5% quality loss, the average performance and energy gains are 10% and 14%, respectively. Thus, ABDTR is able to navigate the trade-off between quality loss and performance-energy improvement based on the user requirements.

Comparing Fig. 6A–C shows that the benefits strongly correlate with the quality loss. Every application, with more quality loss, benefits more of the speedup, energy saving and network average delay reduction from ABDTR. This strong correlation suggests that ABDTR is able to significantly improve both performance and energy consumption by flow regulation through approximating traffic data.

3.4.3 Cost analysis

The hardware cost of ABDTR is mainly due to the controller, throttler and approximator. One ABDTR occupies an area of 883 μm^2 and consumes 1.8 mW of power, according to the synthesis result reported by Synopsys Design Compiler and PrimePower targeting a 45-nm TSMC library. Using DSENT with a 45-nm CMOS technology, a router with three VCs and 16-flit depth FIFO has a total power of 125 mW (under a random traffic trace). and a total area of 67,686 μm^2. Each router configures an ABDTR device. Therefore, the ABDTR circuit's power and area overheads are 1.44% and 1.30%, respectively, of those of an 8 × 8 NoC.

4. Approximate bufferless network-on-chip

The NoC serving as an effective interconnection fabric connects many on-chip components. It provides better scalability and higher bandwidth compared to traditional interconnections such as the bus and crossbar

[48–50]. However, NoCs consume a significant amount of power in CMPs, that is, 40% of the tile power consumption in the 16-tile MIT RAW chip [51], 28% in the 80-tile Intel TeraFLOPS chip [22], and 19% in the 36-tile SCORPIO chip [52]. Buffers consume a large portion of network power and area [22, 23]. This motivates the bufferless design in low-power NoC architecture.

In the past, several bufferless NoCs have been proposed and achieve great power and area savings at a cost of lower throughput compared with conventional buffered networks [23, 53–55]. For example, the CHIPPER [53], a misrouting-based bufferless NoC, reduces the network energy by 55% and achieves 36% area savings by eliminating buffers, but it increases the average runtime of benchmarks by 13.6%. We compared the average latency of bufferless and buffered NoCs as shown in Fig. 7. The SCARAB [54] and CHIPPER [53] are two state-of-the-art bufferless designs. The former is an optimized NACK-based network that sends NACK messages to trigger retransmissions of dropped packets in the event of a collision, and the latter enables NACK-less operations by misrouting the conflicting flits. Compared with the buffered NoC, the SCARAB network shows a lower bandwidth, and the CHIPPER NoC runs with a higher average latency and this becomes worse as the injection rate increases. Therefore, bufferless NoCs are only suitable for low injection rates. However, in the big data era, it is essential to design high-performance bufferless NoC that achieves low latency and high bandwidth while maintaining the energy and area benefit of bufferless design.

Instead of in-router buffering, NACK-based bufferless NoCs discard contending packets in the event of a collision and then retransmit them from the sources. We evaluate the ratio of retransmitted packets under uniform random workloads. All the injections are single-flit packets. Fig. 8 shows that the percentage of retransmitted packets increases as the injection rate increases. When the injection rate is greater than 0.2, more than 50% packets are retransmitted, and about half of them are served more than once. Thus, we intend to reduce packet retransmissions for throughput improvement (part of retransmitted packets are dropped instead of being reinjected). Fig. 9 shows the average latency in bufferless network with different retransmission reduction ratios. Obviously, reducing retransmissions results in a great latency reduction and a bandwidth improvement. And the greater the retransmission reduction, the greater the return is. Therefore, the main goal of this paper is to reduce network retransmissions with low quality loss and achieve a performance improvement in bufferless NoC.

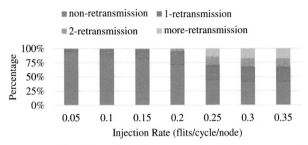

Fig. 8 The percentage of packets with different retransmissions under uniform random workloads.

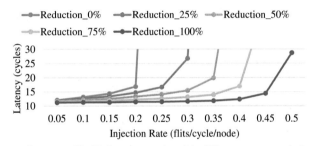

Fig. 9 Average latency of bufferless networks with different retransmission reduction ratios under single-flit random workloads.

In this work, we design the ABNoC, an approximate bufferless NoC. It relaxes the transmission accuracy to reduce packet retransmissions and increase network throughput. The ABNoC is a lossy NACK-based bufferless design. It can discard conflicting approximable flits without retransmission and recover them after packet transmission, thereby reducing packet retransmissions and improving bufferless network performance.

4.1 Approximation framework

4.1.1 Approximation design

In this paper, flits are delivered independently in ABNoC. In each router, any incoming flit performs routing and port allocation. Fig. 10 shows the principle of ABNoC. In case of a routing conflict, only one flit can be allocated successfully and cross the router while all the conflicting flits being dropped. The AAM triggers a NACK message for packet retransmission when a nonapproximable flit is dropped. For the approximable ones, they will be discarded directly and be recovered at the destination node through packet approximation method after the packet transmission. This design will

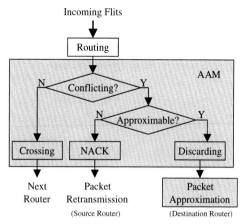

Fig. 10 ABNoC operation flowchart.

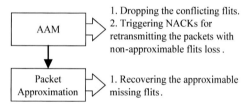

Fig. 11 Approximation framework overview.

reduces the retransmission of packets with approximable flits. In ABNoC, not all nonapproximable flit drops will trigger NACKs immediately in the network conflicts. Some NACKs are triggers in the destination routers when nonapproximate flits in a packet are found missing. The approximate transmission will be expanded in the next subsection.

Another important design is the packet approximation method that works for recovering the approximable missing flits. The packet approximation in ABNoC includes two parts: approximate encoding before packet injection and approximate decoding after packet transmission. We encodes all the approximable flits of a multiflit packet into a nonapproximable flit and transmit it with the multiflit packet. The nonapproximable flits can be preserved in ABNoC. Thus, the missing approximable flits can be decoded from the encoded flit. This approximate method only works for multiflit packet transmission. All the single-flit packets in our design are non-approximable and can achieve lossless transmissions in ABNoC.

Fig. 11 shows the overview of the approximate design. In each router the AAM drops the conflicting flits and triggers NACKs for retransmitting the packets with nonapproximable flits loss. Thereby, the conflicting approximable flits are discarded without retransmission. And ABNoC results in the packet retransmission reduction and network conflicts mitigation. After a packet transmission, the missing approximable flits in the packet will be recovered through the packet approximation method. The ABNoC provides an in-network traffic reduction can accurately capture the network congestions. In ABNoC, only the conflicting approximable flits are discarded and approximated, while others still have lossless delivery. Therefore, ABNoC could improve bufferless network performance while maintaining low application error.

4.1.2 Approximate NACK-based bufferless transmission

In ABNoC, the flit transmissions and ACK/NACK feedbacks are separated. Each packet is transmitted with an exclusive NACK channel for ACK/NACK message transfer. For single-flit packet, the single flit configures the NACK channel while it is transmitted over a router. For multiflit packet, only the head flit is transmitted with a configured NACK channel, while others enable NACK-free transmission and travel independently in ABNoC. There are no buffers at the switch ports in ABNoC, so flits cannot be stopped. In every cycle, flits that arrive at the input ports contend for the output ports. When two or more flits contend for the same output port, only one can be transferred through the output channel, and others are dropped. If the single-flit packet or the head flit of a multiflit packet fails in any allocation or there are no remaining NACK channels, it will be dropped. Then, an NACK message travels along the preconfigured NACK channel back to the source to trigger the whole packet retransmission. Other flits are free to transmit and perform independent routing and port allocation in each router. If they are dropped in a conflict, the NACK message will not be generated immediately. After a multiflit packet finishes its transmission,[a] it will be checked whether all the nonapproximable flits have been received.[b] If there is any non-approximable flit being lost, the NACK channel corresponding to this packet will signal an NACK message to its source node for packet retransmission, and the received flits of the packet will be discarded. For other missing flits,

---

[a] When the head flit and the last flit of a multiflit packet reach the destination or the limited waiting time of the head flit at destination node is over, the packet transmission is considered complete.

[b] The approximable flits' quantity of a multiflit packet is stored in its head flit, and the approximable status is also stored in each flit. Thus, if the number of received nonapproximable flits does not equal the total flits number minus the approximable flits' quantity, there is at least one nonapproximable flit being dropped.

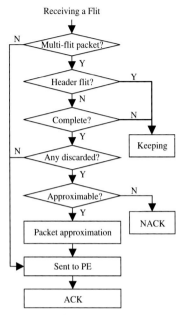

Fig. 12 Destination node operation flowchart.

they will be recovered through the packet approximation method, while an ACK signal will be triggered for releasing the exclusive NACK channel.

Fig. 12 shows the overall flow in the destination node. When a flit successfully reaches its destination, if it is a single-flit packet, it can be sent directly to the processing element (PE) and then an ACK message will be signaled back. Or if it is the head flit of a multiflit packet, it will be kept in the destination node, waiting for the remaining flits of the packet. All the received flits of a multiflit packet can be kept in destination node until the packet transmission is over. After a multiflit packet transfer is completed, the packet will be checked whether there is any flit being discarded. The loss of nonapproximable flits will trigger NACK messages and the approximable missing ones can be approximated. If no nonapproximable flits are lost, an ACK message will be fed back along the exclusive NACK path.

4.2 Architecture design for ABNoC
4.2.1 Multiplane architecture
The ABNoC router separates packet transmission and ACK/NACK rollback by a multiplane architecture to avoid traffic collisions caused by ACK/NACK messages. The ABNoC provides a two-plane architecture:

one for data transfer named data network and the other for ACK/NACK rollback named NACK network. Each packet transmitting in the data network obtains an exclusive channel of the NACK network to the ACK/NACK message rollback. In the NACK network, the exclusive channel is allocated based on the allocation results of the data network. Thus, the NACK network acts as a preconfigured circuit-switched network. Note that each packet is delivered with only one exclusive channel of the NACK network. For a multiflit packet, only the head flit shares allocation results with paired routers in the NACK network and is transmitted with an exclusive NACK channel. Moreover, the head flit or a single-flit packet can be successfully allocated in each router only if the productive output is obtained and a free NACK channel exists along its output path.

Fig. 13 shows the high-level architectural depiction of the ABNoC router. The ABNoC connects the processing elements (PEs). In ABNoC, each router consists of two interrelated but physically separate components: the data router and NACK router. The data router is a fully functional bufferless router, while the NACK router does not need routing computation, and it is configured with the allocation results from the data router. Each data router also contains an AAM that can discard the conflicting approximable flits without NACK messages.

4.2.2 Bufferless data network
4.2.2.1 Bufferless router
Fig. 14 illustrates the design of the bufferless router in the data network. Each router consists of five multiplexers: four cardinal neighbor connections and one local connection. There are no buffers needed for packet transmission in a router. Each cardinal neighbor connection is only equipped with a latch for

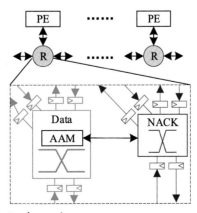

Fig. 13 ABNoC architectural overview.

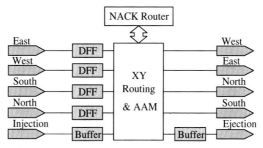

Fig. 14 Bufferless data router architecture.

5-bit	1-bit	3-bit	3-bit	3-bit	3-bit	3-bit	128-bit
Priority	S/H	$X_{destination}$	$Y_{destination}$	X_{source}	Y_{source}	Flit_ID	Data

Header

Fig. 15 The flit packetization.

storing a flit that may come in from the output port of neighbor direction at any cycle. And all arriving flits contend for output ports and are forced to be routed out in the next cycle, except the flits in injection port. Injection can only occur when the required output slot is free. Therefore, some injection buffers are needed to store the flits waiting to be injected. Some ejection buffers are also necessary for receiving the arrived packets. But the injection and ejection buffers can also be replaced with the miss status handling registers (MSHRs) as in previous bufferless designs [53, 54].

The bufferless data router supports XY routing and AAM. The XY routing and AAM are combined together and can quickly determine whether the incoming flits are being served or discarded without retransmission. Due to the lossy transmission, the data router simplifies the complex conflict prevention design in previous bufferless NoCs. It is with a lightweight router architecture which consumes very little area and power and reduces the pipeline stages for single-cycle delivery in a router. In Fig. 14, the assignments of input ports relative to output ports is "twisted." The reason is that straight-through router traversals ($N \Leftrightarrow S$ and $E \Leftrightarrow W$) are more common than turns.

In the data router, each flit in a packet starts with some header bits for independent routing and port arbitration. Fig. 15 shows the flit packetization in an 8×8 mesh network. The header bits are encoded as follows: a 5-bit priority, 1-bit single/header indicator, 6-bit destination address, 6-bit source address and 3-bit flit ID. The 128-bit data bits are used for storing some int or float values. In ABNoC, the XY routing and AAM is a priority-based design.

The nonapproximable flits will have a higher priority than approximable ones. And the priority of nonapproximable flits is incremented with retransmissions. The first 4 bits of the 5-bit priority are used to record the number of retransmissions of nonapproximable flit. The last 1 bit indicates the approximable info of flit (1 for nonapproximable, 0 for approximable). The priority of an approximable flit is always 0. The priority and destination info is used for the XY routing and port allocation. The 1-bit single/header indicator determines whether a flits needs to configure an NACK channel during the transmission. Only the single-flit packet and the head flit of multiflit packet is transmitted with a NACK channel and shares the allocation results with a paired NACK router. The 3-bit flit ID has two functions. It indicates the place of a flit in an 8-flit data packet. And in the encoded head flit, it stores the number of approximable flits in a 8-flit data packet. Based on the source, ID and flit approximable info, at the destination node, we can determine whether the transmission of a multiflit packet is complete and whether there is a nonapproximable flit loss.

4.2.2.2 Routing
The bufferless data network employs XY routing. The productive output port of each flit is based on its destination address, which is stored in its header bits as X and Y values. Routing computation in each router determines the correct output port by comparing the destination address with the local address. XY routing provides each packet with a unique transmission path, so the flits in a multiflit packet are routed independently but with the same output. Therefore, the order of flits in a multiflit packet will not be changed with the bufferless transmission.

4.2.2.3 AAM
The AAM has three functions: (1) allocating the output port to flits based on the routing result, (2) dropping contending flits with low priority, and (3) triggering an NACK message when a single-flit packet or a head flit is dropped. With XY routing, we fix the arbitration scheme for each output port. All the input ports are hardcoded with a default priority: north > south > west > east > injection, which means if flits with the same priority from two different input ports, one always take precedence based on this default priority. This design allows all the incoming flits to quickly be either directed to their preferred output ports or dropped in a conflict. The fixed arbitration follows a rule that a flit going straight has a higher priority than a flit that is turning. The injection port always has the lowest priority because the flits

can be buffered first in the injection port. If a flit in the injection port does not succeed in arbitration for its output port, we can try again in the next cycle until it is injected into the network for transmission.

4.2.2.4 Implementation

The required signals for routing and allocation are obtained from the header bits of the incoming flits at each input port. In Fig. 16, the $V_{direction}$, $P_{direction}$, $S/H_{direction}$, $X_{direction}$, $Y_{direction}$ correspond to the valid bit, priority, single or head flit, destination X and Y values, respectively. Once a flit arrives at an input port, the valid bit will be activated, which indicates a valid flit is waiting at the input port. X_r and Y_r are the local address of the working router. $NACK_{direction}$ represents whether there is at least one NACK wire not being assigned at the output port direction in the NACK network.

Fig. 16A shows the east and west output routing and allocation. Based on the XY routing, the east and west direction output only needs to consider the latches of their opposing input ports and the injection port, regardless of the flit priority. Any time that a flit arrives at either the east or west input port with a nonzero X value, it will be guaranteed to be forwarded straight since it has the highest priority in our fixed arbitration scheme. Routing and arbitration for the north and south output ports are more complicated since they need to consider flits turning.

Fig. 16B shows that the arbitration follows the default hardcoded priority. The multiplexers always check the opposing input port first and then determine if the flit in west input port is turning, followed by the east input port, and finally the injection port, if all the input flits have the same priority. Otherwise, the flit with higher priority will be preferred. Note that a flit traverses at most three 2-to-1 multiplexers from its input port to its output port, which keeps the critical path delay low.

The logic circuit in the ejection port, shown in Fig. 16C, is similar to that of the north and south output ports. Four channels are selected by three multiplexers, and the incoming flits are ranked based on their flit priorities and the default hardcoded rule. The source address of a packet can never be the same as its destination address, thus eliminating the need to connect the ejection port to the injection port. To determine if a packet is destined for the ejection port, both the X and Y values need to be zero.

In addition, each output port is equipped with a selector. If the output flit of the outermost multiplexer is a head flit or a single-flit packet, the selector will work based on the $NACK_{direction}$ signal. Otherwise, the flit will go directly through the selector. The selector is used to ensure that each head

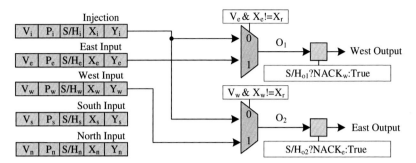

(a) Structure for East and West output

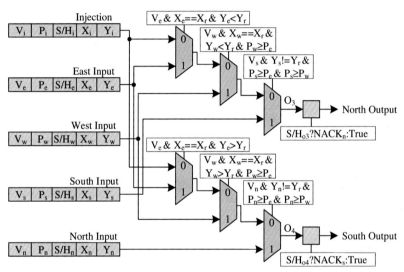

(b) Structure for North and South output

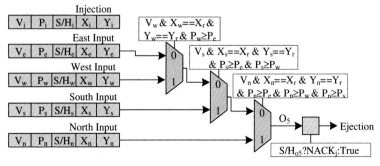

(c) Structure for Ejection

Fig. 16 Routing and AAM.

flit or single-flit packet traversing the router configures a 1-bit NACK path. If the $NACK_{direction}$ signal is false, the head flit or single-flit packet will be dropped and an NACK signal will travel along the corresponding NACK path to trigger retransmission. Moreover, the design performs routing and AAM together to allow flits to traverse each router in a single cycle.

4.2.2.5 Deadlock and Livelock
The XY routing and AAM greatly simplify route computation and port arbitration. Our design can quickly direct flits from the input port to their productive output ports and determines which flits to be dropped with or without the NACK signal in the event of a conflict. All the conflicting flits are dropped instead of being blocked in ABNoC. The bufferless data network is deadlock-free in nature due to the XY routing and AAM design.

To prevent the livelock problem caused by saturating priority, we provide 4-bit priority increment, which supports up to 15 retransmissions before saturation. This can satisfy most packet retransmission requests. Even for the worst case in which a packet is retransmitted up to 15 times, we will stop injecting other packets with 15 retransmissions until this packet finishes transmission. This design will ensure that there is at most one 15-priority packet that is transmitted in the network and avoids livelock.

There is another situation that may lead to a livelock. Flits perform independently in each router, so a multiflit packet injection may be interrupted by transmissions of other flits due to the lowest priority of the injection port. The interruption will result in a long injection interval between two flits of the packet. In addition, for the multiflit packet, the injection period from the head flit to the last flit is the same as its ejection period due to no network congestion. After the ejection space is filled, if the multiflit packet has not completed the transmission, a livelock will be generated. To avoid the livelock caused by the limited ejection space being filled, the packet injection period needs to be limited. In our design, each router allocates a counter of fixed E cycles in the injection port. When the head flit of a multiflit packet is injected, the counter starts to decrease with the number of cycles until it reaches zero or the last flit of the packet is injected. When the count is over, the remaining flits of the multiflit packet will be dropped. Then, the packet will be stored into MSHR and wait for retransmission. Each router injects one flit per cycle. Thus, the $E + 1$ must be larger than the size of the data packet, where the 1 is for the head flit. The waiting time of head flit in destination node will not exceed E cycles, since it is the same as the packet injection period.

4.2.3 NACK Network

4.2.3.1 Multichannel circuit-switched network

The NACK network in the ABNoC operates as a small 5×5 mux-based switch, and each port is equipped with multiple 1-bit channels, which is similar to the design in SCARAB [54]. Fig. 17 illustrates the overview of an NACK router. There are no buffers and routing and arbitrator micro-architecture. The NACK router works with only some logic to control the crossbar for establishing or releasing a 1-bit pathway to transfer the acknowledgment (ACK) and negative acknowledgment (NACK) messages. The routing and allocation information comes from its paired data router. When all channels of a port are assigned, an $NACK_{direction}$ signal will be activated and fed back to the data router. Compared with the traditional NACK-based bufferless NoC that needs to send a full packet to request retransmission, the 1-bit channel NACK network results in a very energy efficient means of signaling feedback.

Moreover, messages in the NACK network take two cycles per hop. Any message sending out on an even clock cycle can only be transferred by the next router in a later even cycle. This enables us to use time division multiplexing (TDM) technology to halve the number of inter-router NACK channels and sustain network bandwidth. This design reduces the number of NACK channels for area and power saving in ABNoC. The TDM allows each inter-router NACK circuit channel to transfer two messages that are differentiated by even and odd cycles. The circuit path setup and message transmission are separate in

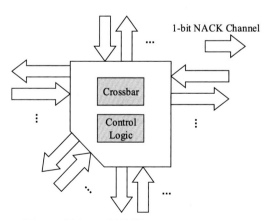

Fig. 17 Overview of the multichannel NACK router.

the NACK network. Thus, any packet transmitted in the data network needs to be NACKed or ACKed after an even period, which drives the maximum injection period (E) to be an even value.

4.2.3.2 Path establishment and release

The NACK network works as a general circuit-switch network, where a dedicated circuit channel must be set up before data transmission. Every packet traverses a router with configuring a 1-bit channel of the NACK network. This NACK channel is connected along the packet's entire path, such that the ACK or NACK message is transmitted along the preconfigured circuit path. In the NACK network, each feedback message transfers each preconfigured hop with two cycles, which creates a deterministic delay of message transmission of $2 \times N$ cycles, where N is the number of hops of the preconfigured circuit path.

Unlike the SCARAB design that allows NACK channels to be reallocated after the implicitly ACKed transmission delay, the ABNoC releases a circuit path with the NACK and ACK message. When a router in the NACK network receives a feedback message, the router releases the corresponding configured connection with the previous node. This means that each established path is only used once for transferring the NACK or ACK message before being released. Moreover, the NACK network in ABNoC is more energy efficient in comparison to that in the SCARAB due to not needing counters to record the implicit ACKed epochs.

In the ABNoC design, a feedback message can be triggered in three situations: (1) An NACK signal is triggered when the head flit or single-flit packet is dropped during its transmission; (2) signaling NACK at the destination node since a nonapproximable flit is not received after the packet transmission; and (3) an ACK signal travels from destination node to the source when the packet is delivered successfully, which means all the nonapproximable flits are received at the destination node. As the transmission delay in both the NACK network and data network are proportional to the delivery hops ($2\times$ hops), the time of the transmission delay plus the NACK/ACK message feedback time will be four times the hops. At the destination node, the ejection period is the same as injection period which is less than E cycles. The maximum waiting time of head flit is E cycles. Thus, the window of time that a packet could be NACKed or ACKed after initial transmission is deterministically bounded, as shown in Eq. (4),

$$L = 4 \times \left(|X_s - X_d| + |Y_s - Y_d| + 1 \right) + E \qquad (4)$$

where L is the maximum delay, E is the maximum injection period, (X_s, Y_s) and (X_d, Y_d) are the source address and the destination address, respectively.

4.3 Packet approximation

Previous approximation designs in NoC focus on the reduction of data injection [24, 25, 56]. They discard some approximable data before being injected into the network. However, the injected flits in these designs still need to face network conflicts and wait, which results in long latency of contending flits. These end-to-end traffic reduction schemes cannot response to network conflicts on the spot. Therefore, we propose an in-network traffic reduction scheme which can accurately capture the network workload. In ABNoC, contending flits are discarded instead of waiting, and the approximable ones can be safely approximated after packet transmission. Packet approximation is a stage for recovering the lost approximable flits. In this paper, we propose a new value approximation method for approximating the missing flits with low quality degradation.

In packet approximation, each multiflit packet starts with a nonapproximable head flit. All the approximable flits are encoded into the head flit. ABNoC ensures that the nonapproximable flit will not be lost in packet transmission. So that, the discarded approximable flits can be decoded from the head flit at the destination node.

Fig. 18 shows a packet approximation example with a 4-flit packet transfer. We assume that the first three flits are approximable and the last one is

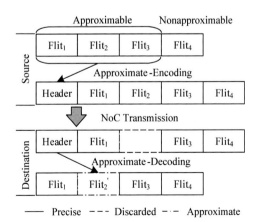

Fig. 18 Approximate encoding and decoding of a 4-flit packet.

nonapproximable. Note that the number and the place of approximable flits have no effect on our design. All the approximable flits can be encoded into a head flit that is signed nonapproximable. The head flit and the existing 4 flits comprise a new 5-flit packet and are injected into the ABNoC. During the packet transmission, any approximable flit could be discarded in a network conflict. In the example, we assume the approximable $flit_2$ is discarded. After the packet reaches the destination router, the $flit_2$ is recovered as the $flit_2'$ by decoding the head flit.

In this work, we focus on 32-bit integer and floating-point value approximation. The 128-bit data of a flit consists of 4 integer or floating-point values. Each flit recovery is to approximate the four values. Note that a flit is annotated as approximable only when the flit stores the words of a same type and they are all approximable. The approximate encoding and decoding will be elaborated in the following subsection.

4.3.1 Approximate encoding

The approximate encoder works by using the 128-bit data space of head flit to encode the words in all approximable flits. The number of approximable flits in a packet is 0 to 8. If the number is 0, we will set the last flit as approximable. This will not cause a loss of quality to the packet, since the head flit can duplicate the words in the approximable flit in a one-approximable-flit packet Thus, there is at least one approximable flit in any multiflit packet. In approximate encoding, the 128-bit data space of head flit is divided according to the number of approximable flits. Then each divided part is used to encode a 4-value approximable flit. The number of approximable flits (1–8) can be stored in the 3-bit ID of head flit.

Algorithm 1 shows the encoding process. In approximate encoding, the 128-bit data space of head flit can be divided into 1, 2, 4 or 8 equal parts, shown as the Fig. 19. Specifically, 1 is for one approximable flit, 2 for two, while 4 is for three and four approximable flits, and 8 is for five, six, seven or eight. Then all the approximate flits are encoded into the divided parts in order.

Encoding a flit is to approximately store the four 32-bit integers or floating-point values of the flit. First, we encode each 32-bit value as a 16-bit data. Fig. 20 shows the integer and floating-point value encoding. The most significant bit of the 16-bit data is used to indicate the value type, 0 for integer and 1 for floating-point. For the integer value, we shift data to

Input: $n \leftarrow$ the number of approximable flits

$flit_0, ..., flit_{n-1} \leftarrow$ the data bits in approximable flits

Output: $H \leftarrow$ the data bits in head flit

1 **if** $n == 1$ **then**

2 $\quad$ $k \leftarrow 1;$ // k is number of equal parts.

3 **else if** $n == 2$ **then**

4 $\quad$ $k \leftarrow 2;$

5 **else if** $n == 3, 4$ **then**

6 $\quad$ $k \leftarrow 4;$

7 **else**

8 $\quad$ $k \leftarrow 8;$

9 **end**

10 $s \leftarrow \frac{128}{k};$ // The data space is divided according to k.

11 **for** $i = 0; i ; n ; i++$ **do**

12 $\quad$ $\{value\} \leftarrow flit_i;$ //$\{value\}$ is the data set stored in $flit_i$.

13 $\quad$ **for** $j = 0; j \leqslant \frac{s}{16}; j++$ **do**

14 $\quad\quad$ //Storing encoded data into H

15 $\quad\quad$ $H \leftarrow H << 16$

16 $\quad\quad$ $H \leftarrow H|(encoding(value_j));$

17 $\quad$ **end**

18 **end**

19 **return** H

the right by n digits for converting an integer value to a maximum 10-bit value (A-value). The shifted number n is stored in the 16-bit data with the 10-bit value. For example, an integer value $+445566789$ will be shifted to the right by 20 digits and stored as $+0 \times 1A8$ (A-value). The coding result is $0 \times 51A8$. The floating-point value is represented as Eq. (5),

$$float = (-1)^S \times (1 + .mantissa) \times 2^{exponent} \qquad (5)$$

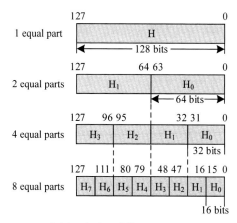

Fig. 19 128-bit data space division in head flit.

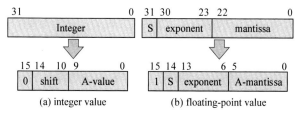

(a) integer value   (b) floating-point value

Fig. 20 Integer and floating-point value approximate encoding.

where S is the sign bit. The mantissa and exponent are stored in a 23-bit and an 8-bit spaces separately, as shown in Fig. 20B We propose to approximate only the mantissa field of the floating-point value. We encode the 23-bit mantissa in a 6-bit A-mantissa by removing the 17-bit LSB and the sign bit and exponent bits are reserved.

Second, the encoded values are stored into a divided part (H_i). Different divisions result in different sizes to store these values. When the divided part is full, the rest encoded values that can not be stored will be discarded. For example, if there are three approximable flits, each flit will be encoded into a 32-bit space. Only the first 2 values of the approximable flit can be encoded and stored in a divided space. The remaining 2 values of the flit will be discarded.

4.3.2 Approximate decoding

Decoding is to recover the missing flits. When a multiflit packet finishes transmission, the decoder will first determine the divisions of the data bits in the head flit by the number of approximable flits. Then, each missing flit will be determined which division it encodes. Finally, we can recover the data in the missing flit.

Algorithm 2 shows the decoding process. If there is only one approximable flit, the values stored in the head flit will be applied directly to recovering the approximable flit. For more than one approximable flit, the data space in the head flit is divided into 2 or 4 or 8 parts based on the approximable quantity. Each part can recover an approximable flit. Different divisions result in different sizes being used for recovering a flit. For example, two divisions will result in each part have 64 bits to recover a flit, while for 8-division scenarios, there are only 16 bits in each division. Each 16 bits of a division can recover a 32-bit data in the flit. The 4 data of a flit will require 64 bits to recover. If there are not enough bits to recover the flit, the remaining data in the flit can be restored to the last decoded data. The decoding that recovers a 32-bit data from 16 bits in a division is the opposite process of encoding. It pads the missing bits with zeros.

4.4 Methodology

We evaluate the effectiveness of the ABNoC in comparison with the outstanding NACK-based bufferless network (SCARAB) [54] and a compression-based approximate design (APPROX-NoC) [25]. The network configuration parameters are listed in Table 3.

4.4.1 NoC configurations

We compare ABNoC against a nonapproximate bufferless NoC and an approximate bufferless NoC. The configurations are listed below:

- **SACRAB**: In this configuration, the NoC is the nonapproximate bufferless SCARAB. The workload is uncompressed and achieves losslessly delivery. All the data packets are filled with 8 flits. As detailed in Ref. [54], each flit needs to perform two stages (switch traversal and link delay) in each SCARAB router. For purposes of fair comparison, the SCARAB network referenced in this paper is designed with priorities but without the opportunistic buffering, as the opportunistic buffering needs router-side buffers and could likewise be applied to the other bufferless design for reducing retransmission latency.

- **SACRAB_APPROX**: There is no approximate design in bufferless NoC so far. We apply the approximation framework APPROX-NoC [25] to the SACRAB network. In this configuration, the network is the nonapproximate SCARAB, but the injection packets are compressed and damaged. Following the APPROX-NoC design, all the data packets

Input: $n \leftarrow$ the number of approximable flits

  $H_j \leftarrow$ the missing flit

  $H \leftarrow$ the data bits in head flit

Output: $V' = \{value_0', ..., value_3'\} \leftarrow$ the data set in the missing flit

1 **if** $n == 1$ **then**

2   $\quad k \leftarrow 1;$ // k is number of equal parts.

3 **else if** $n == 2$ **then**

4   $\quad k \leftarrow 2;$

5 **else if** $n == 3, 4$ **then**

6   $\quad k \leftarrow 4;$

7 **else**

8   $\quad k \leftarrow 8;$

9 **end**

10 $s \leftarrow \frac{128}{k};$

11 $H_j \leftarrow H >> (n - j - 1) \times s;$

12 **for** $i = 0;\ i \leqslant 3;\ i{+}{+}$ **do**

13   $\quad m \leftarrow \frac{s}{16};$

14   $\quad$ **if** $i <\leqslant m$ **then**

15     $\quad\quad$ //recovering a value from 16 bits

16     $\quad\quad$ ecoding $value_i' \leftarrow decoding(0xFFFF\&(H_j >> (m - i - 1) \times 16))$

17   $\quad$ **else**

18     $\quad\quad$ //repeat the last value to recover the rest values in $flit_i$

19     $\quad\quad value_i' \leftarrow value_{i-1}'$

20   $\quad$ **end**

21 **end**

22 **return** V'

Table 3 NoC configurations.

	SCARAB	ABNoC
Channel width	128 bits	149 bits
Routing algorithm	Adaptive	XY
Data packet	8 flits	9 flits
Control packet	1 flits	
Pipeline stages	2	
Network size	8 × 8	

are compressed with the frequent-pattern approximate compression technique which is called FP-VAXX. Its error threshold is set as 10%. The compression latency is three cycles and the decompression latency is two cycles, which is the same as the default configuration in APPROX-NoC.

- **ABNoC**: This is our proposed approximate bufferless NoC. All the data packets are equipped with an extra encoded head flit, but the data are uncompressed. The flits in ABNoC contain more 21-bit header bits for routing and port allocation and routers have the same channel width as the flit size.

4.4.2 Synthetic traffic

We evaluate the average latency and average number of retransmissions in NoCs under several synthetic traffic patterns. We use a cycle-accurate, in-house NoC simulator. All injected packets are multiflit packets, and we randomly select 50% of the packets to be approximable. A varying percentage of approximable flits are also studied to show their impact on the performance of ABNoC. As stated in Ref. [25], the APPROX-NoC enhances the compression ratio by up to 41% and increases the nonapproximate compression ratio by 30%. Therefore, we reduce the approximable packet by 3 flits and the nonapproximable packet by 2 flits. The approximable data packet is compressed into a 5-flit packet, while the nonapproximable ones become 6-flit packets.

We also evaluate the arrival rate in ABNoC to show the percentage of received flits. Its value is the ratio of the number of received flits (not including decoding flits) to the number of input flits. The arrival rate determines the output quality. A high arrival rate means that only a small amount of flits will be approximated. Bandwidth is another important parameter for the NoCs.

It is the injection rate where the network latency becomes excessive. A large bandwidth means low packet latency at high injection rates. In this paper, we evaluate the bandwidth as the maximum injection rate where network latency is less than 100 cycles.

4.4.3 Full-system simulation

To evaluate the impact of our ABNoC mechanism on the overall application output error, we utilize the Pin tool, a dynamic binary instrumentation framework [57]. We hand-annotate the benchmarks mentioned below, in a similar fashion to the load value approximation [5], to identify the data regions that can be approximated. The ABNoC only focuses on integer and floating-point value approximation. An important consideration while hand-annotating approximable data regions is the data type of the variables being determined to be approximable. We annotate the flits as approximable only if their words are approximable and have the same type (integer or floating-point). In ABNoC, there is at least one approximable flit in each data packet due to the encoding and decoding algorithm. If all the flits of a data packet are nonapproximable, the last flit will be annotated as approximable. For a full-system evaluation, we use a modified event-driven many-core simulator [47]. Table 4 lists the full-system simulation parameters.

4.4.4 Benchmarks

The benchmarks used for performance evaluation are selected from the PARSEC [58]. We configure each application with 64 threads and run the benchmarks for 100 million instructions with small input size. All the simulation benchmarks and their quality metrics are listed in Table 5

Table 4 Full–system simulation system configurations.

Number of cores	64 (x86 ISA)
L1 D cache (private)	128 KB, 64–B lines, 4 cycles
L1 I cache (private)	128 KB, 64–B lines, 4 cycles
L2 cache (shared)	512 KB slice/node, 64–B lines, 8 cycles
Caching protocol	MESI
DRAM controllers	4 (on the node 0, 16, 32, 48)
DRAM size	2 GB, 45 cycles

Table 5 Benchmarks.

Name	Quality metric
blackscholes	Percentage of results with error greater than 1%
bodytrack	Average error of the output vectors
canneal	Relative error of the final routing
fluidanimate	Percentage of particles with location error greater than 0.2%
raytrace	Average error of the output vector
streamcluster	Percentage of points that are not accurately clustered
swaptions	Average error of the output prices
vips	Average relative error of final output RGB image's pixel values
x264	The peak signal-to-noise ratio

4.5 Experimental results

4.5.1 Design parameters exploration

4.5.1.1 Injection period

In ABNoC, packet injection period needs to be limited in E cycles. Fig. 21 shows the average packet latency and the arrival rate in ABNoC by varying the limited injection period (E). All the packets are injected randomly. At low injection rates, the average packet latencies with different E values have hardly changed. When the injection rate exceeds 0.2, with E decreases, the average latency and the arrival rate weakly decreases. The reason is that the reduction of E results in more flits to be discarded before injection, thereby the waiting time before injection will be reduced. At a low injection rate, there is little waiting time before injection and the reduction of E reduces the arrival rate and achieves little latency benefit. In ABNoC, considering the arrival rate and the latency benefit, the default value of E is set as 16.

4.5.1.2 NACK Channels

Generally, multiple NACK channels are to meet the high injection rate requirements. We evaluate an 8×8 network with 4, 8, 12, and 16 logical NACK channels. Fig. 22 shows the percentage of failed transmissions due to the lack of the NACK channel. The workload is random single-flit traffic, so that each flit applies a dedicated NACK channel. As the injection rate increases, the NACK channel competition becomes increasingly more intense. At a low injection rate, a small number of NACK channels can satisfy NACK channel contention, whereas it results in a large amount of

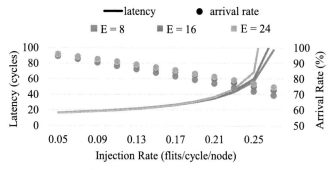

Fig. 21 Maximum count of injection counter sensitivity analysis.

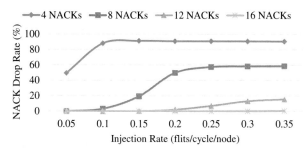

Fig. 22 Percentage of flits dropped due to lack of the NACK channel under a random single-flit traffic workload.

packet drops due to NACK channel contention at a high injection rate. In our tests, when the NACK network is equipped with 16 logical channels in each port, there is little packet drop due to lack of NACK channel. Therefore, we choose the 16 logical NACK channels in ABNoC. In physical design, there are 8 NACK channels per port due to the TDM design.

4.5.2 Performance analysis

In this section, we present the NoC level performance evaluation of the ABNoC using simulated workloads with different patterns.

4.5.2.1 Average Latency and arrival rate

We evaluate the average latency of ABNoC on a random and tornado traffic pattern. Fig. 23A and B shows the results comparing with the SCARAB and SCARAB_APPROX. ABNoC shows a great decrease in average latency compared with SCARAB and SCARAB_APPROX. ABNoC achieves

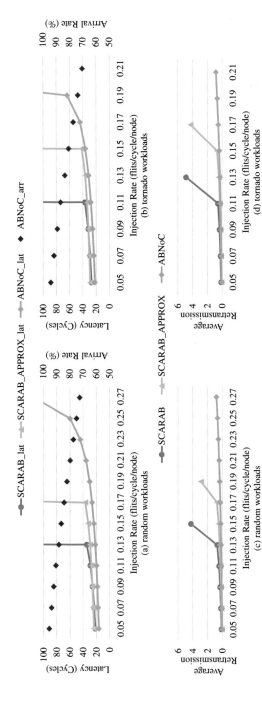

Fig. 23 Throughput and retransmission analysis under random and tornado traffic patterns.

the lowest average packet latency and the highest bandwidth. The average packet latency of SCARAB_APPROX is higher than that of SCARAB when the injection rate is very low. It is due to the SCARAB_APPROX needs more time for compression and decompression packets. At a high injection rate, SCARAB_APPROX reduces the queuing time at source node and slightly reduces network conflicts due to fewer injections. The ABNoC is dedicated to reducing network conflicts by allowing the approximable collision flits to be discarded without retransmission. Therefore, as the injection rate increases, the network conflict intensifies, ABNoC achieves great benefits in latency reduction. Under random workloads, ABNoC achieves an approximate 1.92× and 1.47× bandwidth increase compared with SCARAB and SCARAB_APPROX. For the tornado traffic pattern, ABNoC increases the bandwidth by 1.73× with respect to SCARAB and 1.27× compared with SCARAB_APPROX.

Fig. 23A and B also shows the arrival rate in the ABNoC under random and tornado traffic patterns. Across the simulation workloads, the arrival rate is always higher than 70% even when the network saturates. This means that most of the flits are delivered without loss, even for the approximable flits.

Packet Retransmission is an important parameter affecting network performance in NACK-based bufferless NoCs. The retransmitted packets could exacerbate network conflicts. Fig. 23C and D shows the results of the average number of retransmissions. It shows that the ABNoC can greatly reduce packet retransmissions. The average number of retransmissions in ABNoC is very small even when network throughput has been saturated. ABNoC can greatly reduce network conflicts, thereby most of the packets travel through ABNoC are without retransmission. In addition, due to reducing network injections, the average number of retransmissions is also reduced by the SCARAB_APPROX. But the reductions are much smaller than that in ABNoC. The reason is that the end-to-end traffic reduction cannot immediately react to network congestions, while the in-router AAM of ABNoC can directly discard the conflicting flits.

4.5.2.2 Sensitivity

We show the sensitivity of the percentage of approximable flits. Fig. 24 shows the average packet latency and the arrival rate of the ABNoC as the percentage of approximable flits is varied by 25%, 50% (default), and 75%. The arrival rates are plotted until network saturation. Obviously, the throughput increases and the arrival rate decreases as the percentage of approximable packets increases. The reason is that a high approximable ratio increases the chances

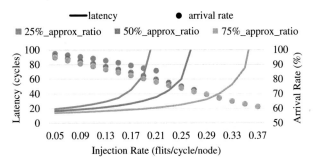

Fig. 24 Approximable flits ratio sensitivity analysis.

of flit discarding without retransmission. Furthermore, for the arrival rates of the 50% and 75% approximable ratios, there is little difference between them. It is due to that at the 50% approximable ratio, network conflicts have been greatly reduced in ABNoC. The arrival rate will be mainly determined by the injection rate and the workload pattern. At a higher approximable ratio, the arrival rate in ABNoC changes very little.

4.5.3 Full system simulation

In this section, we evaluate the performance of ABNoC in a full-system simulation. We present the average packet latency, average number of retransmissions, system runtime, and application error on a range of PARSEC benchmarks.

4.5.3.1 Overall system performance

We analyze the impact of ABNoC on the average network latency and overall system speedup. Fig. 25A shows the normalized average latency for benchmarks compared against the SCARAB and the SCARAB_APPROX. The average latency is normalized to SCARAB. Across the benchmarks, ABNoC reduces the latency by average 34.6% with respect to SCARAB_APPROX and by average 46.7% to SCARAB. This is mainly because the approximate transmission of ABNoC brings more reduction in the number of packet retransmissions, leading to latency benefits. Fig. 25B shows the average number of retransmissions of benchmarks. Obviously, ABNoC greatly reduces packet retransmissions for all the benchmarks. On average, ABNoC reduces packet retransmissions by 83.6% with respect to SCARAB and 81.3% to SCARAB_APPROX. The SCARAB_APPROX also contributes a minor reduction in the average retransmission. Thus, the SCARAB_APPROX also achieves an average 18.5% reduction in latency

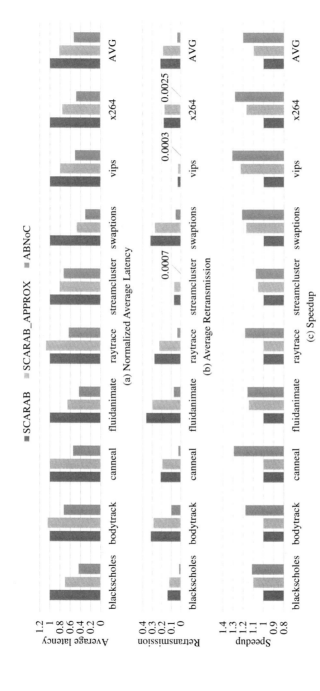

Fig. 25 Network latency, retransmission, and system performance for full-system analysis.

compared to SCARAB. However, for *bodytrack* and *raytrace*, SCARAB_APPROX slightly increases the latency, and for *canneal*, it only achieves moderate improvement. The reason is that the latency benefit of traffic reduction is offset by the compression/decompression overheads in SCARAB_APPROX.

Fig. 25C shows how ABNoC affects overall system runtime. All the results are normalized to SCARAB. The evaluation shows that ABNoC and SCARAB_APPROX achieve an average 1.20× and 1.09× speedup compared with SCARAB, respectively. And for all the benchmarks, ABNoC achieves a higher speedup than SCARAB_APPROX. This is due to the higher reduction of ABNoC in packet latency.

4.5.3.2 Arrival rate and error

The arrival rates and the result errors in ABNoC is shown in Fig. 26. The average arrival rate is over 92%. This means that ABNoC delivers most of the flits without loss. Besides, we analyze the relative error of the encoding algorithm in ABNoC. For integer value and floating-point value, their relative error can be calculated by Eqs. (6) and (7).

$$
error_{int} =
\begin{cases}
\dfrac{\sum\limits_{k=0}^{n} X_k 2^k}{\sum\limits_{k=0}^{n+7} X_k 2^k + 2^{n+8}}, & int < -512, int \geqslant 512 \\[4mm]
0, & -512 \leqslant int < 512
\end{cases}
\tag{6}
$$

where the $error_{int}$ is the relative error of a integer value; the X_k is the kth bit value of the integer value and $X_k \in \{0, 1\}$; n is the shifted bits ($1 \leq n \leq 22$).

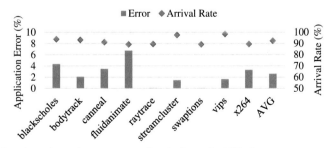

Fig. 26 Benchmark result accuracy and arrival rate in ABNoC.

$$error_{float} = \frac{\displaystyle\sum_{k=7}^{23} X_k 2^{-k}}{1 + \displaystyle\sum_{k=1}^{23} X_k 2^{-k}} \tag{7}$$

where the $error_{float}$ is the relative error of a floating-point value; the X_k is the kth bit value of the mantissa and $X_k \in \{0, 1\}$. Therefore, according to the summation of geometric progression,[c] the maximum error of a integer value must be less than 0.39% and the maximum relative error of a floating-point value encoding is less than 1.56%.[d] These relative errors are very low. In addition, the data discarded in encoding are adjacent to the encoded data in the cache. They could be very similar. Thus, they could be recovered by the encoded data with low error. Besides, most of the flits does not need to be approximated due to the high arrival rate in ABNoC. Evaluations shows that all applications have an error of less than 7%.

4.5.4 Power consumption and area overhead
In this section, we evaluate the effect of ABNoC on the network power consumption and area overhead. To show the variation in dynamic power consumption among SCARAB, SCARAB_APPROX and ABNoC, we depict the dynamic power of different benchmarks in Fig. 27 using the Orion power simulator [59]. All the results are normalized to SCARAB.

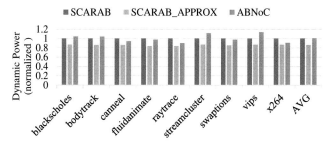

Fig. 27 Normalized dynamic power consumption analysis.

---

[c] $\sum_{k=1}^{n} q^{k-1} = (1 - q^n)/(1 - q)$, where n is the number of terms, and q is the common ratio in the sequence.

[d] $error_{int} < 2^{-8}$, $error_{float} < 2^{-6}$.

In ABNoC, some benchmarks (*canneal, fluidanimate, raytrace, swaptions,* *×264*) consume less dynamic power than that in SCARAB. This is primarily attributed to the reduction in packet retransmissions. However, the others increase their dynamic power consumptions. The main reason is that flits in ABNoC contain more 21-bit bits than that in SCARAB and need more energy to pass a router. When the increase in power consumption is larger than the power savings of retransmission reduction, the dynamic power will increase. For an average comparison, ABNoC has a similar dynamic power consumption to SCARAB. SCARAB_APPROX achieves dynamic power reduction for all the benchmarks compared with SCARAB, It is due to the reduction in the number of injected flits.

To ensure accurate hardware modeling, the ABNoC router is implemented in RTL based on the open source RTL router design [60]. We use the Synopsys design compiler with TSMC 65 nm technology to evaluate the power and area. The results of SCARAB router power and area are cited from Ref. [54]. For purposes of fair comparisons with SCARAB, the injection and ejection buffers are also designed to use MSHRs in the ABNoC router, which is the same as the SCARAB design. We also compare with a regular buffered router that equips with five ports (128-bit channel width), four virtual channels (4-flit each) per port, and a 3-stage pipeline based on the XY routing. All the routers work at a 1.9 GHz frequency (same as the SCARAB). And we assume a constant uniform load on all input ports (same as the SCARAB). The results of router power and area are listed in Table 6. Compared with SCARAB, the ABNoC router has slightly more consumption in terms of power and area. It incurs 5.0% more power consumption and 14.1% more area overhead. It is due to the more bits of channel width in ABNoC and the additional packet approximation design. However, these increments are much smaller relative to the area and power overhead in the buffered design. ABNoC reduces 76.2% power consumption and 88.9% area overhead compared to the buffered router. ABNoC still maintains the area and power benefit of bufferless design.

Table 6 Router power and area comparison.

	Power	Area
SCARAB	46.2 mW	34.1 Kμm^2
ABNoC	48.5 mW	38.9 Kμm^2
Buffered	194.1 mW	307.3 Kμm^2

5. Approximate multiplane network-on-chip

Reducing the power of the NoC while increasing performance is essential for scaling up to larger systems for future CMP designs. Minimizing power consumption requires more efficient use of network resources. Multiplane NoCs have shown their efficiency in total bandwidth usage [23, 61]. Furthermore, multiplane NoCs can be designed with heterogeneous physical subnetworks; as a result, messages are injected into different subnetworks to satisfy different transmission properties. For many approximation-enabled applications, not all messages need to be delivered without loss; that is, approximable data can be transmitted with loss to simplify the NoC architecture and achieve a relatively small area overhead and reduced power consumption. This observation intuitively suggests that instead of having one interconnected plane serving all the on-chip traffic, the NoC may be physically split into two planes: a lossless plane for transmitting nonapproximable traffic and an area- and power-efficient plane that operates with lossy delivery and serves approximable traffic.

In this work, we propose an approximate multiplane NoC (AMNoC). The proposed AMNoC includes a lightweight, low-latency, lossy-transmission NoC (the approximate subnetwork, approx-subnet) and a conventional NoC without traffic loss (the lossless subnetwork, lossless-subnet). The approx-subnet is bufferless with a lightweight router architecture that consumes very little area and power. The lossy transmission in the approx-subnet allows flits to be discarded in the event of a conflict and recovers the missing flits after packet transmission; therefore, there is no congestion in the approx-subnet, and packets are transferred with low latency. In addition, the lossless-subnet provides guarantees that nonapproximate flits can arrive at their destinations without loss.

5.1 Architecture design for AMNoC

5.1.1 Overview

5.1.1.1 Multiplane architecture

In this section, we present the AMNoC design, which consists of a two-plane architecture. Fig. 28 shows the high-level architectural depiction of the AMNoC router. In the AMNoC, which connects the processing elements (PEs), there are two physically separate subnetworks: lossless-subnet and approx-subnet. There is no connection between these two subnetworks, which run independently. The lossless-subnet is a full-functional

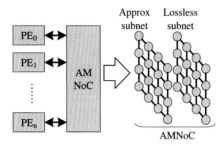

Fig. 28 AMB-NoC architecture overview.

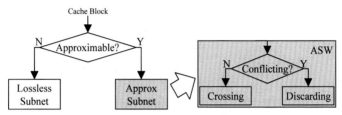

Fig. 29 Approx-subnet operation flowchart.

buffered NoC, while the approx-subnet is a lightweight bufferless NoC that relaxes the transmission accuracy to provide low-latency packet delivery with low power and low area overhead. This two-plane architecture reduces the power consumption of approximable packets while maintaining the lossless transmission of nonapproximable packets.

5.1.1.2 Approximate design

Bufferless NoCs have been shown to be very effective at conserving power and area under low network loads; in contrast, due to retransmission and misrouting, bufferless NoCs incur a significant performance penalty under high network loads [53, 54]. In the AMNoC, all approximable packets are transmitted by the bufferless approx-subnet, which is designed based on an approximate switch (ASW) design. As a result, instead of being retransmitted or misrouted, the contending flits are discarded in a network conflict. Fig. 29 shows the workflow in the AMNoC. There are no buffers at the switch ports in the approx-subnet, so any incoming flits cannot be stopped. In every cycle, flits that arrive at the input ports contend for the output ports. When two or more flits contend for the same output port, only one can be transferred through the output channel, while all other conflicting flits are discarded without being collected. In the approx-subnet, after an approximable packet finishes its transmission, the missing flits will be recovered based on the received flits. Based on the lossy design, the approx-subnet transmits packets without congestion. Through simplified routing and port allocation, the approx-subnet also achieves single-cycle hop delivery, and packets in the approx-subnet can be transmitted with ultra-low latency.

5.1.1.3 Multiplane packet transmission

As an approximate multiplane design, approximable packets are transmitted through the approx-subnet, while nonapproximable packets are injected into the lossless-subnet. However, some critical latency-sensitive messages are non-approximable, and their transmission latencies greatly influence the overall application performance [23, 61]. In particular, the control packets are all single-flit and nonapproximable. In the AMNoC, we also inject these critical messages into the approx-subnet to provide an opportunity for the ultra-fast delivery of these messages; however, these critical messages are also injected into the lossless-subnet to guarantee the lossless delivery of nonapproximable data. Hence, any critical messages dropped by the approx-subnet will still arrive at its destination through the lossless-subnet. Therefore, the dropped critical messages in the approx-subnet do not need to be recovered. Critical messages include all control packets and the first flit in each data packet since the first flit of a data packet contains the initially requested word. When the first flit arrives at the destination node, the processor can continue executing with the received word before the rest of the packet has arrived. In addition, the first flit of an approximable packet is injected into the lossless-subnet; this prevents all the flits in an approximable packet from being discarded in the approx-subnet, which may cause the data request or synchronization to fail. At least one flit exists in each approximable packet, which can reach the destination node. In the AMNoC, the approx-subnet carries approximable data packets, control packets, and the first flits of nonapproximable data packets, while the loss-subnet transmits control packets, nonapproximable data packets, and the first flits of approximable data packets, as shown in Fig. 30. Although some messages are transmitted twice, requiring more energy, the overall

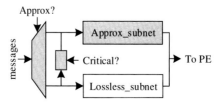

Fig. 30 Multiplane packet transmission in AMNoC.

power overhead is decreased. The reason for this decrease is that packets are transmitted with a much lower power in the bufferless approx-subnet than in regular buffered transmission, and the approximate transmission also reduces energy consumption. The AMNoC therefore provides low-latency transfer for critical messages and minimizes the power consumption of approximable messages.

5.1.2 Approx-subnet design
5.1.2.1 Bufferless router

The AMNoC is designed as a 2D mesh topology, which is commonplace in modern systems. Fig. 31 shows the architecture of a bufferless router in the approx-subnet. Each router consists of five input and output ports: one for each of the four cardinal neighbor connections and one for the local connection. In each neighbor connection, only a latch is used to store an incoming flit from the neighbor direction at any cycle. There are no buffers in any of the four neighbor connections, so incoming flits cannot be stopped. All arriving flits contend for output ports and are forced to be routed out in the next cycle.[e] The injection and ejection port is equipped with some buffers since the packets in local routers can be injected into the approx-subnet only when their ideal output slot is free. Hence, packets in injection ports can be buffered instead of being discarded in allocation conflicts. In addition, ejection buffers are necessary for receiving the arrived packets. In the approx-subnet, the local injection is also controlled by a timer. Multiflit packet injection could be interrupted by flits from other input ports. This would incur additional complexity in receiving the packets to determine the completion of multiflit packet transfer. The timer works to limit the

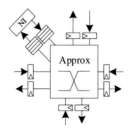

Fig. 31 Bufferless approximate router architecture.

---

[e] If an incoming flit is successfully allocated to an output port, it will be transferred to its neighbor router or ejection buffer; otherwise, it will be discarded.

injection time of each multiflit packet. When the first flit of a multiflit packet is injected, the timer is turned on. Each of the remaining flits must be injected in one cycle; otherwise, the remaining flits will be discarded. All multiflit packets in the approx-subnet are approximable, and the discarded flits can be recovered in the destination node. In addition, the bufferless router supports the ASW design, and the ASW quickly determines whether the incoming flits are to be transferred or discarded.

5.1.2.2 ASW
In the bufferless router, each flit contains some header bits for independent routing and port arbitration. In the approx-subnet, flits are transmitted based on the ASW design, which employs XY routing and simplified port allocation. XY routing computes the productive output port of each flit based only on the destination address, which is stored in the header bits as X and Y values, and provides each packet with a unique transmission path; thus, the flits in a multiflit packet are routed independently but without disorder at the destination node.

In the ASW design, arbitration is based only on a default priority and the approximable notation. The default priority follows a rule that a flit traveling straight has a higher priority than a flit that is turning. The injection port has the lowest priority since the flits can be buffered first in the injection port. Therefore, the default priority is north > south > west > east > injection port. The approximable notation is also stored in the header bits as "1" for approximable and "0" for nonapproximable. Moreover, approximable flits have a higher priority than nonapproximable flits since discarding approximable flits may incur errors in the result, while nonapproximable flits dropped in the approx-subnet are still transmitted by the lossless-subnet.

Fig. 32 shows the ASW design. The required signals are denoted by $V_{direction}$, $A_{direction}$, $X_{direction}$, and $Y_{direction}$, which correspond to the valid bit ("1" for an incoming flit and "0" for none), the approximable notation, and the destination X and Y values, respectively. The east and west output structure, north and south output structure, and ejection port structure are shown in Fig. 32A–C, respectively. Flits are transmitted first along the west (if $X_{direction} > X_r$) or east (if $X_{direction} < X_r$) direction and then along the north (if $Y_{direction} < Y_r$) or south (if $Y_{direction} > Y_r$) direction. Approximate flits always have a higher priority in 2-to-1 multiplexers; the default priority is further hardcoded by three 2-to-1 multiplexers in Fig. 32B and C. Moreover, Fig. 32 shows that any flit traverses at most three 2-to-1 multiplexers from the input port to its productive output port, which reduces

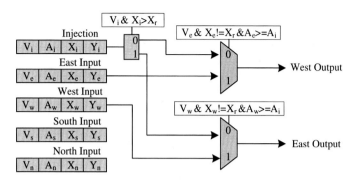

(a) Structure for East and West output

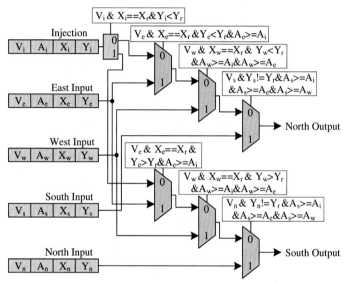

(b) Structure for North and South output

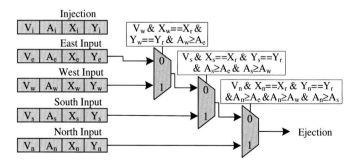

(c) Structure for Ejection

Fig. 32 ASW design.

the critical path delay. Additionally, the routing, allocation, and traversal in the router are combined, and thus, flits pass through each hop in only a single step. Therefore, the approx-subnet achieves single-cycle hop delivery without congestion.

The ASW design greatly simplifies route computation and port arbitration. Packets are transmitted through a single-cycle latency pipeline, and all conflicting flits are dropped instead of creating congestion. The approx-subnet is thus naturally free of deadlocks.

5.1.3 Packet recovery

In previous approximate designs, some annotation frameworks have been proposed that label sections of the approximable data [5, 45]. We manually annotate benchmarks in a fashion similar to these methods. In the AMNoC, a multiflit data packet is annotated as approximable only when the packet stores words of the same type (integer or floating point) and only if those words are all approximable. At the destination node, packet recovery is implemented to approximate the missing data in an approximable data packet. Two stages are required to recover a data packet: (1) determine the missing flits of a data packet and (2) approximate the values in the missing flits.

In the approx-subnet, flits of an approximable data packet are injected (or dropped) within a single-cycle interval. Due to the ASW design, these flits are transmitted in succession. Therefore, at the destination node, flits from the same approximable packet can be easily gathered together. The location of a flit in the packet is also stored in its header bits and transmitted along with the flit. After an approximable packet is transmitted,[f] the missing flits can be determined based on the locations of the received flits.

Previous studies have proposed many value approximation designs, such as last value, stride, FCM [37], and VTAGE [38]. Reducing the complexity and enhancing the accuracy are the two main challenges of value approximation. In this context, the above-mentioned methods are either insufficiently accurate [37] or excessively complex, thereby incurring a high power consumption and a high overhead [38]. We choose linear interpolation due to its low complexity and reasonable accuracy. Linear interpolation requires only one addition to perform value approximation. The data in an

---

[f] The transmission of an approximable packet is considered complete when the last flit of the packet reaches the destination, when the maximum waiting time (<packet size) of the first flit received at the destination node is reached, or when the first flit transmitted by the lossless-subnet has been received.

approximable packet are generally fetched from successive memory blocks; hence, the data in a packet (e.g., the adjacent pixels in an image) are considerably similar. The data of dropped flits are most relevant to the preceding and following flits. Therefore, the linear interpolation error is small. If the dropped flit is the first flit or last flit, its approximation is to copy the received flit that is the closest to it.

5.2 Methodology

5.2.1 NoC configurations

We evaluate the effectiveness of the AMNoC in comparison with a baseline single-plane buffered network.

The configurations are listed below:

- **Baseline**: In this configuration, the NoC is a single lossless buffered NoC with 16-byte channels, as in Table 7.
- **AMNoC**: This NoC is composed of a lossy bufferless plane (approx-subnet) and a regular buffered plane (lossless-subnet). The approx-subnet has 10-byte channels (8 bytes for data and 2 bytes for header bits), while the lossless-subnet operates with 8-byte channels. The total channel width of the AMNoC is 18 bytes, but it uses 16 bytes for data transfer, which is the same as the baseline NoC. The approximate multiflit data packets and critical single-flit packets are injected into the approx-subnet, while the nonapproximate multiflit data packets and single-flit control packets are carried by the lossless-subnet. Data packets in the buffered NoC must be packaged with a head flit for routing, while a head flit is not required in the bufferless NoC. Thus, the data packet size in the approx-subnet is 1 flit smaller than that in the lossless-subnet.

Table 7 NoC configurations.

	Baseline	Lossless-subnet	Approx-subnet
Channel width	16 bytes	8 bytes	10 bytes
Virtual channels	4 × 4	4 × 4	none
Data packet	5 flits	9 flits	8 flits
Control packet	1 flits	1 flits	1 flits
Pipeline stages	3	3	1
Injection/ejection		16 flits	
Routing algorithm		XY	
Network size		8 × 8	

Full-system simulation For a full-system evaluation, we use a modified event-driven many-core simulator [47]. Table 4 lists the full-system simulation parameters.

Benchmarks The benchmarks used for performance evaluation are selected from the PARSEC [58]. We configure each application with 64 threads and run the benchmarks for 100 million instructions with small input size. All the simulation benchmarks and their quality metrics are listed in Table 5

5.3 Experimental results

5.3.1 Evaluation with synthetic traces

5.3.1.1 Average latency

We use synthetic traces to evaluate the communication latency with varying traffic loads. We generate traces such that there are 30 K data messages in the NoCs. The synthetic traces are then studied with different traffic patterns and different injection rates. In the baseline NoC, each data message contains 5 flits. In the AMNoC, a nonapproximable message is injected into the lossless-subnet as a 9-flit packet, while an approximable message is injected into the approx-subnet as an 8-flit packet, and its first flit is also injected into the lossless-subnet. We randomly select 50% of the data messages to be approximable. Fig. 33 shows the average latency for the baseline NoC and

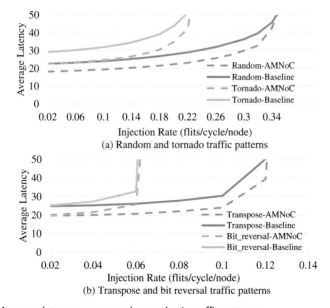

Fig. 33 Average latency curves under synthetic traffic patterns.

AMNoC under different synthetic traffic patterns. The AMNoC shows a significant decrease in average latency compared with the baseline NoC; this is because the approximable messages in the approx-subnet are transmitted with low latency. The AMNoC and baseline NoC saturate around the same injection rate under different traffic patterns; the reason for this phenomenon is that the lossless-subnet and the baseline network have a similar architecture, and the selection of 50% approximable packets generates the same injection rate in both the lossless-subnet and the baseline network.

5.3.1.2 Dropped flit ratio
Some flits are dropped in the AMNoC, and the drop rate will influence the output quality of applications. Here, we evaluate the dropped flit ratio, which is the percentage of dropped flits among all transmitted flits, as shown in Fig. 34. The dropped flit ratio is plotted until network saturation is reached. As illustrated, all the dropped flit ratios are less than 14%. This means that most data can be transmitted without approximation.

5.4 Evaluation with full-system analysis
5.4.1 Latency and speedup
In this section, we evaluate the average packet latency, system runtime, and dropped flit ratio on a variety of PARSEC benchmarks. Fig. 35 shows the normalized average latency and speedup for the benchmarks compared against the baseline NoC with 50% approximable data packets. For all the benchmarks, the AMNoC reduces the latency by an average of 41.9% with respect to the baseline NoC and achieves an average 1.12× speedup.

5.4.2 Sensitivity
Next, varying approximable packet ratios are studied to investigate their impacts on the AMNoC performance. Fig. 36 shows the normalized packet

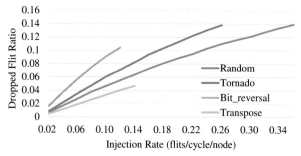

Fig. 34 Dropped flit ratios in the AMNoC.

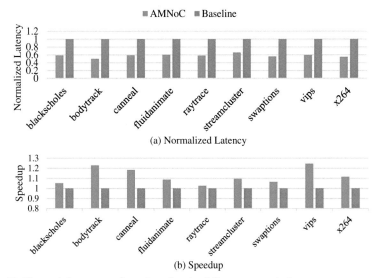

(a) Normalized Latency

(b) Speedup

Fig. 35 Network latency and runtime for the full-system analysis.

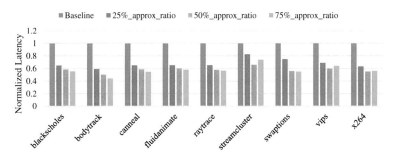

Fig. 36 Approximable packet ratio sensitivity analysis.

latency as the percentage of approximable flits is varied by 25%, 50%, and 75%. Evidently, in most applications, the average latency decreases as the percentage of approximable packets increases; this relationship is observed because a high approximable packet ratio increases the percentage of packets transmitting in the approx–subnet with low latency. The more approximable packets there are the greater the improvement in the average latency. For "streamcluster" "vips" "×264," the average latency decreases when approximable flit ratio increases from 25% to 50% but increases when approximable flit ratio increases from 50% to 75%. This is due to that 75% approximable flits brings more workload to the approx–subnet and cause more waiting time before packet injection. When the increased waiting time exceeds the benefits of transmitting time, the packet latency becomes even greater. However, it is clear that the average latency has improved significantly in all experiments, which shows the effectiveness of the AMNoC design.

5.4.3 Approximated data ratio

The data that are recovered in destination node is called approximated date. Fig. 37 shows the percentage of approximated data under different approximable packet ratios. When the approximable packet ratio is 25%, the percentage of approximated data can be less than 0.1%. For a 50% approximable packet ratio, the proportion of approximated data in all applications is less than 5%. Even when applications are annotated with 75% approximable data packets, their approximated data ratios can still be less than 10%. These results show that only a very small amount of data needs to be recovered and most data can be transmitted losslessly in AMNoC. This is due to the low dropped flit ratio in the AMNoC.

5.5 Power and area

We model the power consumption and area overhead using the design space exploration for network tool (DSENT) [62] with a 45-nm complementary metal oxide semiconductor (CMOS). The results are shown in Table 8. We use the average injection rate obtained in our full-system simulation across all benchmarks to simulate the dynamic power consumption. The AMNoC achieves 48.6% power savings and reduces the area overhead by 25.7%.

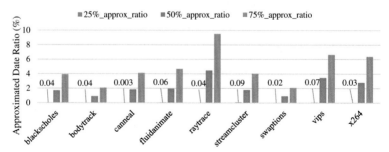

Fig. 37 The percentage of recovered data in benchmarks.

Table 8 Router power and area comparison.

	Total power	Total area
AMNoC	19.11 mW	0.084 mm^2
Basline	37.15 mW	0.113 mm^2

References

[1] S. Mittal, A survey of techniques for approximate computing, ACM Comput. Surv. 48 (4) (2016) 1–33.

[2] A. Raha, H. Jayakumar, S. Sutar, V. Raghunathan, Quality-aware data allocation in approximate DRAM, in: Proceedings of the 2015 International Conference on Compilers, Architecture and Synthesis for Embedded Systems, IEEE, 2015, pp. 89–98.

[3] A. Sampson, J. Nelson, K. Strauss, L. Ceze, Approximate storage in solid-state memories, ACM Trans. Comput. Syst. 32 (3) (2014) 9:1–9:23.

[4] Y. Luo, S. Govindan, B. Sharma, M. Santaniello, J. Meza, A. Kansal, J. Liu, B. Khessib, K. Vaid, O. Mutlu, Characterizing application memory error vulnerability to optimize datacenter cost via heterogeneous-reliability memory, in: Proceedings of the 44th Annual IEEE/IFIP International Conference on Dependable Systems and Networks, IEEE, 2014, pp. 467–478.

[5] J.S. Miguel, M. Badr, N.E. Jerger, Load value approximation, in: Proceedings of the 47th Annual IEEE/ACM International Symposium on Microarchitecture, IEEE Computer Society, 2014, pp. 127–139.

[6] R.S.t. Amant, A. Yazdanbakhsh, J. Park, B. Thwaites, H. Esmaeilzadeh, A. Hassibi, L. Ceze, D. Burger, General-purpose code acceleration with limited-precision analog computation, ACM SIGARCH Comput. Archit. News 42 (3) (2014) 505–516.

[7] A. Yazdanbakhsh, G. Pekhimenko, B. Thwaites, H. Esmaeilzadeh, O. Mutlu, T.C. Mowry, RFVP: rollback-free value prediction with safe-to-approximate loads, ACM Trans. Archit. Code Opt. 12 (4) (2016) 62:1–62:26.

[8] J. Sartori, R. Kumar, Branch and data herding: reducing control and memory divergence for error-tolerant GPU applications, IEEE Trans. Multimedia 15 (2) (2013) 279–290.

[9] L. Renganarayana, V. Srinivasan, R. Nair, D. Prener, Programming with relaxed synchronization, in: Proceedings of the 2012 ACM Workshop on Relaxing Synchronization for Multicore and Manycore Scalability, ACM, 2012, pp. 41–50.

[10] Y. Wang, J. Deng, Y. Fang, H. Li, X. Li, Resilience-aware frequency tuning for neural-network-based approximate computing chips, IEEE Transactions on Very Large Scale Integration (VLSI) Systems 25 (2017) 2736–2748. Article Number: 10. https://doi.org/10.1109/TVLSI.2017.2682885.

[11] H. Esmaeilzadeh, A. Sampson, L. Ceze, D. Burger, Architecture support for disciplined approximate programming, in: Proceedings of the Seventeenth International Conference on Architectural Support for Programming Languages and Operating Systems, IEEE, 2012, pp. 301–312.

[12] S. Venkataramani, V.K. Chippa, S.T. Chakradhar, K. Roy, A. Raghunathan, Quality programmable vector processors for approximate computing, in: Proceedings of the 46th Annual IEEE/ACM International Symposium on Microarchitecture, IEEE, 2013, pp. 1–12.

[13] A. Chandrasekharan, D. Große, R. Drechsler, ProAct: a processor for high performance on-demand approximate computing, in: Proceedings of the 2017 on Great Lakes Symposium on VLSI, ACM, 2017, pp. 463–466.

[14] G. Ndour, T.T. Jost, A. Molnos, Y. Durand, A. Tisserand, Evaluation of variable bit-width units in a RISC-V processor for approximate computing, in: Proceedings of the 16th ACM International Conference on Computing Frontiers, 2019, pp. 344–349.

[15] C.-C. Hsiao, S.-L. Chu, C.-Y. Chen, Energy-aware hybrid precision selection framework for mobile GPUs, Comput. Graph. 37 (5) (2013) 431–444.

[16] D. Peroni, M. Imani, H. Nejatollahi, N. Dutt, T. Rosing, ARGA: approximate reuse for GPGPU acceleration, in: Proceedings of the 56th ACM/IEEE Design Automation Conference, IEEE, 2019, pp. 1–6.

[17] M. Imani, R. Garcia, S. Gupta, T. Rosing, Rmac: runtime configurable floating point multiplier for approximate computing, in: Proceedings of the International Symposium on Low Power Electronics and Design, ACM, 2018, pp. 1–6.

[18] R.R. Osorio, G. Rodríguez, Truncated SIMD multiplier architecture for approximate computing in low-power programmable processors, IEEE Access 7 (2019) 56353–56366.

[19] C.K. Jha, J. Mekie, Seda-single exact dual approximate adders for approximate processors, in: Proceedings of the 56th Annual Design Automation Conference, ACM, 2019, pp. 1–2.

[20] T. Moreau, M. Wyse, J. Nelson, A. Sampson, H. Esmaeilzadeh, L. Ceze, M. Oskin, SNNAP: approximate computing on programmable socs via neural acceleration, in: Proceedings of the 21st IEEE International Symposium on High Performance Computer Architecture, IEEE, 2015, pp. 603–614.

[21] H. Esmaeilzadeh, A. Sampson, L. Ceze, D. Burger, Neural acceleration for general-purpose approximate programs, Commun. ACM 58 (1) (2014) 105–115.

[22] Y. Hoskote, S. Vangal, A. Singh, N. Borkar, S. Borkar, A 5-GHz mesh interconnect for a teraflops processor, IEEE Micro 27 (5) (2007) 51–61.

[23] Z. Li, J. San Miguel, N.E. Jerger, The runahead network-on-chip, in: Proceedings of the 2016 IEEE International Symposium on High Performance Computer Architecture, IEEE, 2016, pp. 333–344.

[24] V.Y. Raparti, S. Pasricha, DAPPER: data aware approximate NoC for GPGPU architectures, in: Proceedings of the Twelfth IEEE/ACM International Symposium on Networks-on-Chip, IEEE Press, 2018, pp. 7:1–7:8.

[25] R. Boyapati, J. Huang, P. Majumder, K.H. Yum, E.J. Kim, APPROX-NoC: a data approximation framework for network-on-chip architectures, in: Proceedings of the 44th Annual International Symposium on Computer Architecture, ACM, 2017, pp. 666–677.

[26] Y. Chen, M.F. Reza, A. Louri, DEC-NoC: an approximate framework based on dynamic error control with applications to energy-efficient NoCs, in: Proceedings of the 36th IEEE International Conference on Computer Design, IEEE, 2018, pp. 480–487.

[27] A.B. Ahmed, D. Fujiki, H. Matsutani, M. Koibuchi, H. Amano, AxNoC: low-power approximate network-on-chips using critical-path isolation, in: Proceedings of the Twelfth IEEE/ACM International Symposium on Networks-on-Chip, IEEE/ACM, 2018, pp. 6:1–6:8.

[28] G. Ascia, V. Catania, S. Monteleone, M. Palesi, D. Patti, J. Jose, Improving energy consumption of NoC based architectures through approximate communication, in: Proceedings of the 7th Mediterranean Conference on Embedded Computing, IEEE, 2018, pp. 1–4.

[29] G. Ascia, V. Catania, S. Monteleone, M. Palesi, D. Patti, J. Jose, Approximate wireless networks-on-chip, in: Proceedings of the 2018 Conference on Design of Circuits and Integrated Systems, IEEE, 2018, pp. 1–6.

[30] F. Betzel, K. Khatamifard, H. Suresh, D.J. Lilja, J. Sartori, U. Karpuzcu, Approximate communication: techniques for reducing communication bottlenecks in large-scale parallel systems, ACM Comput. Surv. 51 (1) (2018) 1–32.

[31] M.F. Reza, P. Ampadu, Approximate communication strategies for energy-efficient and high performance NoC: opportunities and challenges, in: Proceedings of the 2019 on Great Lakes Symposium on VLSI, 2019, pp. 399–404.

[32] M. Li, Q.-A. Zeng, W.-B. Jone, DyXY: a proximity congestion-aware deadlock-free dynamic routing method for network on chip, in: Proceedings of the 43rd Annual Design Automation Conference, ACM, 2006, pp. 849–852.

[33] P. Lotfi-Kamran, M. Daneshtalab, C. Lucas, Z. Navabi, BARP-a dynamic routing protocol for balanced distribution of traffic in NoCs, in: Proceedings of the 2008 Conference on Design, Automation and Test in Europe, ACM, 2008, pp. 1408–1413.

[34] L.P. Tedesco, T. Rosa, F. Clermidy, N. Calazans, F.G. Moraes, Implementation and evaluation of a congestion aware routing algorithm for networks-on-chip, in: Proceedings of the 23rd Symposium on Integrated Circuits and System Design, ACM, 2010, pp. 91–96.

[35] Y. Sazeides, J.E. Smith, The predictability of data values, in: Proceedings of the 30th Annual ACM/IEEE International Symposium on Microarchitecture, IEEE Computer Society, 1997, pp. 248–258.

[36] P. Marcuello, J. Tubella, A. González, Value prediction for speculative multithreaded architectures, in: Proceedings of the 32nd annual ACM/IEEE international symposium on Microarchitecture, IEEE Computer Society, 1999, pp. 230–236.

[37] B. Goeman, H. Vandierendonck, K. De Bosschere, Differential FCM: increasing value prediction accuracy by improving table usage efficiency, in: Proceedings of the Seventh IEEE International Symposium on High-Performance Computer Architecture, IEEE, 2001, pp. 207–216.

[38] A. Perais, A. Seznec, Practical data value speculation for future high-end processors, in: Proceedings of the 20th IEEE International Symposium on High Performance Computer Architecture, IEEE, 2014, pp. 428–439.

[39] A. Perais, A. Seznec, BeBoP: a cost effective predictor infrastructure for superscalar value prediction, in: Proceedings of the 21st IEEE International Symposium on High Performance Computer Architecture, IEEE, 2015, pp. 13–25.

[40] A. Perais, A. Seznec, EOLE: combining static and dynamic scheduling through value prediction to reduce complexity and increase performance, ACM Trans. Comput. Syst. 34 (2) (2016) 4:1–4:33.

[41] M. Carbin, S. Misailovic, M.C. Rinard, Verifying quantitative reliability for programs that execute on unreliable hardware, Commun. ACM 59 (8) (2016) 83–91.

[42] H. Esmaeilzadeh, A. Sampson, L. Ceze, D. Burger, ACM, Architecture support for disciplined approximate programming, ACM SIGPLAN Notices 47 (4) (2012) 301–312.

[43] W. Baek, T.M. Chilimbi, Green: a framework for supporting energy-conscious programming using controlled approximation, in: Proceedings of the 31st ACM SIGPLAN Conference on Programming Language Design and Implementation, 2010, pp. 198–209.

[44] J. Park, H. Esmaeilzadeh, X. Zhang, M. Naik, W. Harris, Flexjava: language support for safe and modular approximate programming, in: Proceedings of the 2015 10th Joint Meeting on Foundations of Software Engineering, ACM, 2015, pp. 745–757.

[45] A. Sampson, W. Dietl, E. Fortuna, D. Gnanapragasam, L. Ceze, D. Grossman, EnerJ: approximate data types for safe and general low-power computation, in: Proceedings of the 32nd ACM SIGPLAN Conference on Programming Language Design and Implementation, 2011, pp. 164–174.

[46] C. Bienia, Benchmarking modern multiprocessors, Ph.D. thesis, Princeton University 2011, january.

[47] X. Wang, M. Yang, Y. Jiang, P. Liu, M. Daneshtalab, M. Palesi, T. Mak, On self-tuning networks-on-chip for dynamic network-flow dominance adaptation, ACM Trans. Embed. Comput. Syst. 13 (2s) (2014) 1–21.

[48] S. Bell, B. Edwards, J. Amann, R. Conlin, K. Joyce, V. Leung, J. MacKay, M. Reif, L. Bao, J. Brown, Tile64-processor: a 64-core soc with mesh interconnect, in: Proceedings of the 2008 IEEE International Symposium on Solid-State Circuits Conference, IEEE, 2008, pp. 88–598.

[49] A. Agarwal, C. Iskander, R. Shankar, Survey of network on chip (noc) architectures & contributions, J. Eng. Comput. Archit. 3 (1) (2009) 21–27.

[50] S. Borkar, Future of interconnect fabric: a contrarian view, in: Proceedings of the 12th ACM/IEEE International Workshop on System Level Interconnect Prediction, ACM, 2010, pp. 1–2.

[51] M.B. Taylor, W. Lee, J. Miller, D. Wentzlaff, I. Bratt, B. Greenwald, H. Hoffmann, P. Johnson, J. Kim, J. Psota, Evaluation of the Raw microprocessor: an exposed-wire-delay architecture for ILP and streams, ACM SIGARCH Comput. Archit. News 32 (2) (2004) 2.

[52] B.K. Daya, C.-H.O. Chen, S. Subramanian, W.-C. Kwon, S. Park, T. Krishna, J. Holt, A.P. Chandrakasan, L.-S. Peh, SCORPIO: a 36-core research chip demonstrating snoopy coherence on a scalable mesh NoC with in-network ordering, in: Proceedings of the 41st ACM/IEEE International Symposium on Computer Architecture, IEEE, 2014, pp. 25–36.

[53] C. Fallin, C. Craik, O. Mutlu, CHIPPER: a low-complexity bufferless deflection router, in: Proceedings of the 2011 IEEE International Symposium on High Performance Computer Architecture, IEEE, 2011, pp. 144–155.

[54] M. Hayenga, N.E. Jerger, M. Lipasti, Scarab: a single cycle adaptive routing and bufferless network, in: Proceedings of the 42nd Annual IEEE/ACM International Symposium on Microarchitecture, IEEE, 2009, pp. 244–254.

[55] T. Moscibroda, O. Mutlu, A case for bufferless routing in on-chip networks, in: Proceedings of the 36th Annual International Symposium on Computer Architecture, ACM, 2009, pp. 196–207.

[56] L. Wang, X. Wang, Y. Wang, ABDTR: approximation-based dynamic traffic regulation for networks-on-chip systems, in: Proceedings of the 2017 IEEE International Conference on Computer Design, IEEE, 2017, pp. 153–160.

[57] C.-K. Luk, R. Cohn, R. Muth, H. Patil, A. Klauser, G. Lowney, S. Wallace, V.J. Reddi, K. Hazelwood, Pin: building customized program analysis tools with dynamic instrumentation, in: Proceedings of the 2005 ACM SIGPLAN Conference on Programming Language Design and Implementation, ACM, 2005, pp. 190–200.

[58] C. Bienia, S. Kumar, J.P. Singh, K. Li, The PARSEC benchmark suite: characterization and architectural implications, in: Proceedings of the 17th International Conference on Parallel Architectures and Compilation Techniques, ACM, 2008, pp. 72–81.

[59] H.-S. Wang, X. Zhu, L.-S. Peh, S. Malik, Orion: a power-performance simulator for interconnection networks, in: Proceedings of the 35th Annual ACM/IEEE International Symposium on Microarchitecture, IEEE Computer Society Press, 2002, pp. 294–305.

[60] D.U. Becker, Efficient microarchitecture for network-on-chip routers, Ph.D. thesis, Stanford University, Palo Alto, 2012.

[61] A.K. Abousamra, R.G. Melhem, A.K. Jones, Deja vu switching for multiplane nocs, in: Proceedings of the Sixth IEEE/ACM International Symposium on Networks-on-Chip, IEEE, 2012, pp. 11–18.

[62] C. Sun, C.-H.O. Chen, G. Kurian, L. Wei, J. Miller, A. Agarwal, L.-S. Peh, V. Stojanovic, DSENT-a tool connecting emerging photonics with electronics for opto-electronic networks-on-chip modeling, in: Proceedings of the Sixth IEEE/ACM International Symposium on Networks-on-Chip, IEEE, 2012, pp. 201–210.

About the authors

Ling Wang received the B.S. degree in monitoring and control technology from the Harbin University of Science and Technology, China, in 2010, and the M.S. degree in biomedical engineering from the Harbin Institute of Technology, China, in 2012, where he also get his Ph.D. degree in computer applied technology in 2021. He is currently a lecturer at Xidian University. His research interests include high-performance many-core architecture, network on chip and AI accelerator.

Xiaohang Wang received the B.Eng. and Ph.D. degrees in communication and electronic engineering from Zhejiang University, in 2006 and 2011, respectively. He is currently an Associate Professor with the South China University of Technology. His research interests include many-core architecture, power efficient architectures, optimal control, and NoC-based systems.

CHAPTER SEVEN

Power-efficient network-on-chip design by partial topology reconfiguration

Mehdi Modarressi[a,b] and S. Hossein SeyyedAghaei Rezaei[a]
[a]School of Electrical and Computer Engineering, Faculty of Engineering, University of Tehran, Tehran, Iran
[b]School of Computer Science, Institute for Research in Fundamental Sciences (IPM), Tehran, Iran

Contents

Advances in Computers, Volume 124
ISSN 0065-2458
https://doi.org/10.1016/bs.adcom.2021.11.001

217

Abstract

Network topology greatly affects the energy consumption, performance, cost, and design time/effort of the Networks-on-Chip (NoC) employed in today's Systems-on-Chip (SoCs). Most existing NoC architectures adopt either standard regular or customized topologies. In this chapter, a new architecture is proposed which is capable to establish virtual adaptive links between non-adjacent network nodes of a regular mesh-based NoC to shorten the distance between them. These virtual adaptive links are short-cut paths created by bypassing the intermediate routers along the path. The proposed adaptive links use part of the bit-width of the regular mesh links to create short-cut paths, so they customized irregular topologies on a regular mesh while imposing negligible area overhead. Consequently, the proposed architecture benefits from the regularity, desirable design time/effort, and scalability of a regular mesh, as well as the superior power/performance of customized irregular topologies. The reconfigurability, by which the adaptive links can be set up or torn down dynamically at run time, is the key advantage of the proposed architecture over the existing topology customization methods. The adaptive links can be constructed either at run time or design time. The run-time mechanism of creating virtual adaptive links consists of two light and fast procedures: on-chip traffic monitoring and virtual link reconfiguration. The former gathers network traffic statistics during a given time interval and the latter reconfigures adaptive link paths based on the current on-chip traffic pattern and in favor of heavy traffic flows. Experimental results show significant improvements can be achieved by the proposed architecture over some state-of-the-art NoC designs in terms of flexibility and energy-efficiency, with negligible area overhead.

1. Introduction

Networks-on-Chip (NoCs) are now considered as a promising solution to address the increasing communication demands of large-scale multicore systems [1–5]. A survey in Ref. [6] reveals that many SoC developers use NoC in their recent commercial products.

Traditional industry-standard solutions for on-chip communication are the bus-based architectures, which suffer from poor scalability in large-scale systems, and custom dedicated links, which have large area overhead. Besides the large area imposed by custom dedicated links, they suffer from lack of reusability (due to their ad hoc nature), low resource utilization, and poor predictability of electrical parameters (prior to the place and route design stages). However, the regular structure of standard NoC topologies makes the circuit-level design and implementation a much easier task. NoCs are also reusable which further reduces the NoC-based system design time/ effort [7]. Moreover, in packet-switched NoCs, packets are buffered at each node and then forwarded to the next node after a sequence of arbitration and routing operations. By storing and forwarding packets in a distributed manner, packet-switched NoCs feature high scalability and resource utilization.

Among different on-chip communication schemes, the ideal communication in terms of power and latency can be often obtained by establishing dedicated links between any pairs of communicating nodes. Packet-switched NoCs, although solve the scalability and area overhead problems of dedicated links, increase these ideal values due to the delay and power imposed by the routers [8,9].

Adding circuit-switching [10,11] or more advanced switching mechanisms [8,12–15] as a second switching method to the baseline packet-switching to compensate for its drawbacks have been proposed in some previous works. In addition, some recent researches [16–18] proposed a reconfigurable topology to share idle resources dynamically and gaining higher performance but these methods still suffer from more power and area consumption.

Targeting this problem, a packet-switched NoC with hybrid topology is employed in this chapter. The proposed architecture, in addition to the regular links of the baseline topology, can provide extra virtual ad-hoc links between every two remote nodes. These long links are dynamic and configured based on the current application's traffic pattern. The virtual adaptive links (or VALs, for short) are low-latency, low-power paths constructed by bypassing router's pipeline stages along a path.

The proposed NoC architecture is realized by two sub-networks: the base sub-network (BS) which features the baseline NoC topology (a regular mesh topology), and the adaptive sub-network (AS) which has a dynamic topology that can be customized for the current on-chip traffic pattern. Although we used mesh as the underlying topology, our proposal can be applied to any other topology, such as the ones introduced in Ref. [19].

To keep the cost of our proposed architecture unaffected, the AS and BS sub-networks share resources of the original network using Spatial Division Multiplexing (SDM). SDM is an alternative for the well-known Time-Division Multiplexing (TDM). TDM is a resource division scheme in circuit-switched NoCs that shares the bandwidth of NoC links among several circuits. In TDM, during each time slot, the whole wires of a link are allocated for a circuit, whereas in SDM, the wires of a link are spatially divided into several sub-links and each sub-link is dedicated to one circuit. The positive impact of SDM on the performance of NoCs is shown in Ref. [20].

The routers in this proposal can then be considered as higher radix routers with 16 half-sized (n/2-bit wide) ports (8 input ports and 8 output ports, ignoring the local ports to the core). Eight ports of each router are fixed and connect two adjacent routers, in much the same way as the links of a regular mesh topology, and the other eight ports implement an ad hoc topology using VALs. In this work, VALs are established in the AS part in favor of heavy communication flows to shorten the distance between their end-point nodes and improve the average performance and power consumption of the network.

The most popular topology proposed for regular NoCs is the mesh which provides well-controlled electrical parameters and low cost structured global interconnections with reduced power consumption on global wires. Despite such interesting advantages of meshes for on-chip implementation, there are some drawbacks associated with them including long communication latencies between two remotely located nodes due to lack of short (low hop count) paths between them. In addition to long zero load latency, the probability of message blocking is increased because there are many hops between two remotely communicating nodes; it makes predicting packet latencies and guaranteeing service quality almost impossible. Fully customized irregular topologies, on the other hand, can achieve superior performance, but at the expense of altering the regularity of mesh and losing reusability. The proposed NoC architecture can take the best from both worlds; it can achieve the superior performance of application-customized topologies by using the AS subnetwork, and as the AS subnetwork is constructed over regular structures, it can also fully exploit the reusability and predictability (of electrical parameters) of regular NoCs.

Several previous works try to find a fully or partially customized topology for NoCs [21–23]. In Ref. [22], authors add some long-range links to a regular mesh to adapt it to a target application. The links are fabricated at design time for a single application. Thus, the physical design and optimization procedures are repeated for each application. This also involves using routers with different port counts. The lack of reusability is the main drawback of this method. The key advantage of our method is the reconfiguration capability which makes the NoC reusable. The router architecture of our proposal is also the same in all nodes. Some other works present fully reconfigurable topology for NoCs [24–27], but our method can exploit the appealing properties of mesh (to support local traffic) and good performance of irregular topologies (to support long-distance traffic) in the same platform by making minor modifications to a conventional router architecture.

The systems that utilize NoCs can be broadly classified into two types: application-specific multiprocessor systems-on-chip (MPSoC) and Chip Multi-Processor (CMP). In MPSoCs, the design usually targets a single or a fixed set of applications. The communication between various cores is known at design time and the interconnect architecture can be tailored to suit the application traffic characteristics. In general-purpose CMPs, however, identical processor cores are used to run the tasks of different applications, often unknown at design time. Hence, the communication demands between cores cannot be easily pre-characterized. Our NoC can be used in both architectures. For MPSoCs, a suitable AS topology is calculated for each target application in advance and is created in the NoC when the application starts running. For CMPs, on the other hand, we reconfigure the adaptive links when on-chip traffic pattern changes at run-time. To this end, we divide time into some fixed intervals and reconfigure the set of VALs at the beginning of an interval time based on the traffic behavior in the previous interval. The reconfiguration algorithm considers heavy traffic flows and finds a set of VALs that covers as many flows as possible. The established VALs then live until the next interval when a new set of VALs are constructed. This reconfigurability would be particularly advantageous for emerging NoC-based many core deep learning accelerators [28–31], where the communication pattern changes significantly not only across different deep learning models but also across the layers of a single model.

In the dynamic VAL construction scheme, it takes some time to find the new set of VALs, but the construction procedure is done by a very fast, light-weight, and infrequently-invoked background process that runs in parallel with the normal NoC operation. The packets also keep on traveling using the previous NoC configuration. So, the setup time of adaptive links does not influence the network performance. This is contrary to some related architectures, for example, traditional circuit-switched NoCs, where circuit setup latency greatly affects the total performance. In circuit-switching, data transfer starts once a circuit is established; this waiting period would be long if the network is highly loaded and we must wait for current circuits to release their recourses. Even if the network is not under high load, sending probes/ACKs to construct circuits in circuit-switched NoCs makes a long setup time that directly increases network latency. The benefits of VALs over circuit-switching are twofold. First, VALs collaborate with a packet-switched network to handle the on-chip traffic. So, when a VAL is not found for a communication flow, it is directed to the packet-switched network and doesn't wait for resources to become available. Second, VAL path calculation is done in parallel with network normal operation, and hence VALs have zero setup latency from the packet perspective.

Our early results [32,33] showed the effectiveness of this approach in reducing the NoC performance/power consumption. In this work, we improve the architecture of the previously proposed NoC and algorithm of adaptive link construction. We present a more efficient algorithm to observe on-chip traffic patterns and reconfigure VALs in order to decrease the overall NoC power consumption. We also design a fast and small hardware module to reconfigure VALs at run-time to further boost the VAL reconfiguration process. The proposed method is compared with some state-of-the-art related methods and explores different trade-offs between area overhead and performance gain of VALs.

The rest of this chapter is organized as follows. Section 2 introduces some related work; Section 3 considers the proposed router architecture, followed by the virtual adaptive link construction algorithm in Section 4. Section 5 presents the experimental results, and finally, Section 6 concludes the chapter.

2. Related work
2.1 Improving packet-switching performance

Most NoCs generally use packet-switching. Circuit switching is also used to reduce on-chip communication latency and power. However, this switching method has some drawbacks such as poor resource utilization due to the need for reserving network resources for circuits. It also degrades performance due to circuit setup delay. During the past years, several NoC proposals focused on integrating a second switching method into a packet-switched NoC to improve its power and performance [8,12,13,22,33].

In our previous work [32], we proposed an early architecture and design flow for the virtual adaptive links. It monitors the traffic flows of the network at run-time and once the communication flow between two nodes exceeds a predefined threshold rate, it immediately starts building a virtual link. The new virtual link may tear down old virtual links, provided that the sum of the communication rates passing over the torn down virtual links is not higher than the rate of the new one. In this work, we improve both the architecture and routing algorithm of adaptive links. The new algorithm finds a better set of virtual links since it limits considering VAL construction request to some specific times only. It then considers all requesting flows at once and can make a better decision based on the global view of the path and weight of requesting flows. We also modified the router architecture to allow more packets use VALs to shorten their path length.

A similar approach has been applied in the hybrid circuit-packet switched NoC in Ref. [33]. This work uses SDM to integrate a packet-switched and a circuit-switched network into a single fabric. In this work, which is one of the bases of the current proposal, the circuits (which are equivalent to our adaptive links) are set up per packet in a single cycle and are held throughout the packet life time in the network. This involves a fast control network made by low-latency on-chip wires which is capable to set up the circuits within a single router pipeline stage. Our method constructs the adaptive links for a communication flow, instead of individual packets, to reduce the frequency of circuit construction procedure activation and it also removes the need for fast setup network.

In Ref. [13], a hybrid packet-circuit switched architecture is presented. The main task of the packet-switched part of this network is to carry out the circuit setup procedure instead of transferring the application data. The path of a circuit is determined by a packet-switched network, using local information at each node and, like our previous work [32], a new circuit can tear down old circuits.

The authors in Ref. [8] propose to add some bypass paths, called Express Virtual Channels (EVC), between some fixed source-destination pairs in a conventional mesh NoC. In this design, packets are allowed to virtually bypass the pipeline stages of intermediate routers along their paths. EVS is made of a set of virtual channels and may start from specific nodes or every node. EVCs are established statically independent of the running application and are limited to connect the nodes along a single dimension. Our proposed adaptive links, however, are configured based on the traffic pattern of the current application to shorten the distance between any two nodes with high communication demands.

A semi-regular topology is presented in Ref. [22] which is obtained by inserting so-called Long Range Links (LRL) between every two nodes that will communicate at a high rate. The LRL insertion procedure is static and performed at design time. The customized topology is especially realized for a single application and is not beneficial to run applications with different traffic patterns. Although the authors report significant benefits for the specific application for which the topology is generated, this method cannot be employed in modern multi-core SoCs and CMPs that target several different applications with different traffic patterns, mostly unknown at design time. Our proposed adaptive links are dynamic and can adapt to the current application traffic pattern. As LRLs are placed at design time, the procedure of physical design and optimization of this semi-regular NoC should be repeated for each SoC. This increases the design time/effort of the NoC and is one of the disadvantages of this work, in comparison to our work. LRLs are implemented by inserting extra ports on some routers, so routers are not identical, while our routers are all identical and similar to the routers of a conventional mesh NoC.

Several customized irregular topologies can also be found in the literature [21,23]. Again, reconfigurability and reusability is the advantage of our work over them.

Virtual Point-to-point (VIP) connections, introduced in Ref. [12], are added as a second switching mechanism to a packet-switched NoC to improve its performance. VIP connections are constructed over one virtual channel (which bypasses the entire router pipeline) at each physical channel of routers. They are constructed and torn down dynamically by a power- and latency-aware online algorithm. VIPs use virtual channels to share the link bandwidth among different packets, while our approach uses SDM for this purpose. Besides, VIPs are dedicated connections between the source and destination of a flow, whereas adaptive links can be used by any packets; both the packets of the flows for which the links are constructed and packets that find adaptive links useful to shorten their paths toward the destination.

2.2 Resource partitioning

We have used SDM for resource sharing in our proposal. A detailed study on the impact of SDM on NoC performance is done in Ref. [20]. In Ref. [34] a comparison has been made between the SDM and TDM and it has been shown that removing the scheduling complexity and memory demand for storing the configurations of switches is the advantage of SDM over TDM. Besides, the power that TDM consumes at each cycle for reconfiguring switches is omitted with SDM. The SDM capability in providing specific levels of QoS [35] and throughput [36] has been shown in some previous work.

2.3 Dynamic path finding in NoCs

Given a set of communication flows between corresponding source-destination pairs, we should find a set of adaptive links in the AS sub-network that maximizes the number of flows provided by adaptive links. This is an instance of an NP-hard problem [37]. Consequently, we adapt a fast and efficient heuristic to solve this problem. Several heuristic algorithms are proposed to solve problems with similar objectives and constraints including routing in FPGAs [38], slot allocation in TDM-based circuit-switching [39], NoC topology reconfiguration [40], and circuit setup in NoCs [12]. Applying the well-established graph-embedding [41] and flow network problems [42] in graph theory to allocate limited network paths to candidate connections often gives results with high quality. However, they are often complicated and have a long execution time, hence are not suitable for online network reconfiguration.

VAL path calculation is done based on the traffic rate and number of distinct paths, or routing flexibility, between communicating NoC nodes. Routing flexibility criteria are used in some previous work to solve the mapping and routing [37] and static power-aware routing [43] problems in NoCs.

3. NOC architecture
3.1 Resource partitioning policy

In the proposed architecture, the NoC is split into an adaptable sub-network (AS) and a base sub-network (BS) using SDM. This NoC handles the traffic between remotely located nodes by establishing VALs. These links collaborate with the normal mesh links to handle the entire on-chip traffic.

We use SDM to get the benefits of adaptive links without considerable cost overhead. SDM divides the links into two narrower sub-channels; it is desirable as it increases distinct paths between adjacent nodes, but it affects network latency by increasing the packet serialization latency. This negative effect of SDM is more noticeable in low traffic rates when the contention on shared resources occurs infrequently, hence the head of line (HOL) blocking is not important. However, in medium and high traffic loads, the increased path diversity in the same direction reduces the HoL problem and not only can compensate the latency overhead of SDM but also can improve the average packet latency and throughput of the network.

In Ref. [20], the effect of dividing wide NoC channel links into narrower sub-channels on the network latency and throughput is analyzed. Dividing the original 512-bit wide NoC links into a varying number of parallel sub-links with equal widths, the paper reports that dividing the links into two equal sub-links gives the best performance improvement with 50–60% increase in throughput. This configuration, however, shows around 10% negative effect on zero-load packet latency.

In this work, the low-latency VALs of the AS sub-network is capable to mitigate the performance degradation of SDM across most traffic loads and even improve latency and throughput. From the energy point of view, our proposed network has a great advantage over the conventional network across all traffic loads, as it provides power-efficient paths for a portion of the traffic. Consequently, energy reduction with no negative impact on performance and area is the main achievement of this work.

3.2 Overall VAL-based NoC architecture

Fig. 1 shows the overall structure of a VAL-based NoC. Assume that the bold red line is the only VAL constructed in AS. It makes a short cut path between nodes X and Y by skipping over some intermediate nodes. The flits enter a register in each node of the VAL and continue in the next cycle, allowing them to pass the VAL in a pipelined manner. Packets enter the VAL from node Z; the East output port of Z is the VAL entry. Packets that arrive at different ports of Z can request for the AS part of the East output port to continue over the VAL. The AS arbiter of the East port of Z selects one of the requesters' packets and allows it to enjoy the low-power/latency VAL.

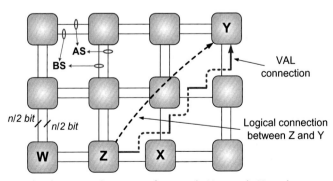

Fig. 1 A VAL that bypasses the routers from node X to node Y. packets enter the VAL from the east port of node Z. from the viewpoint of node Z, there is a direct connection from its east port to node Y.

Consequently, all packets that arrive at Z, including the packets injected into the network from Z, can use VALs. To get the most out of VALs, they are constructed between network nodes with heavy communication demand.

3.3 Structure of routers

Fig. 2 shows the architecture of the router used in our design. As shown in the figure, the proposed router architecture is similar to the architecture of a conventional packet-switched router [44] with some minor modifications.

Every n-bit wide network data-path component (link, buffer, and crossbar switch) is divided into two parallel n/2-bit sub-components. The number of buffers slots is doubled, but each is half-sized, resulting in no change in buffering capacity of each port. Switch allocator is also divided into two allocators; one acts as a conventional allocator handling the BS sub-links and other handles the AS sub-links.

The output allocator in AS sub-network is simply implemented as a register to indicate whether the corresponding output port is a part of an adaptive link; when it is a part of an adaptive link, this register specifies which input port should be directly connected to it. This register is set when the adaptive link is constructed. This allocator and the multiplexers of the input ports (see Section 3.4) form a virtual link between any two nodes by chaining the crossbars and links along the path.

In this NoC, packets belonging to the BS sub-network are buffered at each hop and then, have to go through the router pipeline stages.

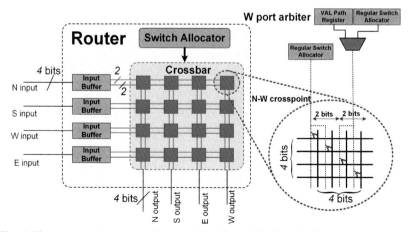

Fig. 2 The proposed router architecture. The bit-width of the NoC is 4 and each sub-network is 2 bits wide.

AS packets, on the other hand, travel on VALs and are not required to go through power-hungry buffering and time-consuming routing and arbitration pipeline stages of intermediate routers. They just pass through the crossbars and links which form the physical distance between the two end point nodes of adaptive links. The packets may switch between the two subnetworks several times before arriving at their destinations.

The other router components including the routing function and crossbar are not changed. We can also use virtual channels. In this case, each virtual channel is horizontally split into two AS and BS parts (just like the division scheme used for a single VC); hence both sub-networks have the same virtual channel count. Fig. 2 shows a 2-bit wide router with a single VC per port.

When the AS part of a physical link between two routers is used by no virtual adaptive link and is free, it is used as a regular link between the two routers. In this way, the network presents two parallel n/2 links between adjacent routers and some performance improvement can be achieved. To handle this case, the regular AS allocator controls the AS part of the crossbar. The regular AS allocator is then used in two cases: when the AS link is used as a regular mesh link and when a VAL starts from the output port (Node Z in Fig. 1). The other allocator is used for the VAL intermediate nodes.

3.4 Structure of input ports

The input port of the routers, depicted in Fig. 3, controls the flow of packets over the two sub-networks and also manages the movement of packets from one sub-network to another. It can support several data-paths that are established by the multiplexers of Fig. 3.

Case 1 Packets of the BS sub-network go to the BS buffer and after passing the router pipeline stages continues traveling either over the BS or AS sub-networks. A packet can continue over BS by passing through

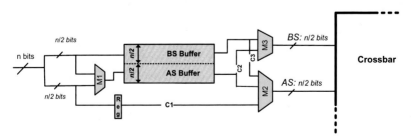

Fig. 3 The structure of the input ports in the proposed architecture.

multiplexer M3. If there is a VAL originated at the current router in the AS sub-network, the packet can switch to the AS sub-network and will be forwarded toward its destination over the VAL. In this case, the packet takes line C3 and multiplexer M2. Obviously, this occurs when the routing function finds it useful to forward the packet through the VAL.

Case 2: When the AS part of the input is a part of a VAL and this router is an intermediate hop of the VAL, arrived packets skip pipeline stages of the router (e.g., buffering, routing, switch and VC allocation, and related flow control tasks) and go to the next router through line C1 and multiplexer M2 in Fig. 3. REG in Fig. 3 is a 1-flit deep register that allows flits of a packet to get buffered when they arrive and enter the crossbar at the next cycle. By this register, flits travel over virtual links in a pipelined manner. Flits may also get buffered after crossbar traversal if the link and crossbar traversal cannot be done within a single clock cycle.

Case 3: when the AS part of the input port is a part of an established VAL and this router is the end point of that VAL, packets are buffered in the AS buffer. At this node, packets enter the routing procedure and may remain in the AS sub-network (if there is a VAL originating at the current router and router decides to use it) or switch to the BS sub-network through line C2 and multiplexer M3.

Case 4: When the AS part of a port is used by a VAL, the corresponding AS buffer is not used and is skipped by AS packets; this unused buffer space is added to the neighboring BS part to increase the BS buffering capacity for further performance improvement. BS packets can enter either BS buffer or AS buffer through multiplexer M1. Both buffers are controlled by the BS input controller. In this case, the buffer width will remain n/2 bits, but buffer depth is doubled.

3.5 Dynamic subnetwork bit-width

An interesting point about the proposed NoC is that by some minor changes in the router architecture we can have a vast variety of VAL-enabled NoCs. Having subnetworks with unequal bit-widths is one possible option. The bit-widths can be set either dynamically (based on the dynamic traffic behavior) or statically. By setting the bit-width of the AS to zero, we obtain a conventional packet-switched network.

Dynamic partitioning can be achieved by a bit vector that determines the link-width assigned to each sub-network.

Setting a bit in the bit-vector to 1 indicates that the corresponding bit-wire in the NoC link is assigned to the BS sub-network, otherwise, it belongs to the AS sub-network. In this case, the logics that generate control signals are slightly modified to direct control signals to appropriate bit-wires. For example, Fig. 4 shows the revised structure of crossbar in Fig. 2 to support dynamic sub-network bit-width. The same modification is done for the input ports to support dynamic partitioning.

Please note that as the extra logic applies to router control paths (and not datapaths), adding the capability of dynamic bit-width selection has no negative impact on the datapath capacitance switched per data transfer, and hence imposes negligible negative impact on NoC power consumption.

To reduce the partitioning bit-vector size, the link wires can be considered as a bundle of b wires and each bundle is assigned a bit in the vector. Sub-network sizes in this case should be a multiple of the bundle size.

Once the bit-width of each sun-network is set, the width of flits of each sub-network should be set accordingly. This can be easily achieved by shift registers [34]. Obviously, the flit width should be long enough to guarantee that the routing and other required information can be embedded into one flit (the header flit).

We can also consider NoCs with multiple AS and BS subnetworks that work in parallel. This is especially viable for NoCs with wider links which may become commonplace in the future [20].

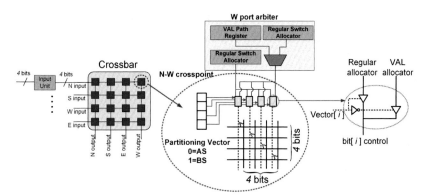

Fig. 4 The structure of a cross-point in the crossbar with dynamic bit-width selection. The bit-width of the NoC is 4 and the partitioning vector determines the bit-width of each sub-network.

3.6 Packet routing

Once flits are buffered in AS buffer (Path 3) or BS buffer (Path 1), they should go through pipeline stages of the current router. The proper output port for a packet is determined based on a minimal routing scheme, so a packet is only allowed to take the ports along one of the shortest paths toward the destination (at most two directions and four n/2-bit ports). The algorithm checks the related free ports and allocates one of them to the packet. Also, in particular, our algorithm considers whether there exists a free VAL that begins from the current router and ends at some router along the path toward the packet's destination. If there is such a VAL, it will be prioritized over the BS ports and used to shorten packet's journey. If more than one VAL is found, the longer VAL is considered first. This involves maintaining the end point address of each VAL in the VAL's source router to use for making decision. Thus, as mentioned before, packets in our network may switch between the AS and BS sub-networks several times to reach their destinations.

The network guarantees deadlock freedom by employing the well-known escape channel concept [45]. One of the virtual channels of the BS ports is used as an escape channel and adopts the dimension-order XY deadlock-free routing algorithm. To guarantee deadlock freedom, once a packet enters an escape channel, it cannot switch again to adaptive VCs and the AS sub-network. If there is a single VC per port, the entire BS sub-network uses the XY algorithm and acts as an escape channel for the AS. For flow control over VALs, we use the methods presented in Ref. [12].

4. VAL construction algorithm

In this section, an algorithm to establish VALs based on the current on-chip traffic pattern is presented. Constructing VALs is done in favor of high-volume traffic flows to improve performance and power consumption of the network. The reason behind selecting source and destination nodes of heavy flows is that by constructing VALs for such nodes, more packets can use these VALs and this can give more power and performance gains over a conventional network.

4.1 VAL construction and reconfiguration- general approach

The proposed NoC architecture can be used in systems with both dynamic (which is often the case in homogenous CMPs) and static predictable (which is often the case in heterogeneous multicore SoCs) traffics. In the latter, it is assumed that each input application is spatially partitioned into several tasks, each of which is assigned to a processing unit. On-chip traffic pattern remains relatively static as each core performs a fixed task. Many NoC customization methods for this kind of application can be found in the literature [21–24].

In such systems, the traffic flows of each application are modeled by a *Communication Task Graph* (CTG) that outlines the average bandwidth demands between any two nodes and can be obtained through code analyzing or program profiling. The CTG is a directed graph $G(V,E)$, where each v_i represents a task, and a directed edge $e_{i,j}$ represents a communication flow from v_i to v_j. $t(ei,j)$ is the label on $e_{i,j}$ that characterizes its communication rate.

For such applications, the paths of AS links are calculated a priori. The calculated VAL paths are maintained and then loaded into the network when the corresponding application starts running. The VAL path information that should be maintained includes the value of the multiplexer select lines of the input units and the VAL allocator values for all involved nodes along the VAL path.

In systems with dynamic traffic, on the other hand, we should monitor the on-chip traffic to construct the application CTG online. Afterward, the obtained CTG is fed to the algorithm to adapt AS links to the traffic at run-time. This procedure must be repeated periodically to handle dynamic traffic patterns.

NoCs reconfigure the VALs when a new application replaces an old one (in the static case) or new VAL paths are calculated at run-time (in the dynamic case). In either case, the existing VALs are destroyed and the new configuration is set up. This is done by sending configuration data to all NoC nodes. The configuration data includes values of AS allocators together with values of the select lines of the multiplexers at input ports. Old VALs are destroyed when all in-transit packets are delivered to the destination nodes. Then the new VALs are set up.

The run-time VAL path finding is carried out in parallel with the normal network operation. Therefore, packets do not wait for a new configuration to be set up; rather they keep traveling on old NoC configuration and unlike traditional circuit-switched networks, network performance is not degraded due to the VAL setup latency. Therefore, there is no need to have a real-time path finding algorithm for dynamic VALs. Nonetheless, the proposed algorithm is simple and fast and can react in real-time.

4.2 Dynamic VAL reconfiguration

For NoCs with static traffic, the CTG is directly fed to the algorithm to calculate VAL paths. For NoCs with dynamic traffic, however, the CTG should be first constructed at run-time to feed the algorithm with it. In this case, the system life time is divided into intervals during which the traffic information of each flow of the network is gathered. At the beginning of the next interval, all existent VALs are destroyed and a new set of VALs are set up based on the traffic pattern of the last interval. Then, the procedure of gathering information for the new interval starts again. The dynamic algorithm, as a result, needs two more phases, i.e., traffic monitoring, and delivering the traffic information to some manager (or root) node to calculate the VAL paths.

To monitor the network traffic, each node, during each time interval, keeps track of its communication flows by counting the packets it sends to each destination. At the end of the interval, the m most significant bits of the register storing the traffic volume of each flow are considered as the rate of the flow. This rate is used as the flow weight in the VAL construction algorithm. m is a design parameter that makes a trade-off between data precision (hence the algorithm accuracy) and reconfiguration power/bandwidth overhead. At the end of each time interval, the gathered traffic information is sent to some root node.

The dynamic version of the VAL construction algorithm runs on one of the NoC nodes, which is referred to as the root node hereinafter. In most SoCs, there is a central configuration processor that manages and configures the system [10,12]. We give the special responsibility of collecting NoC state, finding a new set of VALs, and sending the information required to construct VALs at every time interval back to NoC nodes to the root node. The root can be any network node, but we select the middle node of the mesh, i.e., node $(n/2, m/2)$ in an $n \times m$ mesh, to balance the control traffic across the network. Each node sends information of its traffic flows (the flow end-point nodes and current rate) to the root node. Once the root node collects all traffic flow information, it starts calculating the new VAL configuration.

In some previous work [13,33], a dedicated network is used to transfer such control messages. In Ref. [12] it has been shown that for small and infrequent control messages, using the main data network to transfer the control messages can hardly have negative effect on NoC performance. So, as the reconfiguration procedure occurs infrequently and each reconfiguration involves sending several small flow weights to the root node, we use the main data network for the control message transfer.

To reduce the communication and computation overhead of VAL reconfiguration procedure, the flows are filtered out based on a predefined communication rate threshold. Each node only sends the flows with the communication rate higher than the threshold to the root processor and discards low-rate flows.

4.3 Problem formulation

The VAL path calculation algorithm takes the CTG of the application as input and customizes the VALs for it. The algorithm aims to find as many VALs as possible to maximize the amount of traffic covered by VALs. As stated in Section 2, this is an instance of an NP-hard problem; hence we use a heuristic algorithm to solve it.

The algorithm maintains a set called *CFS* of communication flows that are the CTG edges with communication rates higher than a predefined threshold.

Formally stated, given an $n \times m$ mesh-connected NoC and the current set of communication flows *CFS*, the algorithm aims to find a new configuration for VALs in such a way that:

$$MAX\left\{ \sum_{\forall cf_{i,j} \in CFS} R\left(cf_{i,j}\right) \times VAL\_Func\left(cf_{i,j}\right) \right\}$$

where $cf_{i,j} \in CFS$ represents a traffic flow with weight $R(cf_{i,j})$ between nodes x and y and the *VAL_Func* function is defined as

$$VAL\_Func\left(cf_{i,j}\right) = \left\{ \begin{array}{ll} 1 & , \quad if\ VAL_{i,j} \in VAL\_Set \\ 0 & \quad otherwise \end{array} \right\}$$

Taking $cf_{i,j} \in CFS$ as input, the function returns *true* if there is a VAL between nodes i and j, otherwise it returns *false*. In this function, *VAL_Set* is the set of all existing VALs of the NoC.

In addition to weight, routing flexibility is another attribute of each communication flow. As there may exist multiple shortest paths between any two nodes in a mesh, the *routing flexibility* for $cf_{i,j} \in CFS$, represented by $RF(cf_{i,j})$, is defined as the number of different shortest paths from nodes i to j. Although *routing flexibility* is obtained by a more complex relation, we use a simpler approximation for this parameter as follows:

$$RF\left(cf_{i,j}\right) = \left|i_x - j_x\right| \times \left|i_y - j_y\right|$$

where i_x and i_y are the x and y coordinates of node i, respectively. We cast the RF value to m bits to have the same range as the flow rate.

The weight of a flow that will be used by some steps of the algorithm is an appropriate mix of the flow rate and routing flexibility and defined as

$$W\left(\mathit{f}_{i,j}\right) = \alpha \times R\left(\mathit{f}_{i,j}\right) + \beta \times \left(MaxRF - RF\left(\mathit{f}_{i,j}\right)\right)$$

where $MaxRF$ is the maximum RF a flow can have in the topology. α and β are the weights of the two terms. The first term of this relation is the communication rate-sensitive term, while the second term accounts for flexibility. This weight function improves with decreasing routing flexibility and increasing rate. The reason is that a flow with lower routing flexibility has fewer options for path finding than a flow with higher flexibility and should be prioritized over it during the path finding algorithm. Higher rate should also increase the chance of VAL construction, which in turn, increases the chance of covering more portion of traffic by VALs. Investigating several CTGs, we found out that emphasizing on flow rate results in higher quality answers. So, we set $\alpha = 2$ and $\beta = 1$.

4.4 The algorithm

The algorithm is composed of two steps: a congestion estimation step which calculates the cost of using each link for VAL construction based on the demand placed on that link by other flows and a flow routing, which finds the path for one flow at a time using a shortest-path algorithm. Algorithm 1 outlines the algorithm steps.

Step 1. In this step, each flow places a demand on the links on one of the shortest paths between its two end-point nodes. The shortest path is selected in a dimension-order manner, i.e., the path, starting from the upstream node, follows the XY routing algorithm to reach the downstream node. This path is referred to as the initial path of each flow, henceforth. In this step, individual links can be requested by more than one flow. The cost of a link is then calculated as the cumulative weight of the flows that place a demand on it. We examined some other routing algorithms (YX, Zigzag) for the initial link demanding in this step, but none of them led to considerable improvement in the final solution.

Step 2. Once the approximation of the initial demand for each link is obtained, the second step of the algorithm finds a route for every flow through the links with lower demand. Considering the network as a graph and the cost of each link as edge weight, a shortest path algorithm for weighted directed graphs is used to route the flows.

To this end, the flows are selected in the order of their weights and the shortest path algorithm finds the path with minimum cost between the flow source and destination nodes. The cost of a path is the cumulative cost of the edges it contains. Upon fixing a path, the algorithm assigns a cost of infinity to the path links to prevent consequent flows using it. The flow then removes the demand from the links of its initial path by decreasing its weight from the links' cost.

Algorithm 1. VAL path finding

```
For each cf_{i,j} ∈ CFS do
  For each link l_i ∈ the XY route from i to j do
    Cost(l_i)+=W(cf_{i,j});
  End For
End For
For each cf_{i,j} ∈ CFS do
  Select the cf_{i,j} with maximum weight and remove it from CFS;
  SP_{i,j} =path with minimum cost between the cf_{i,j} endpoint nodes;
  For each l_i ∈ the XY route from i to j do
    Cost(l_i)-=W(cf_{i,j});
  End For
  If (cost(SP_{i,j})=∞)
    Break; //failure! Continue with the next flow
  End if
  Save SP_{i,j};
  For each l_i ∈ SP_{i,j} do
    Cost(l_i)=∞;
  End For
End For;
```

We apply Dijkstra's algorithm to find a path with minimum cost for the VALs. The canonical Dijkstra's algorithm finds a path with minimum cost, but there is no guarantee that a path with minimum cost is also a topological minimal path (when considering the hop count). The VAL paths, however, must be one of the topological shortest paths between the flow source and destination. To address this problem, we modify Dijkstra's algorithm in such a way that it is restricted to use only the links of shortest paths between source and destination nodes. The algorithm is also restricted to use the links directing toward the destination node. For example, if the source address is (0,5) and the destination address is (2,2), considering North and East as positive directions, the algorithm can only use links with the West and North directions inside the rectangular area formed by (0,2) and (0,5).

The algorithm may meet the same initial path of a flow for more than one time. In this case, the weight of the flow should be added to the path cost only once.

4.5 Enhancing AS resource utilization

Partial VALs. When the main algorithm terminates, we apply two techniques to fully utilize the AS sub-network resources. First, the flows which did not succeed in finding a VAL are considered for constructing a partial VAL; a VAL from the source node to some node on one of the shortest paths toward the destination node (and not the destination node itself). This partial VAL can deliver the packets to an intermediate node from which the packets can switch to the BS subnetwork or take another VAL to move toward its destination.

In this complementary step of the algorithm, the remaining flows are considered in order of their weight and Dijkstra's algorithm is used again. As in this case, all flows have removed their demand for initial XY-paths, the weight of all links is either 0 or infinity. When Dijkstra's algorithm terminates, it calculates the shortest path from the source node to all nodes of the network located on flow source-destination shortest paths. We select one of the nodes with finite weight that has the longest topological distance from the source node and construct a VAL from the source to that node.

Unused AS links. The second enhancement, which is done at the final stage of VAL construction procedure, is to use the AS links not used by any VAL as a regular mesh link between two adjacent nodes. The multiplexers of the input ports and allocators support this case.

4.6 Implementing the parallel algorithm

We design a hardware functional unit to efficiently implement the run-time VAL path finding algorithm. The hardware has some simple processing cells, each corresponding to an NoC node, arranged and connected in much the same way as the main network nodes are connected (2D mesh in this work). There is also controller logic that manages the functional unit through some control links. Fig. 5 shows the structure of this module. Each cell has four

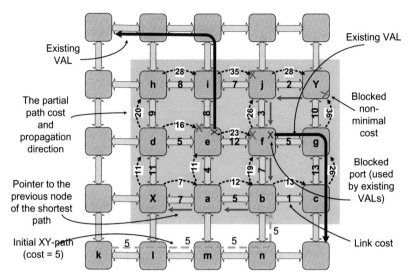

Fig. 5 The general structure of the hardware module is used to find VAL paths. Each square represents a cell. The numbers on the links are the cost of the outgoing links of the cells. The dashed lines show the direction and value of cost propagation. The shaded region is the shortest path region of cells *X* and *Y*. A chain of pointers (represented by arrows) keeps the path with minimum cost in a distributed manner.

incoming and four outgoing ports to neighboring cells. Ports are memory spaces to which the neighboring nodes write their data. It has one arithmetic unit and four sets of registers, each dedicated to one outgoing connection to calculate the total communication volume of the flows that request it.

The controller of the functional unit sends the flow information (weight, source, and destination of each flow) to the cells corresponding to each flow source node. Each cell, during the steps of this algorithm, is responsible for maintaining and updating the cost of the outgoing links of its corresponding node in the NoC. The cells then implement the VAL path routing algorithm in a distributed manner in two stages; initial path request and VAL path finding.

The flow weights are *m* bits wide. However, the link cost bit width (which determines the width of inter-cell links and arithmetic units) is wider to prevent weight overflow. The extra precision added to the cost width is determined by the network size.

Initial path finding. This step aims to calculate the cost of each link, which is the sum of the weights of the requesting flows, and maintain it in the cell corresponding to the upstream node of the link in the NoC. To place a demand on the links of the initial path of the flows, the following steps are done by each cell.

1. Source cell of each flow first passes the x coordinate and y coordinate of the destination address, as well as the weight of the flow, to the neighboring cells in the x dimension along the appropriate direction (X^+ if $x_d > x_s$, X^- if $x_d < x_s$, Y^+ or Y^- if $x_d = x_s$). These three data items are sent in parallel along the direction toward the destination cell. Sending data to a neighbor node involves writing the data to its appropriate registers. To prevent overwriting a waiting flow, each cell has a flag to the neighboring cells to signal them if it still contains the previous data. Once it passed all its flow data to next nodes, it clears the flag to allow them to pass new data. In particular, each cell first sends its data and then allows other data pass across it to go to their destination nodes.

2. The data advances along the selected direction until it reaches the same coordinate as the destination in that dimension. Each intermediate cell receives the flow information, checks the x coordinate of the data, and decides whether the packet should be passed to the next cell. If it should be passed, the cell then:

   a. Adds the flow weight to the cost it keeps for the corresponding output link. It also maintains the source, destination, and weight of each flow that places a demand on the link.

   b. Sends the data to the next cell along the direction

3. When the destination column is encountered, the data is routed in the vertical direction along the direction of the destination until it reaches the destination cell. The same procedure is carried out in the y dimension, but only the y coordinate and flow weight is sent.

4. After the end of this step, the initial demand of each link is calculated. *Finding VAL paths with minimum cost.* To calculate a VAL path with minimum cost for a flow, the partial path cost is propagated from the source cell toward the destination cell on all of the shortest paths between them. Partial path is defined as an incomplete path between the source cell and some intermediate cell. Without loss of generality, we follow the path finding algorithm from cell *x* to cell *y* in Fig. 5. Fig. 6 shows the behavioral description of the task of each cell in this step.

Cell *x* sends the cost of its E and N ports as the initial path cost to its neighboring cells along the shortest path toward the destination, that are cells *d* and *a* in this example. These two cells add the received partial cost to the cost of their E and N ports and propagate the result to the next cells toward the destination. This procedure continues until the final cost is received at

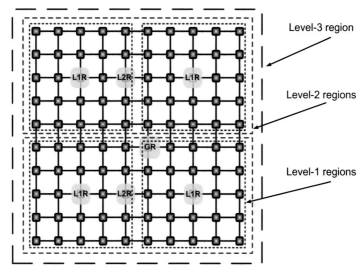

Fig. 6 Hierarchical VAL construction on a 10×10 NoC partitioned into four 5×5 regions. (L_iR = level-i local root, GR = global root).

the destination cell. Some intermediate cells, e for example, receive two costs from the neighboring cells (a and d in this example) at the same cycle. Each cell selects the minimum cost of the two to propagate. Each cell maintains a pointer to the cell from which the minimum cost is received. When the final cost is delivered to the destination cell, the path is maintained by the chain of the pointers that starts from the destination cell and ends at the source cell.

In this case, the VAL is established in the network by appropriately configuring the multiplexers and allocators along the path. The cell ports corresponding to the NoC links on which the VAL is constructed is disabled in the algorithm to prevent other VALs to use them during their path finding process (port E of cell f and port N of cell e). This is equivalent to setting their cost to infinity in Dijkstra's algorithm.

A path for a flow cannot be found if all partial paths are blocked up by disabled ports (equivalent to already used NoC links) before reaching the destination cell. In this case, the algorithm fails and continues with the next flow.

For the cells to send the partial cost in the right direction, the controller logic determines the cell ports that are allowed to propagate the received cost. First, only the cells within the shortest path region between the source and destination cells are involved in the algorithm (the shaded region in Fig. 5). Second, all ports of the cells inside the region with the direction toward the destination cell are legal (ports E and W in Fig. 5), excepting those that end to a cell outside the region (the N and E ports of the cells at the east and north boundaries of the shortest path region in Fig. 5). As mentioned before, reusing the ports used by already created VALs are also prohibited.

Each controller also maintains a list of initial paths that contribute to a partial path cost. If a partial path meets an initial path for more than one time, the respective initial path cost is not added to the partial path cost again. Otherwise, the algorithm may overestimate the cost of some paths. For example, if multiple addition of the weight of a single initial path is allowed, the cost of establishing a VAL between cells k and n through cells k, l, m, and n, is 5×3. However, the actual cost of this path is 5 that is achieved by taking the place of the initial path between cells k and b.

After finishing with all flows, the complementary partial VAL construction is done for unsuccessful flows. This is done by the same procedure of complete VAL path finding, but returns the cell with minimum weight with the farther distance from the source to build a partial VAL.

Hardware implementation. We have designed this hardware module in VHDL and synthesized it for the Spartan-3 Xilinx FPGA family. For an 8×8 cell array with 4-bit weight precision and 8-bit links and registers, it consumes 1018 logic cells which is a reasonable area overhead, especially when considering that it is up to $42 \times$ faster than the software implementation of the algorithm (on a CPU with an Intel Core i3 processor working at 2.0 GHz). Simple structure, which leads to very high clock frequencies, and parallel nature of this hardware module are the main sources of its high-speed operation. Further, it is a customized hardware implementation of the algorithm and eliminates common software overheads.

4.7 Addressing the scalability issue of the algorithm

As we will show in Section 5, the VALs can provide appropriate power/performance improvement in moderate-sized NoCs (of a few tens of nodes). For NoCs with several hundreds of nodes, we propose to use a hierarchical VAL construction approach. To this end, the NoC is partitioned into several disjoint regions with a manageable size. Every region has a root node that handles VAL construction inside the region. Inter-region VALs are handled by a global root node. This partitioning scheme distributes the processing load of the VAL construction and reconfiguration algorithm across the roots. In addition, the average node-to-root distance is kept short in the hierarchical approach. This improves the speed of the method, as VAL construction request messages that target the root and SCP paths that are sent by the root incur a smaller latency when the root is located at a shorter distance.

Fig. 6 shows a 10×10 NoC that is partitioned into four 5×5 regions. Local root nodes, which act as level-1 root nodes (marked by L1R in Fig. 6), are responsible to receive requests and construct VALs in their regions, whereas level-2 roots (marked by L2R in Fig. 6) handle VAL requests that span across two adjacent regions. Finally, the global root (marked by GR) handles remaining inter-region requests. In this case, the global root is activated first to construct longer inter-region VALs. Then it activates local roots to use the remaining resources to construct SCPs inside regions. Higher-level roots send the path of constructed VLAs to corresponding nodes, as well as to the corresponding lower-level roots of the node to keep the local roots updated.

As the cost of the proposed VAL construction hardware unit is very small, replicating the unit comes with negligible cost in terms of area overhead.

4.8 Supporting uniform traffic

Up to now, we have considered supporting applications with non-uniform spatial traffic distribution. The main target of such an approach is common SoC applications where the communication occurs along with some limited communication flows and the communication pattern often remains unchanged during the application life time. It is also suitable for those CMP applications that have temporal high-volume communication flows between source-destination pairs. In either case, we select source-destination pairs with high communication demand and establish a VAL for them.

However, applications that exhibit uniform traffic pattern, in which each source is equally likely to send data to each destination, cannot take full advantage of VALs. In this case, as there is no distinction among different flows, VALs can establish regular connections among nodes to generate a topology with shorter diameter, and hence potentially reduce average packet latency.

5. Experimental results

5.1 Simulation setup

For the simulations, we use BookSim [46], a cycle-accurate simulator for NoCs. To evaluate the power consumption of the NoCs we add the

Orion III power library [47] to BookSim. The simulator is configured to have speculative pipelined routers with a per-hop latency of 4 cycles. The network bit-width is 64. Each port has two virtual channels with 8-flit buffers per VC. The power consumption results are reported based on an NoC implementation in 65 nm technology working at 1 GHz.

In our experiments, we use a set of synthetic traffic patterns, as well as Splash-II [48] to evaluate the dynamic VAL construction and embedded multi-task applications to evaluate the static design-time VAL construction.

In all simulation experiments, the threshold weight of each node is set to the average weight of its communication flows and the VAL requests are done for the flows (or CTG edges) weighting higher than this threshold.

The evaluation includes comparing the proposed NoC with a reconfigurable NoC [24], NoCs with Long Range Links [22], NoCs with Express Virtual Channels [8], and a conventional packet-switched NoC.

Done in parallel with NoC normal operation, the dynamic VAL path calculation has no impact on NoC performance. However, the effect of the VAL path finding algorithm is included in the energy consumption results. This energy consumption is estimated by figuring out the required algorithm operations and estimating the energy consumption of each operation using the Orion library elements. We also take the control traffic among the NoC nodes and the root node into account in our simulations.

5.2 Synthetic traffic

We use the *n-hot flow* traffic patterns presented in Ref. [12] to generate synthetic traffic loads. This synthetic traffic pattern takes into account the temporal and spatial locality that often exists in common SoC and scientific CMP workloads [13]. Each node in n-hot flow synthetic traffic pattern sends a considerable portion of the generated packets (50% in our experiments) to exactly n destination nodes (hot destinations) and the remaining traffic to other randomly chosen nodes (with uniform distribution). Each node selects its hot destinations independently and the hot destination nodes change periodically. The time when hot destinations are changed is determined according to a poison distribution with a mean of 500,000 cycles. The size of each packet is 5 flits and each simulation runs for 10,000,000 cycles. The network reconfigures the VALs every 500,000 cycles. The network size is 6×6. The results are also compared to the Express Virtual Channels (EVS) [8]. Here, we use the static EVCs of size 2.

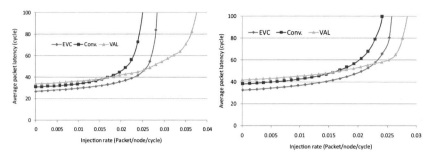

Fig. 7 Average packet latency of a proposed, EVC-based, and conventional NoC in a 6 × 6 mesh under 1-hot flow (left) and 3-hot flow (right) traffic patterns.

Latency. Fig. 7 shows the latency vs offered traffic load for the VAL-based, EVC-based, and conventional NoCs under 1-hot and 3-hot traffic patterns. As we expect, the latency of the proposed network performs a little worse in low traffic load. In moderate load, the effect of VALs overcomes the negative effect of packet size and in heavy load, VALs provide substantial performance improvements over other considered NoC architectures. VALs also improve the maximum network throughput.

Comparing the network behavior under 1-hot flow and 3-hot flow traffic patterns reveals that increasing the number of favored destinations of each source node in 3-hot flow pattern, which spreads the traffic more evenly across the communication flows, results in decreasing the effect of VALs. This is because the number of heavy flows is increased while the network resources that can be used by VALs are not. Consequently, NoC provides fewer flows with VALs and larger portion of the traffic uses the BS sub-network. Fig. 8 shows the portion of the on-chip traffic that travels over VALs for the two traffic patterns. This portion decreases when the number of hot flows increases. However, even in 3-hot traffic pattern, VAL connections can handle a fair portion of on-chip traffic and improve performance/energy.

Energy. We select two points in the injection rate range to compare energy results. The selected injection rates are one rate from the low injection rate region, and the rate just before the saturation point of the conventional NoC. Fig. 9 shows the normalized energy per packet results in these points. The numbers are normalized with respect to the energy per packet of the conventional NoC. As figures show, VALs provide energy reduction over conventional NoC for the full range of network load. For 1-hot flow traffic pattern, VALs always outperform EVCs.

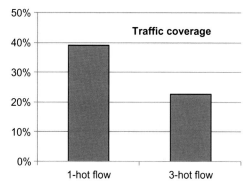

Fig. 8 The portion of the packets travel over VALs for 1-hot flow and 3-hot flow traffic patterns.

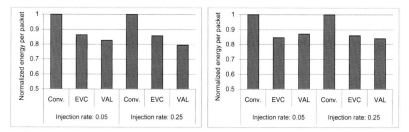

Fig. 9 Average energy per packet of a proposed, EVC-based, and conventional NoC in a 6 × 6 mesh under 1-hot flow (left) and 3-hot flow (right) traffic patterns. The bars are normalized to the conventional NoC energy.

In conclusion, EVCs are a better choice when the traffic pattern is uniform, while VALs are more appropriate when there is non-uniformity in the traffic load.

5.3 Splash workloads

The effectiveness of the dynamic VALs is evaluated under the traffic traces generated from the SPLASH-2 programs. We set the period of the VAL reconfiguration procedure execution to 500,000 cycles, while each simulation is running for 5,000,000 cycles. Fig. 10 compares the latency and energy per packet values obtained by the proposed NoC architecture and a conventional NoC. The NoC size is 7 × 7, but the other simulation parameters are the same as the synthetic workloads.

As Fig. 10 shows, the proposed NoC improves average packet latency and energy consumption of the conventional NoC by 18% and 33%,

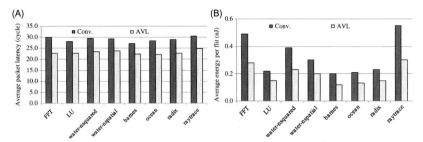

Fig. 10 The average packet latency (A) and energy per flit (B) results of a VAL-enabled and a conventional NoC for the SPLASH-2 programs.

respectively. As mentioned before, energy consumption improvement is more than latency improvement, due to the negative effect of SDM on latency.

In both energy and latency results, the improvements come from the fact that the intermediate routers of VLAs are bypassed. For latency, some of the improvements are neutralized by the negative effect of increased packet size, but the power gain of the packets traveling on VALs is directly translated to total NoC power reduction.

In these experiments, the network is rather heavily loaded (but still works well below the saturation point) to increase the NoC resource utilization and fully exploit NoC bandwidth. We explored the behavior of the proposed NoC across the full network load range in the previous subsection.

5.4 The effect of reconfiguration frequency

An important parameter in the proposed dynamic VAL reconfiguration procedure is the frequency of its execution. Obviously, higher frequencies can capture the changes of the traffic pattern better and give more adaption to on–chip traffic patterns. On the other hand, this would increase the overhead of the algorithm in terms of processing power/load on the root processor and bandwidth usage of the control data. Fig. 11 shows the effect of changing VAL construction period from 500,000 cycles to 250,000 cycles. As the execution time of the dynamic path finding algorithm is in the order of 10,000–20,000 cycles, this reconfiguration period is well longer than the time required for reconfiguration calculations. We observed that by further increasing in the reconfiguration frequency (less than 250k period), no improvement is obtained due to the overhead of reconfiguration. We used the power-delay product of NoCs as a comparison metric to show the combined results of power and latency in a single plot.

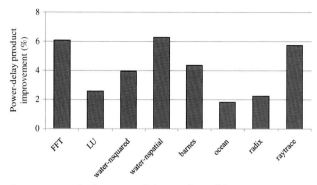

Fig. 11 The improvement in the power-delay product of the SPLASH-2 programs for the reconfiguration period of 250,000 cycles over the period of 500,000 cycles.

As figure indicates, up to 7% improvement (average of 4%) can be achieved by doubling the frequency. Obviously, the suitable frequency of reconfiguration depends on the behavior of each individual program; the more intense the program traffic varies over time, the more improvement is achieved with higher frequencies.

5.5 System-on-chip workload

The static design-time VAL construction algorithm targets systems whose applications are known a priori. To evaluate the performance improvements that can be achieved by VALs for such applications, we select GSM encoder, an embedded multi-task application as benchmark. The application CTG can be found in Ref. [49]. The tasks of the application have been mapped onto a 6×8 mesh. In the simulations, packets are generated with an exponential distribution with the communication rates proportional to the CTG edge weights. The simulation parameters are set as those in Section 5.2.

We compare our results to the reconfigurable NoCs (ReNoC) [24], and Long-Range Links (LRL) [22] which target the same class of applications. The LRLs have an extra link budget equivalent to 40 regular mesh links. The original 6×8 NoC has 164 links.

Fig. 12 shows the latency and energy per flit consumption of the three networks under the GSM encoder benchmark. As the figure shows, LRLs give the best performance, followed by ReNoC. As the latency of the conventional NoC under this benchmark is 38 cycles, ReNoC, LRL, and VAL bring about 24%, 19%, and 7% improvement in latency. The main reason for the higher improvement of ReNoC and LRL is the more area they use, but

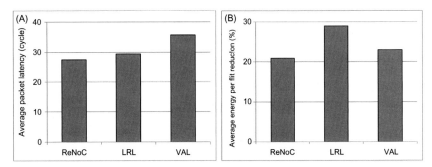

Fig. 12 The average packet latency in cycles (A) and percentage of the energy per flit reduction with respect to a conventional NoC (B) for the three considered NoCs under the GSM encoder benchmark.

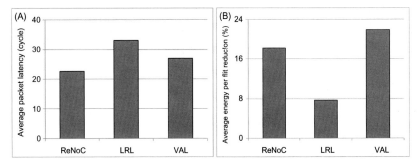

Fig. 13 The average packet latency in cycles (A) and percentage of the energy per packet reduction (B) for the three considered NoCs under the GSM decoder benchmark. The application uses the LRLs of the GSM encoder.

the VALs are restricted to have the same area as a conventional NoC. In energy consumption, however, ReNoC, LRL, and VAL give 21%, 29%, and 25% reduction compared to a conventional NoC, respectively.

Adaptability. As mentioned before, adaptability is the key advantage of our work over LRLs. To evaluate this, we assume that a new application, *GSM decoder,* is added to the system after chip synthesis. The new application runs on the same MPSoC designed for GSM-encoder. It uses the same set of cores as the GSM encoder but with a different inter-core communication pattern. Fig. 13 shows the latency and energy consumption of the three networks.

As the figure indicates, the adaptable nature of ReNoC and VALs allows them to keep reasonable performance improvement for a new application, while many static LRLs designed based on the old application are not useful

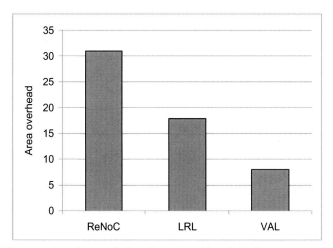

Fig. 14 The area overhead of the three considered NoCs over a conventional packet-switched NoC.

for the newly added communication pattern. Adaption capability without negative impact on design time/effort of the NoC is an interesting property of VALs.

Cost. Both ReNoC and VALs are adaptable, but ReNoC gives better performance than VALs. However, the advantage of VALs over ReNoC is the less area it occupies. Less area, in turn, results in less static power consumption. Fig. 14 shows the area overhead of LRLs, ReNoC and VALs used in this study over a conventional mesh NoC. The area of the NoCs is modeled using an analytical NoC area model presented in Ref. [24]. The VAL area overhead is mainly due to the multiplexers and wiring in the input ports. The right NoC implementation option can be chosen based on flexibility, performance constraints, and area budget.

5.6 Trade-off between area and performance

In this section, we investigate the performance improvement that would be achieved if the n/2-bit sub-network width constraint is relaxed. In Fig. 15, the effect of increasing the link bit-width of two sub-networks on the performance improvement of the proposed architecture is shown. The results are compared to the results given by the original sub-network bit-width of 32 (in a 64-bit wide NoC). When the bit width of sub-networks is 64, the entire network bit-width is in fact doubled. In this case, we can assume that a 64-bit VAL network is added to the main 64-bit mesh network.

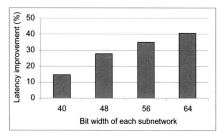 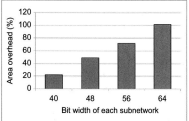

Fig. 15 (Right) The percentage of latency reduction by increasing the subnetwork link width from 32 to 64 bits. The bars are the average of latency reduction of all Splash programs. (Left) The area overhead of VAL-based NoCs when increasing subnetwork bit-width from 32 bits to 64 bits. The figures show the area overhead over an NoC with 32-bit subnetworks.

The network size is 7×7 and the simulation is done for the Splash programs. To keep the figure size manageable, each bar of the figure shows the average latency reduction across all Splash benchmarks.

Fig. 15 also shows the area overhead (over the 32-bit sub-network) imposed by increasing the sub-network bit-width to explore the trade-off between area overhead and performance of VALs.

5.7 Sensitivity to network size

We evaluate the effectiveness of VALs in a larger NoC to explore its scalability. A larger system is obtained by integrating the CTGs of GSM encoder and GSM decoder into a single 120 task CTG. The CTG is then mapped onto a 12×10 mesh NoC. The energy consumption and latency results are shown in Fig. 16. By comparing the improvement reported in Fig. 16 with the results shown in Fig. 15, we can see that although VALs continue to deliver better performance than a conventional NoC (9% reduction in latency and 17% in energy), the performance margin reduces when the size increases. The reason is that larger NoCs have longer VALs, on average. This increases the conflict among VAL paths and therefore, the number of flows covered by VALs reduces. However, VALs still are advantageous in larger NoC sizes.

Another issue with larger network sizes is the algorithm execution time. As mentioned in Section 4.7, a hierarchical approach is used to construct VALs for large networks. Fig. 17 shows the normalized execution time of the algorithms when finding SCP paths in a 12×10 network (2 GSM Encoder + 1 GSM Decoder) with and without the hierarchical scheme.

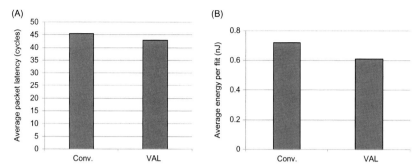

Fig. 16 The average packet latency in cycles (A) and the energy per flit (nJ) (B) for a conventional and VAL NoCs under the GSM decoder + GSM encoder benchmark in a 9 × 10 mesh.

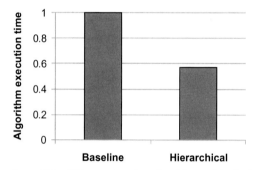

Fig. 17 Execution time of the VAL construction algorithms with the baseline and hierarchical approaches in a 12 × 10 NoC under a mix of GSM applications. The bars are normalized with respect to the baseline.

The results compare the execution time of the algorithms when a central root manages all SCP construction requests with the case when the network is divided into four 6 × 5 regions and five root nodes (four local and one global) perform the SCP construction job. The results are normalized with respect to results of the baseline scheme with a single root. As the figure indicates, using the hierarchical method can keep the execution time of the SCP construction algorithm to a reasonable value when the network size increases.

6. Conclusions

In this work, we proposed to add virtual adaptable links (VALs) to a conventional packet-switched NoC. These links virtually connect two distant nodes and provide them with low-latency low-power connections.

The links borrow the regular NoC links bit-width, rather than using distinct physical links. In this case, a part of the NoC link wires are connected according to the mesh topology, while the other part can implement any ad-hoc topology. Our proposed architecture benefits from the interesting properties of both regular and customized ad-hoc topologies; it can be customized for a target application and at the same time, has a regular structure to take advantage of the reusability and predictability of regular NoCs. Adding VALs to conventional packet-switched NoCs imposes a negligible area overhead of 7%. The key advantage of the proposed architecture is its reconfigurability and dynamic adaption to running application traffic. We developed an algorithm to adapt the virtual links to the current on-chip traffic pattern. The algorithm can work in both run-time and design-time. The experimental results for a set of benchmarks showed that this architecture improves the energy consumption and performance of NoCs, compared to some state-of-the-art designs. Some performance gains of VALs are neutralized by negative impacts of SDM on packet latency. Consequently, VALs can improve energy consumption more than latency.

References

[1] G. De Micheli, Keynote talk: NoCs: a short history of success and a long future, in: 2018 Twelfth IEEE/ACM International Symposium on Networks-on-Chip (NOCS), Turin, 2018, pp. 1–2.
[2] S.H.S. Rezaei, M. Modarressi, R. Ausavarungnirun, M. Sadrosadati, O. Mutlu, M. Daneshtalab, NoM: Network-on-Memory for Inter-Bank Data Transfer in Highly-Banked Memories, IEEE CAL, 2020.
[3] P. Lotfi-Kamran, M. Modarressi, H. Sarbazi-Azad, Near-ideal networks-on-chip for servers, in: 2017 IEEE International Symposium on High Performance Computer Architecture (HPCA), Austin, TX, 2017, pp. 277–288.
[4] A. Hemani, et al., Network on a chip: an architecture for billion transistor era, in: Proc. IEEE NorChip Conf, 2000, pp. 166–173.
[5] J. Yin, et al., Experiences with ML-driven design: a NoC case study, in: 2020 IEEE International Symposium on High Performance Computer Architecture (HPCA), San Diego, CA, USA, 2020, pp. 637–648. .intel.
[6] Implementing Low-Power AI SoCs Using NoC Interconnect Technology, 2012. www.sonics.com.
[7] C. Lin, W. He, Y. Sun, B. Pei, Z. Mao, M. Seok, 25.8 A Near- threshold-voltage network-on-chip with a metastability error detection and correction technique for supporting a quad-voltage/frequency-domain ultra-low-power system-on-a-chip, 2020, in: IEEE International Solid- State Circuits Conference—(ISSCC), San Francisco, CA, USA, 2020, pp. 394–396.
[8] A. Kumar, L.S. Peh, P. Kundu, N.K. Jha, Express virtual channels: toward the ideal interconnection fabric, in: Proc. Int. Symp. Comput. Architect. (ISCA), 2007, pp. 150–161.

[9] H.G. Lee, N. Chang, U.Y. Ogras, R. Marculescu, On-chip communication architecture exploration: a quantitative evaluation of point-to-point, bus, and network-on-chip approaches, ACM Trans. Design Automat. Electron. Syst. 12 (3) (2007) 1–20.

[10] K. Goossens, J. Dielissen, A. Radulescu, The Æthereal network on chip: concepts, architectures, and implementations, IEEE Des. Test Comput. 22 (2005) 414–421.

[11] T. Bjerregaard, J. Sparso, A router architecture for connection-oriented service guarantees in the MANGO clockless network-on-Chip, in: Proceedings of the Conference on Design, Automation and Test in Europe—volume 2 (DATE '05), Vol. 2, IEEE Computer Society, 2005, pp. 1226–1231.

[12] M. Modarressi, A. Tavakkol, H. Sarbazi-Azad, Virtual point-to-point connections for NoCs, IEEE Trans. Comput. Aided Des. Integr. Circuits Syst 29 (6) (2010) 855–868.

[13] N.E. Jerger, M. Lipasti, L. Peh, Circuit-switched coherence, IEEE Comput. Archit. Lett. 6 (2007) 5–8.

[14] A. Kumar, L.S. Peh, N.K. Jha, Token flow control, in: Proceedings of the 41st annual IEEE/ACM International Symposium on Microarchitecture (MICRO 41), 2008, pp. 342–353.

[15] M. Modarressi, M. Asadinia, H. Sarbazi-Azad, Using task migration to improve non-contiguous processor allocation in NoC-based CMPs, Elsevier J. Syst. Archit. 59 (7) (2013) 468–481.

[16] S.H. Seyyedaghaei Rezaei, A. Mazloumi, M. Modarressi, P. Lotfi-Kamran, Dynamic resource sharing for high-performance 3-DNetworks-on-Chip, IEEE Comput. Archit. Lett. 15 (1) (Jan. 2016) 5–8.

[17] S.H. Seyyedaghaei, M. Modarressi, R.Y. Aminabadi, M. Daneshtalab, Fault-Tolerant 3-D Network-on-Chip Deignusing Dynamic Link Sharing, Design, Automation & Test in Europe Conference & Exhibition (DATE), 2016.

[18] S.H.S. Rezaei, M. Modarressi, M. Daneshtalab, S. Roshanise-fat, A three-dimensional networks-on-chip architecture with dynamic buffer sharing, in: 24th Euromicro International Conference on Parallel, Distributed, and Network-Based Processing, IEEE, 2016.

[19] J. Balfour, W.J. Dally, Design tradeoffs for tiled CMP on-chip networks, in: Proc. of the International Conference of Supercomputing, 2006, pp. 178–189.

[20] C. Gomez, M.E.G. Requena, P.L. Rodriguez, J.D. Marin, Exploiting wiring resources on interconnection network: increasing path diversity, in: Proceedings of the 16th Euromicro Conference on Parallel, Distributed and Network-Based Processing (PDP 2008) (PDP '08), IEEE Computer Society, Washington, USA, 2008, pp. 20–29.

[21] J. Hu, R. Marculescu, Energy- and performance-aware mapping for regular NoC architectures, IEEE Trans. Comput. Aided Des. Integr. Circuits Syst. 24 (1) (2005) 551–562.

[22] U.Y. Ogras, R. Marculescu, "It's a small world after all": NoC performance optimization via long-range link insertion, IEEE Trans. Very Large Scale Integr. Syst. 14 (7) (2006) 693–706.

[23] S. Murali, D. Atienza, P. Meloni, S. Carta, Synthesis of predictable networks-on-Chip-based interconnect architectures for Chip multiprocessors, IEEE Trans. Very Large Scale Integr. Syst. 15 (8) (2007) 869–880.

[24] M. Modarressi, A. Tavakkol, H. Sarbazi-Azad, Application-aware topology reconfiguration for on-chip networks, IEEE Trans. Very Large Scale Integr. Syst. 19 (11) (2011) 2010–2022.

[25] M. Stensgaard, J. Sparsø, ReNoC: a network-on-chip architecture with reconfigurable topology, in: Proceedings of International Symposium on Networks-on-Chip (NoCS), 2008, pp. 55–64.

[26] S. Vassiliadis, I. Sourdis, Flux networks: interconnects on demand, in: Proceedings of International Conference on Embedded Computer Systems: Architectures, Modeling and Simulation (IC-SAMOS), 2006, pp. 160–167.

[27] M. Keramati, et al., Thermal management in 3d networks-on-chip using dynamic link sharing, Elsevier J. Microprocess. Microsyst. 52 (2017) 69–79.

[28] R. Hojabr, M. Modarressi, M. Daneshtalab, A. Yasoubi, A. Khonsari, Customizing Clos network-on-chip for neural networks, IEEE Trans. Comput. 66 (11) (2017) 1865–1877.

[29] A. Firuzan, M. Modarressi, M. Daneshtalab, M. Reshadi, Reconfigurable network-on-chip for 3D neural network accelerators, in: 2018 Twelfth IEEE/ACM International Symposium on Networks-on-Chip (NOCS), Turin, 2018, pp. 1–8.

[30] S.M. Nabavinejad, M. Baharloo, K.-C. Chen, M. Palesi, T. Kogel, M. Ebrahimi, An overview of efficient interconnection networks for deep neural network accelerators, IEEE J. Emerging Sel. Top. Circuits Syst. 10 (3) (2020) 268–282.

[31] K. Chen, M. Ebrahimi, T. Wang, and Y. Yang. "NoC-based DNN accelerator: a future design paradigm," In Proceedings of the 13th IEEE/ACM International Symposium on Networks-on-Chip, pp. 1–8.

[32] N. Teimouri, M. Modarressi, H. Sarbazi-Azad, Power and performance efficient partial circuits in packet-switched networks-on-chip, in: 2013 21st Euromicro International Conference on Parallel, Distributed, and Network-Based Processing, 2013, pp. 509–513.

[33] M. Modarressi, H. Sarbazi-Azad, M. Arjomand, An SDM-based hybrid packet-circuit-switched on-Chip network, Design, Automation & Test in Europe Conference & Exhibition (DATE), 2009.

[34] A. Leroy, P. Marchal, A. Shickova, F. Catthoor, F. Robert, D. Verkest, Spatial division multiplexing: a novel approach for guaranteed throughput on NoCs, in: Proceedings of the 3rd IEEE/ACM/IFIP International Conference on Hardware/Software Codesign and System Synthesis (CODES + ISSS '05), ACM, New York, NY, USA, 2005, pp. 81–86.

[35] A. Morgenstein, et al., Link division multiplexing (LDM) for network-on-chip link, in: Proc. of 24th IEEE Convention of Electrical and Electronics Engineers, 2006.

[36] T. Wolkotte, G.J.M. Smit, G.K. Rauwerda, L.T. Smit, An energy-efficient reconfigurable circuit switched network-on-chip, in: Proc. of Reconfigurable Architecture Workshop (RAW), Denver, Colorado, 2005.

[37] J. Hu, R. Marculescu, Exploiting the routing flexibility for energy/performance aware mapping of regular NoC architectures, Design, Automation & Test in Europe Conference & Exhibition (DATE), 2003.

[38] R. Lysecky, F. Vahid, S. Tan, Dynamic FPGA routing for just-in-time FPGA compilation, in: Proceedings of the 41st Annual Design Automation Conference (DAC '04), ACM, New York, NY, USA, 2004, pp. 954–959.

[39] M. Schoeberl, F. Brandner, J. Sparsø, E. Kasapaki, A statically scheduled time-division-multiplexed network-on-chip for real-time systems, in: Proceedings of the 2012 IEEE/ACM Sixth International Symposium on Networks-on-Chip (NOCS '12), IEEE Computer Society, Washington, DC, USA, 2012, pp. 152–160.

[40] M. Modarressi, H. Sarbazi-Azad, A. Tavakkol, An efficient dynamically reconfigurable on-chip network architecture, in: Design Automation Conference (DAC'10), 2010, pp. 310–313.

[41] J. Newsome, D. Song, GEM: graph embedding for routing and data-centric storage in sensor networks without geographic information, in: Proceedings of the 1st International Conference on Embedded Networked Sensor Systems (SenSys '03), ACM, New York, NY, USA, 2003, pp. 76–88.

[42] R. Stefan, K. Goossens, A TDM slot allocation flow based on multipath routing in NoCs, Microprocess. Microsyst 35 (2) (2011) 130–138.

[43] G. Chen, F. Li, M. Kandemir, Compiler-directed channel allocation for saving power in on-chip networks, in: Proc. of ACM Symposium on Principles of Programming Languages, 2006.

[44] W.J. Dally, B. Towles, Principles and Practices of Interconnection Network, Morgan Kaufmann, San Mateo, 2004.
[45] J. Duato, A necessary and sufficient condition for deadlock-free adaptive routing in wormhole networks, IEEE Trans. Parallel Distrib. Syst. 6 (10) (1995) 1055–1067.
[46] Booksim NoC simulator, n.d., http://nocs.stanford.edu/booksim.html.
[47] A. Kahng, B. Li, L. Peh, K. Samadi, ORION 2.0: a fast and accurate NoC power and area model for early-stage design space exploration, DATE, France, 2009, pp. 423–428.
[48] SPLASH-2, n.d., http://www.flash.stanford.edu/apps/SPLASH.
[49] M. Schmitz, Energy Minimization Techniques for Distributed Embedded Systems, Ph.D. thesis, University of Southampton, 2003.

About the authors

S. Hossein SeyyedAghaei Rezaei received the BSc degree from Babol Noshirvani University of Technology, Babol, Iran, in 2012, and the MSc degree from University of Tehran, Tehran, Iran, in 2015, both in Computer Engineering. He is currently working toward the PhD degree in computer architecture in the Department of Electrical and Computer Engineering, University of Tehran, Tehran, Iran. His research interests include memory systems, GPUs, networks-on-chip and Process in Memory (PIM).

Mehdi Modarressi is currently an Assistant Professor with the Department of Electrical and Computer Engineering, College of Engineering, University of Tehran, Tehran, Iran. He received his PhD in Computer Engineering from Sharif University of Technology, Tehran, Iran, in 2011. His research interests include interconnection networks, parallel processing, and hardware architectures for deep, on which he has published more than 100 peer-reviewed papers and one book.

CHAPTER EIGHT

The design of an energy-efficient deflection-based on-chip network

Rachata Ausavarungnirun[b] and Onur Mutlu[a]
[a]SAFARI Research Group, ETH Zürich, Switzerland
[b]King Mongkut's University of Technology North Bangkok, Bangkok, Thailand

Contents

Advances in Computers, Volume 124
ISSN 0065-2458
https://doi.org/10.1016/bs.adcom.2021.12.002

Abstract

As the number of cores scale to tens and hundreds, the energy consumption of routers across various types of on-chip networks in chip muiltiprocessors (CMPs) increases significantly. A major source of this energy consumption comes from the input buffers inside Network-on-Chip (NoC) routers, which are traditionally designed to maximize performance. To mitigate this high energy cost, many works propose bufferless router designs that utilize deflection routing to resolve port contention. While this approach is able to maintain high performance relative to its buffered counterparts at low network traffic, the bufferless router design suffers performance degradation under high network load.

In order to maintain high performance and energy efficiency under *both* low and high network loads, this chapter discusses critical drawbacks of traditional bufferless designs and describes recent research works focusing on two major modifications to improve the overall performance of the traditional bufferless network-on-chip design. The first modification is a minimally-buffered design that introduces limited buffering inside critical parts of the on-chip network in order to reduce the number of deflections. The second modification is a hierarchical bufferless interconnect design that aims to further improve performance by limiting the number of hops each packet needs to travel while in the network. In both approaches, we discuss design tradeoffs and provide evaluation results based on common CMP configurations with various network topologies to show the effectiveness of each proposal.

1. Introduction

As commercial processors incorporate more cores, scalability and energy efficiency demand better interconnection substrates. Different interconnect designs such as a 2D mesh [1–17] or a flattened butterfly [18] have become increasingly popular as scalable, high-performance on-chip networks for chip multiprocessors (CMPs) as the number of core grows to hundreds. Unfortunately, these high-performance Network-on-Chip (NoC) designs are projected to consume significant amount of power. For example, 28% of the chip power is consumed by the NoC in the Intel Terascale

80-core chip [19], 36% in MIT RAW [20], and 36% in Intel 48-core SCC [21]. To reduce the overall power consumption of modern processors, NoC energy efficiency is a critically important design goal [2–4,22–25].

Previous on-chip interconnection network designs commonly assume that each router in the network needs to contain buffers to buffer the packets (or flits) transmitted within the network. Indeed, buffering within each router improves the bandwidth efficiency in the network because buffering reduces the number of dropped or "misrouted" packets [1], i.e., packets that are sent to a less desirable destination port. On the other hand, buffering has several disadvantages. First, buffers consume significant energy/power: dynamic energy when read/written and static energy even when they are not occupied. Second, having buffers increases the complexity of the network design because logic needs to be implemented to place packets into and out of buffers. Third, buffers can consume significant chip area: even with a small number (16) of total buffer entries per node where each entry can store 64 bytes of data, a network with 64 nodes requires 64 KB of buffer storage. In fact, in the TRIPS prototype chip, input buffers of the routers were shown to occupy 75% of the total on-chip network area [26]. Energy consumption and hardware storage cost of buffers increases as future many-core chips incorporate more network nodes.

Mechanisms have been proposed to make conventional input-buffered NoC routers more energy-efficient (i.e., use less energy per unit of performance). For example, bypassing empty input buffers [27,28] reduces some dynamic buffer power, but static power remains. Such bypassing is also less effective when buffers are not frequently empty. Bufferless deflection routers [2] remove router input buffers completely (thereby eliminating their static and dynamic power) to reduce router power. In a conventional bufferless deflection network, flits (several of which make up one packet) are independently routed, unlike most buffered networks, where a packet is the smallest independently-routed unit of traffic. When two flits contend for a single router output, one must be deflected [29] to another output. Thus, a flit never requires a buffer in a router. By controlling which flits are deflected, a bufferless deflection router can ensure that all traffic is eventually delivered. Removing buffers yields simpler and more energy-efficient NoC designs: for example, CHIPPER [4] reduces average network power by 54.9% in a 64-node system compared to a conventional buffered router.

Unfortunately, at high network load, deflection routing reduces performance and efficiency. This is because deflections occur more frequently when many flits contend in the network. Each deflection sends a flit further from its destination, causing unnecessary link and router traversals. Relative to a buffered network, a bufferless network with a high deflection rate wastes energy, and suffers from worse congestion, because of these unproductive network hops. In contrast, a buffered router is able to hold flits (or packets) in its input buffers until the required output port is available, incurring no unnecessary hops. Thus, a buffered network can sustain higher performance at peak load [2], but at the cost of large buffers, which can consume significant power and die area.

The best interconnect design would obtain the energy efficiency of the bufferless approach with the high performance of the buffered approach. To obtain the best of both worlds, a router would contain only a small amount of buffering for flits that actually require it, and the network topology would allow flits to reach their destination using the fewest number of hops.

On top of the bufferless approach, to achieve scalability to many (tens, hundreds, or thousands) of cores, *mesh* or other higher-radix topologies are used. Both the Intel SCC [30] and Terascale [31] CMPs, and several other many-core CMPs (e.g., the MIT RAW [20] prototype, the UT-Austin TRIPS chip [32], several Tilera products [33,34]), and more recently, the Intel Skylake [35], Intel Cascade Lake [36], and Intel Ice Lake [37] server processors exchange packets on a mesh. A mesh topology has good scalability because there is no central structure which needs to scale (unlike a central bus or crossbar-based design), and bisection bandwidth increases as the network grows. However, routers in a 2D mesh can consume significant energy and die area due to overheads in buffering, in routing and flow control, and in the switching elements (crossbars) that connect multiple inputs with multiple outputs.

Mainstream commercial CMPs today most commonly use *ring*-based interconnects. Rings are a well-known network topology [1], and the idea behind a ring topology is very simple: all routers (also called "ring stops") are connected by a loop that carries network traffic. At each router, new traffic can be injected into the ring, and traffic in the ring can be removed from the ring when it reaches its destination. When traffic is traveling on the ring, it continues uninterrupted until it reaches its destination. A ring router thus *needs no in-ring buffering or flow control* because it prioritizes on-ring traffic. In addition, the router's datapath is very simple compared to a mesh router, because the router has fewer inputs and requires no large, power-inefficient crossbars; typically it consists only of several MUXs to allow traffic to enter and leave, and one pipeline register. Its latency is typically only one clock cycle, because no routing decisions or output port allocations are necessary (other than removing traffic from the ring when it arrives). Because of these advantages, several prototype and commercial multicore processors have utilized ring interconnects. Examples of these processors include the Intel Larrabee [38], IBM Cell [39], Intel Sandy Bridge [40], Intel Skylake [35], Intel Coffee Lake [41] and more recently, the Intel Ice Lake [37] and the three interwoven rings in the NVIDIA DGX-1 NVLink [42].

Past work has shown that rings are competitive with meshes up to tens of nodes [43–45]. Unfortunately, rings suffer from a fundamental scaling problem because a ring's bisection bandwidth does not scale with the number of nodes in the network. Building more rings, or a wider ring, serves as a stop-gap measure but increases the cost of every router on the ring in proportion to the bandwidth increase. As commercial CMPs incorporate more cores, a new network design will be needed that balances the simplicity and low overhead of rings with the scalability of more complex topologies.

A hybrid design is possible. In this chapter, we introduce an approach to construct rings in a *hierarchy* such that groups of nodes share a simple ring interconnect, and these "local" rings are joined by one or more "global" rings. Fig. 1 shows an example of such a *hierarchical ring* design [23,24,44–51]. Past works [44–51] propose hierarchical rings as a scalable alternative to single ring and mesh networks. These proposals join rings with *bridge routers*, which reside on multiple rings and transfer traffic between rings. This design was shown to yield good performance and scalability [46]. A state-of-the-art design [46] requires *flow control and buffering* at every node router (ring stop), because a ring transfer can cause one ring to back up and stall when another ring is congested. While this previously proposed hierarchical ring is much more scalable than a single ring [46], the reintroduction of in-ring buffering and flow control nullifies one of the primary advantages of using ring networks in the first place (i.e., the lack of buffering and buffered flow control within each ring).

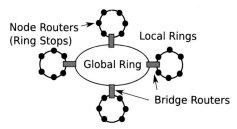

Fig. 1 A traditional hierarchical ring design [46–50] allows "local rings" with simple node routers to scale by connecting to a "global ring" via bridge routers. *Reproduced from R. Ausavarungnirun, et al., Improving energy effciency of hierarchical rings via deflection routing, SAFARI Technical Report TR-2014-002: http://safari.ece.cmu.edu/tr.html (2014).*

To combine the concept of a minimally-buffered router with a hierarchical ring, previous work allows a bridge router with a full buffer to *deflect* packets, called HiRD (i.e., Hierarchical Rings with Deflection [23,24,52]). Rather than requiring buffering and flow control in the ring, packets simply cycle through the network and try again. While deflection-based, bufferless networks have been proposed and evaluated in the past [2–4,6,19,23,24,27,53–65], this minimally-buffered hierarchical ring approach is effectively an elegant hybridization of bufferless (rings) and buffered (bridge routers) styles. To prevent packets from potentially deflecting around a ring arbitrarily many times (i.e., to prevent live-lock), we introduce two new mechanisms, the *injection guarantee* and the *transfer guarantee*, that ensure packet delivery even for adversarial/pathological conditions (as discussed in [23,24,52] and evaluated with worst-case traffic in Section 5.10). This simple hierarchical ring design provides a more scalable network architecture while retaining the key simplicities of ring networks (no buffering or flow control within each ring). Section 5.10 shows that HiRD provides better performance, lower power, and better energy efficiency with respect to various buffered hierarchical ring designs [46] as well as other NoC designs.

2. Bufferless Routing

2.1 Why Bufferless? (and when?)

Bufferless[a] NoC design has recently been evaluated as an alternative to traditional virtual-channel buffered routers [2–4,6,19,23,24,27,53–65]. It is appealing mainly for two reasons: reduced power consumption, and simplicity in design. As core count in modern CMPs continues to increase, the interconnect becomes a more significant component of system power consumption. Several prototype manycore systems point toward this trend: in MIT RAW, interconnect consumes 40% of system power; in the Intel Terascale chip, 30%. Buffers consume a significant portion of this power. A recent work [2], described in Section 2.2 reduces network energy by 40% by eliminating buffers. Furthermore, the complexity reduction of the design at the high level could be substantial: a bufferless router requires only pipeline registers, a crossbar, and arbitration logic. This can translate into reduced system design and verification cost.

---

[a] More precisely, a "bufferless" NoC has no in-router (e.g., virtual channel) buffers, only pipeline latches. Baseline bufferless designs, such as BLESS [2], still require reassembly buffers and injection queues. Another subsequent work CHIPPER [4] eliminates these buffers as well.

Bufferless NoCs present a tradeoff: by eliminating buffers, the peak network throughput is reduced, potentially degrading performance. However, network power is often significantly reduced. For this tradeoff to be effective, the power reduction must outweigh the slowdown's effect on total energy. Moscibroda and Mutlu [2] reported minimal performance reduction with bufferless when NoC is lightly loaded, which constitutes many of the applications they evaluated. Bufferless NoC design thus represents a compelling design point for many systems with low-to-medium network load, eliminating unnecessary capacity for significant savings.

2.2 BLESS: Baseline bufferless deflection routing

Here we briefly introduce bufferless deflection routing in the context of BLESS [2]. BLESS routes flits, the minimal routable units of packets, between nodes in a mesh interconnect. Each flit in a packet contains header bits and can travel independently, although in the best case, all of a packet's flits remain contiguous in the network. Each node contains an injection buffer and a reassembly buffer; there are no buffers within the network, aside from the router pipeline itself. Every cycle, flits that arrive at the input ports contend for the output ports. When two flits contend for one output port, BLESS avoids the need to buffer by misrouting one flit to another port. The flits continue through the network until ejected at their destinations, possibly out of order, where they are reassembled into packets and delivered. Deflection routing is not new: it was first proposed in [29], and is used in optical networks because of the cost of optical buffering [60]. It works because a router has as many output links as input links (in a 2D mesh, 4 for neighbors and 1 for local access). Thus, the flits that arrive in a given cycle can always leave exactly N cycles later, for an N-stage router pipeline. If all flits request unique output links, then a deflection router can grant every flit's requested output. However, if more than one flit contends for the same output, all but one must be deflected to another output that is free.

2.2.1 Livelock freedom

Whenever a flit is deflected, it moves further from its destination. If a flit is deflected continually, it may never reach its destination. Thus, a routing algorithm must explicitly avoid livelock. It is possible to probabilistically bound network latency in a deflection network [66,67]. However, a deterministic bound is more desirable. BLESS [2] uses an Oldest-First prioritization rule to give a deterministic bound on network latency. Flits arbitrate based on packet timestamps. Prioritizing the oldest traffic creates a consistent total order and allows this traffic to make forward progress. Once the oldest packet arrives, another packet becomes oldest. Thus livelock freedom is guaranteed inductively. However, this age-based priority mechanism is expensive [27,55] both in header information and in arbitration critical path. Alternatively, some bufferless routing proposals do not provide or explicitly show deterministic livelock-freedom guarantees [55–57]. This can lead to faster arbitration if it allows for simpler priority schemes. However, a provable guarantee of livelock freedom is necessary to show system correctness in all cases. CHIPPER [4], described in Section 2.3, provides strong livelock freedom guarantees in a bufferless design.

2.2.2 Injection

BLESS guarantees that all flits entering a router can leave it, because there are at least as many output links as input links. However, this does not guarantee that new traffic from the local node (e.g., core or shared cache) can always enter the network. A BLESS router can inject a flit whenever an input link is empty in a given cycle. In other words, BLESS requires a "free slot" in the network in order to insert new traffic. When a flit is injected, it contends for output ports with the other flits in that cycle. Note that the injection decision is purely local: that is, a router can decide whether to inject without coordinating with other routers.

2.2.3 Ejection and packet reassembly

A BLESS router can eject one flit per cycle when that flit reaches its destination. In any bufferless deflection network, flits can arrive in random order; therefore, a packet reassembly buffer is necessary. Once all flits in a packet arrive, the packet can be delivered to the local node. Importantly, this buffer must be managed so that it does not overflow, and in such a way that maintains correctness. BLESS [2] does not consider this problem in detail. Instead, it assumes an infinite reassembly buffer, and reports maximum occupancies for the evaluated workloads.

2.3 Deflection routing complexities

While bufferless deflection routing is conceptually and algorithmically simple, a straightforward hardware implementation leads to numerous complexities. In particular, two types of problem plague baseline bufferless deflection routing: high hardware cost, and unaddressed correctness issues. The hardware cost of a direct implementation of bufferless deflection routing is nontrivial, due to expensive control logic. Just as importantly, correctness issues arise in the reassembly buffers when they have practical (non-worst-case) sizes, and this fundamental problem is unaddressed by current work. Here, we describe the major difficulties: output port allocation, expensive priority arbitration, and reassembly buffer cost and correctness. Prior work cites these weaknesses [2].

To address these drawbacks, CHIPPER [4] proposes a new bufferless router architecture, CHIPPER, that solves these problems through three key insights. First, CHIPPER eliminates the expensive port allocator and the crossbar, and replace both with a permutation network; deflection routing maps naturally to this arrangement, reducing critical path length and power/area cost. Second, CHIPPER provides a strong livelock guarantee through an implicit token passing scheme, eliminating the cost of a traditional priority scheme. Finally, CHIPPER proposes a simple flow control mechanism for correctness with reasonable reassembly buffer sizes, and propose using cache miss buffers (MSHRs [68–70]) as reassembly buffers. CHIPPER [4] shows that at low-to-medium load, the reduced-complexity CHIPPER design performs competitively to a traditional buffered router (in terms of application performance and operational frequency) with significantly reduced network power, and very close to baseline bufferless (BLESS [2]) with a reduced critical path.

For low-to-medium network load, CHIPPER delivers performance close to a conventional buffered network as shown in Fig. 2, because the deflection rate is low: thus, most flits take productive network hops in every cycle, just as in the buffered network. In addition, the bufferless router has significantly reduced power (hence improved energy efficiency), because the buffers in a conventional router consume significant power. However, as network load increases, the deflection rate in a bufferless deflection network also rises, because flits contend with each other more frequently.

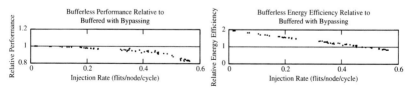

Fig. 2 System performance and energy efficiency (performance per watt) of bufferless deflection routing, relative to conventional input-buffered routing (4 VCs, 4 flits/VC) that employs buffer bypassing, in a 4 × 4 2D mesh. Injection rate (X axis) for each workload is measured in the baseline buffered network. *Reproduced from C. Fallin, et al., MinBD: Minimally-Buffered Deflection Routing for Energy-Effcient Interconnect, NOCS (2012).*

With a higher deflection rate, the dynamic power of a bufferless deflection network rises more quickly with load than dynamic power in an equivalent buffered network, because each deflection incurs some extra work. Hence, bufferless deflection networks lose their energy efficiency advantage at high load. Just as important, the high deflection rate causes each flit to take a longer path to its destination, and this increased latency reduces the network throughput and system performance.

Overall, neither design obtains both good performance and good energy efficiency at all loads. If the system usually experiences low-to-medium network load, then the bufferless design provides adequate performance with low power (hence high energy efficiency). But, if we use a conventional buffered design to obtain high performance, then energy efficiency is poor in the low-load case, and even buffer bypassing does not remove this overhead because buffers consume static power regardless of use. Finally, simply switching between these two extremes at a per-router granularity, as previously proposed [71], does not address the fundamental inefficiencies in the bufferless routing mode, but rather, uses input buffers for all incoming flits at a router when load is too high for the bufferless mode (hence retains the energy-inefficiency of buffered operation at high load).

This chapter (in Section 4) describes MinBD [3,72,73], a minimally-buffered deflection router that combines bufferless and buffered routing in a new way to reduce this overhead.

3. Scalability in mesh-based interconnects

Despite the simplicity advantage of a ring-based network, rings have a fundamental *scalability* limit: as compared to a mesh, a ring stops scaling at fewer nodes because its bisection bandwidth is *constant* (proportional only to link width) and the average hop count (which translates to latency for a packet) increases linearly with the number of nodes. (Intuitively, a packet visits half the network on its way to the destination, in the worst case, and a quarter in the average case, for a bidirectional ring.) In contrast, a mesh has bisection bandwidth proportional to one dimension of the layout (e.g., the square-root of the node count for a 2D mesh) and also has an average hop count that scales only with one dimension. The higher radix, and thus higher connectivity, of the mesh allows for more path diversity and lower hop counts which increases performance.

To demonstrate this problem quantitatively, Fig. 3 shows application performance averaged over a representative set of network-intensive

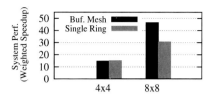

Fig. 3 Performance as mesh and ring networks to 64 nodes.

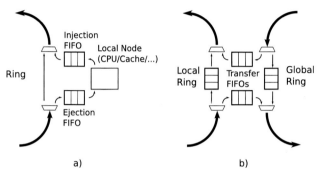

a)   b)

Fig. 4 (A) A ring router. (B) Buffered hierarchical ring detail, as proposed in prior work [46]: one *bridge router* which connects two rings, as proposed in HiRD [23]. *Reproduced from R. Ausavarungnirun, et al., Improving energy effciency of hierarchical rings via deflection routing, SAFARI Technical Report TR-2014-002: http://safari.ece.cmu.edu/tr.html (2014).*

workloads on (i) a single ring network and (ii) a conventional 2D-mesh buffered interconnect. As the plot shows, although the single ring is able to match the mesh's performance at the 16-node design point, it degrades when node count increases to 64. Note that in this case, the ring's bisection bandwidth is kept equivalent to the mesh, so the performance degradation is solely due to higher hop count; considerations of practical ring width might reduce performance further.

3.1 Simplicity in ring-based interconnects

Ring interconnects are attractive in current small-to-medium-scale commercial CMPs because ring routers (ring stops) are simple, which leads to smaller die area and energy overhead. In its simplest form, a ring router needs to perform only two functions: injecting new traffic into the ring, and removing traffic from the ring when it has arrived at its destination. Fig. 4A depicts the router datapath and control logic (at a high level) necessary to implement this functionality. For every incoming *flit*, the router only needs to determine whether the flit stays on the ring or exits at this node. Then, if a flit is waiting to inject, the router checks whether a flit is already present on the ring in this timeslot, and if not, injects the new flit using the in-ring MUX.

Rings achieve very good energy efficiency at low-to-medium core counts [43]. Despite the simplicity advantage of a ring-based network, rings have a fundamental *scalability* limit: a ring stops scaling at fewer nodes because its bisection bandwidth is *constant* (proportional only to link width) and the average hop count (which translates to latency for a packet) increases linearly with the number of nodes. (In the worst case for a bidirectional ring, a packet visits half the network on its way to its destination, and a quarter on average.)

3.2 Hierarchical rings for scalability

Fortunately, past work has observed that *hierarchy* allows for additional scalability in many interconnects: in ring-based designs [46–50], with hierarchical buses [74], and with hierarchical meshes [75]. The state-of-the-art hierarchical ring design [46] in particular reports promising results by combining local rings with one or more levels of higher-level rings, which we refer to as global rings, that connect lower level rings together. Rings of different levels are joined by Inter-Ring Interfaces (IRIs), which we call "bridge routers" in this work. Fig. 4B graphically depicts the details of one bridge router in the previously proposed buffered hierarchical ring network.

Unfortunately, connecting multiple rings via bridge routers introduces a new problem. Recall that injecting new traffic into a ring requires an open slot in that ring. If the ring is completely full with flits (i.e., a router attempting to inject a new flit must instead pass through a flit already on the ring every cycle), then no new traffic will be injected. But, a bridge router must inject transferring flits into those flits' destination ring in exactly the same manner as if they were newly entering the network. If the ring on one end of the bridge router is completely full (cannot accept any new flits), and the transfer FIFO into that ring is also full, then any other flit requiring a transfer must *block* its current ring. In other words, ring transfers create new dependences between adjacent rings, which creates the need for end-to-end flow control. This flow control forces every node router (ring stop) to have an in-ring FIFO and flow control logic, which increases energy and die area overhead and significantly reduces the appeal of a simple ring-based design. Table 1 compares the power consumption for a previously proposed

Table 1 Power and area comparison (16-node network).

Metric	Buffered HRing	Bufferless HRing
Normalized power	1	0.535
Normalized area	1	0.495

Reproduced from R. Ausavarungnirun, et al., Improving energy effciency of hierarchical rings via deflection routing, SAFARI Technical Report TR-2014-002: http://safari.ece.cmu.edu/tr.html (2014).

hierarchical ring design (Buffered HRing) and a bufferless hierarchical ring design on a system with 16 nodes using DSENT 0.91 [76] and a 45 nm technology commercial design library. In the examined idealized bufferless design, each ring has no in-ring buffers, but there are buffers between the hierarchy levels. When a packet needs a buffer that is full, it gets deflected and circles its local ring to try again. Clearly, eliminating in-ring buffers in a hierarchical ring network can reduce power and area significantly.

However, simply removing in-ring buffers can introduce livelock, as a deflected packet may circle its local ring indefinitely. Our goal in this work is to introduce a hierarchical ring design that maintains simplicity and low cost, while ensuring livelock and deadlock freedom for packets.

4. MinBD: Minimally-buffered deflection router design

The MinBD (minimally-buffered deflection) router [3,72,73] is a new router design that combines bufferless deflection routing with a small buffer, which we call the *side buffer*. We start by outlining the key principles we follow to reduce deflection caused inefficiency by using buffering:

1. When a flit would be deflected by a router, it is often better to buffer the flit and arbitrate again in a later cycle. Some buffering can avoid many deflections.
2. However, buffering every flit leads to unnecessary power overhead and buffer requirements, because many flits will be routed productively on the first try. The router should buffer a flit only if necessary.
3. Finally, when a flit arrives at its destination, it should be removed from the network (ejected) quickly, so that it does not continue to contend with other flits.

4.1 Basic high-level operation

The MinBD router does not use input buffers, unlike conventional buffered routers. Instead, a flit that arrives at the router proceeds directly to the routing and arbitration logic. This logic performs deflection routing, so that when two flits contend for an output port, one of the flits is sent to another output instead. However, unlike a bufferless deflection router, the MinBD router can also buffer up to one flit per cycle in a single FIFO-queue side buffer. The router examines all flits at the output of the deflection routing logic, and if any are deflected, one of the deflected flits is removed from the router pipeline and buffered (as long as the buffer is not full). From the side buffer, flits are reinjected into the network by the router, in the same way that new traffic is injected. Thus, some flits that would have been deflected in a bufferless deflection router are removed from the network temporarily into this side buffer, and given a second chance to arbitrate for a productive router output when re-injected. This reduces the network's deflection rate (hence improves performance and energy efficiency) while buffering only a fraction of traffic.

We will describe the operation of the MinBD router in stages. First, Section 4.1 describes the deflection routing logic that computes an initial routing decision for the flits that arrive in every cycle. Then, Section 4.2 describes how the router chooses to buffer some (but not all) flits in the side buffer. Section 4.3 describes how buffered flits and newly-generated flits are injected into the network, and how a flit that arrives at its destination is ejected. Finally, Section 4.4 discusses correctness issues, and describes how MinBD ensures that all flits are eventually delivered.

4.2 Deflection routing

The MinBD router pipeline is shown in Fig. 5 Flits travel through the pipeline from the inputs (on the *left*) to outputs (on the *right*). We first discuss the deflection routing logic, located in the Permute stage on the right. This logic implements deflection routing: it sends each input flit to its preferred output when possible, deflecting to another output otherwise. MinBD uses the deflection logic organization first proposed in CHIPPER [4]. The permutation network in the Permute stage consists of two-input blocks arranged into two stages of two blocks each. This arrangement can send a flit on any input to any output. (Note that it cannot perform all possible permutations of inputs to outputs, but as we will see, it is sufficient for correct operation that at least one flit obtains its preferred output.) In each two-input block, arbitration logic determines which flit has a higher priority, and sends that flit in the direction of its preferred output. The other flit at the two-input block, if any, must take the block's other output. By combining two stages of this

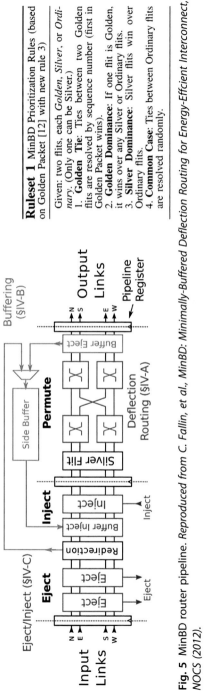

Ruleset 1 MinBD Prioritization Rules (based on Golden Packet [12] with new rule 3)

Given: two flits, each *Golden*, *Silver*, or *Ordinary*. (Only one can be Silver.)
1. **Golden Tie**: Ties between two Golden flits are resolved by sequence number (first in Golden Packet wins).
2. **Golden Dominance**: If one flit is Golden, it wins over any Silver or Ordinary flits.
3. **Silver Dominance**: Silver flits win over Ordinary flits.
4. **Common Case**: Ties between Ordinary flits are resolved randomly.

Fig. 5 MinBD router pipeline. *Reproduced from C. Fallin, et al., MinBD: Minimally-Buffered Deflection Routing for Energy-Efficient Interconnect, NOCS (2012).*

arbitration and routing, deflection arises as a distributed decision: a flit might be deflected in the first stage, or the second stage. Restricting the arbitration and routing to two-flit subproblems reduces complexity and allows for a shorter critical path, as demonstrated in CHIPPER [4]. In order to ensure correct operation, the router must arbitrate between flits so that every flit is eventually delivered, despite deflections. We adapt a modified version of the Golden Packet priority Scheme [4], which solves this livelock-freedom problem. This priority scheme is summarized in Ruleset 1. The basic idea of the Golden Packet priority scheme is that at any given time, at most one packet in the system is golden. The flits of this golden packet, called "golden flits," are prioritized above all other flits in the system (and contention between golden flits is resolved by the flit sequence number). While prioritized, golden flits are never deflected by non-golden flits. The packet is prioritized for a period long enough to guarantee its delivery. Finally, this "golden" status is assigned to one globally-unique packet ID (e.g., source node address concatenated with a request ID), and this assignment rotates through all possible packet IDs such that any packet that is "stuck" will eventually become golden. In this way, all packets will eventually be delivered, and the network is livelock-free. (See CHIPPER [4] for the precise way in which the Golden Packet is determined; we use the same rotation schedule.)

However, although Golden Packet arbitration provides correctness, a performance issue occurs with this priority scheme. Consider that most flits are not golden: the elevated priority status provides worst-case correctness, but does not impact common-case performance (prior work reported over 99% of flits are delivered without becoming golden [4]). However, when no flits are golden and ties are broken randomly, the arbitration decisions in the two permutation network stages are not coordinated. Hence, a flit might win arbitration in the first stage, and cause another flit to be deflected, but then lose arbitration in the second stage, and also be deflected. Thus, unnecessary deflections occur when the two permutation network stages are uncoordinated.

In order to resolve this performance issue, we observe that it is enough to ensure that in every router, at least one flit is prioritized above the others in every cycle. In this way, at least one flit will certainly not be deflected. To ensure this when no golden flits are present, we add a "silver" priority level, which wins arbitration over common-case flits but loses to the golden flits. One silver flit is designated randomly among the set of flits that enter a router at every cycle (this designation is local to the router, and not propagated to other routers). This modification helps to reduce deflection rate.

Prioritizing a silver flit at every router does not impact correctness, because it does not deflect a golden flit if one is present, but it ensures that at least one flit will consistently win arbitration at both stages. Hence, deflection rate is reduced, improving performance.

4.3 Using a small buffer to reduce deflections

The key problem addressed by MinBD is deflection inefficiency at high load. In other words, when the network is highly utilized, contention between flits occurs often. This results in many deflected flits. We observe that adding a small buffer to a deflection router can reduce deflection rate, because the router can choose to buffer rather than deflect a flit when its output port is taken by another flit. Then, at a later time when output ports may be available, the buffered flit can re-try arbitration.

Thus, to reduce deflection rate, MinBD adds a "side buffer" that buffers only some flits that otherwise would be deflected. This buffer is shown in Fig. 5 above the permutation network. In order to make use of this buffer, a "buffer eject" block is placed in the pipeline after the permutation network. At this point, the arbitration and routing logic has determined which flits to deflect. The buffer eject block recognizes flits that have been deflected, and picks up to one such deflected flit per cycle. It removes a deflected flit from the router pipeline, and places this flit in the side buffer, as long as the side buffer is not full. (If the side buffer is full, no flits are removed from the pipeline into the buffer until space is created.) This flit is chosen randomly among deflected flits (except that a golden flit is never chosen: see Section 4.4). In this way, some deflections are avoided. The flits placed in the buffer will later be re-injected into the pipeline, and will re-try arbitration at that time. This re-injection occurs in the same way that new traffic is injected into the network, which we discuss below.

4.4 Injection and ejection

So far, we have considered the flow of flits from router input ports (i.e., arriving from neighbor routers) to router output ports (i.e., to other neighbor routers). A flit must enter and leave the network at some point. To allow traffic to enter (inject) and leave (eject), the MinBD router contains inject and eject blocks in its first pipeline stage (see Fig. 5). When a set of flits arrive on router inputs, these flits first pass through the ejection logic. This logic examines the destination of each flit, and if a flit is addressed to the local router, it is removed from the router pipeline and sent to the local network node.[b] If more than one locally-addressed flit is present, the ejector picks one, according to the same priority scheme used by routing arbitration.

However, ejecting a single flit per cycle can produce a bottleneck and cause unnecessary deflections for flits that could not be ejected. In the workloads we evaluate, at least one flit is eligible to eject 42.8% of the time. Of those cycles, 20.4% of the time, at least two flits are eligible to eject. Hence, in 8.5% of all cycles, a locally-addressed flit would be deflected rather than ejected if only one flit could be ejected per cycle. To avoid this significant deflection-rate penalty, we double the ejection bandwidth. To implement this, a MinBD [3,72,73] router contains two ejector blocks. Each of these blocks is identical, and can eject up to one flit per cycle. Duplicating the ejection logic allows two flits to leave the network per cycle at every node.

After locally-addressed flits are removed from the pipeline, new flits are allowed to enter. There are two injector blocks in the router pipeline shown in Fig. 5: (i) re-injection of flits from the side buffer, and (ii) injection of new flits from the local node. (The "Redirection" block prior to the injector blocks will be discussed in the next section.) Each block operates in the same way. A flit can be injected into the router pipeline whenever one of the four inputs does not have a flit present in a given cycle, i.e., whenever there is an "empty slot" in the network. Each injection block pulls up to one flit per cycle from an injection queue (the side buffer, or the local node's injection queue), and places a new flit in the pipeline when a slot is available. Flits from the side buffer are re-injected before new traffic is injected into the network. However, note that there is no guarantee that a free slot will be available for an injection in any given cycle. We now address this starvation problem for side buffer re-injection.

4.5 Ensuring side buffered flits make progress

When a flit enters the side buffer, it leaves the router pipeline, and must later be re-injected. As we described above, flit reinjection must wait for an empty slot on an input link. It is possible that such a slot will not appear for a long time. In this case, the flits in the side buffer are delayed unfairly while other flits make forward progress.

---

[b] Note that flits are reassembled into packets after ejection. To implement this reassembly, we use the Retransmit-Once scheme, as used by CHIPPER [4], which uses MSHRs (Miss-Status Handling Registers [77], or existing buffers in the cache system) to reassemble packets in place.

To avoid this situation, we implement buffer redirection. The key idea of buffer redirection is that when this side buffer starvation problem is detected, one flit from a randomly-chosen router input is forced to enter the side buffer. Simultaneously, the flit at the head of the side buffer is allowed to inject into the slot created by the forced flit buffering. In other words, one router input is "redirected" into the FIFO buffer for one cycle, in order to allow the buffer to make forward progress. This redirection is enabled for one cycle whenever the side buffer injection is starved (i.e., has a flit to inject, but no free slot allows the injection) for more than some threshold $C_{threshold}$ cycles (in our evaluations, $C_{threshold} = 2$). Finally, note that if a golden flit is present, it is never redirected to the buffer, because this would break the delivery guarantee.

4.6 Livelock and deadlock-free operation

MinBD provides livelock-free delivery of flits using Golden Packet and buffer redirection. If no flit is ever buffered, then Golden Packet [4] ensures livelock freedom (the "silver flit" priority never deflects any golden flit, hence does not break the guarantee). Now, we argue that adding side buffers does not cause livelock. First, the buffering logic never places a golden flit in the side buffer. However, a flit could enter a buffer and then become golden while waiting. Redirection ensures correctness in this case: it provides an upper bound on residence time in a buffer (because the flit at the head of the buffer will leave after a certain threshold time in the worst case). If a flit in a buffer becomes golden, it only needs to remain golden long enough to leave the buffer in the worst case, then progress to its destination. We choose the threshold parameter ($C_{threshold}$) and golden epoch length so that this is always possible. More details can be found in our extended technical report [72].

MinBD achieves deadlock-free operation by using Retransmit-Once [4], which ensures that every node always consumes flits delivered to it by dropping flits when no reassembly/request buffer is available. This avoids packet reassembly deadlock (as described in [4]), as well as protocol level deadlock, because message-class dependencies [78] no longer exist.

5. HiRD: Simple hierarchical rings with deflection

With the design of a minimally-buffered deflection router, minBD [3,72,73], we now describe how a similar concept can be integrated to a deflection-based hierarchical ring interconnect called HiRD [23,24] in order to further improve performance, energy efficiency and scalability of NoCs. HiRD is built on several basic operation principles:

1. Every node (e.g., CPU, cache slice, or memory controller) resides on one *local ring*, and connects to one *node router* on that ring.
2. Node routers operate exactly like routers (ring stops) in a single-ring interconnect: locally-destined flits are removed from the ring, other flits are passed through, and new flits can inject whenever there is a free slot (no flit present in a given cycle). There is no buffering or flow control within any local ring; flits are buffered only in ring pipeline registers. Node routers have a single-cycle latency.
3. Local rings are connected to one or more levels of *global rings* to form a tree hierarchy.
4. Rings are joined via *bridge routers*. A bridge router has a node-router-like interface on each of the two rings it connects, and has a set of transfer FIFOs (one in each direction) between the rings.
5. Bridge routers consume flits that require a transfer whenever the respective transfer FIFO has available space. The head flit in a transfer FIFO can inject into its new ring whenever there is a free slot (exactly as with new flit injections). When a flit requires a transfer but the respective transfer FIFO is full, the flit remains in its current ring. It will circle the ring and try again next time it encounters the correct bridge router (this is a *deflection*).

By using *deflections* rather than buffering and blocking flow control to manage ring transfers, HiRD retains node router simplicity, unlike past hierarchical ring network designs. This change comes at the cost of potential live-lock (if flits are forced to deflect forever). We introduce two mechanisms to provide a deterministic guarantee of livelock-free operation in [23,24].

While deflection-based bufferless routing has been previously proposed and evaluated for a variety of off-chip and on-chip interconnection networks (e.g., [2–4,8,9,29,65]), deflections are trivially implementable in a ring: if deflection occurs, the flit[c] continues circulating in the ring. Contrast this to past deflection-based schemes that operated on mesh networks where multiple incoming flits may need to be deflected among a multitude of possible out-bound ports, leading to much more circuit complexity in the router microarchitecture, as shown by [4,27,55]. Our application of deflection to rings leads to a simple and elegant embodiment of bufferless routing.

---

[c] All operations in the network happen in a flit level similar to previous works [1–17,65,75].

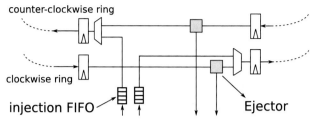

counter-clockwise ring

clockwise ring

injection FIFO

Ejector

Fig. 6 Node router. *Reproduced from R. Ausavarungnirun, et al., Improving energy effciency of hierarchical rings via deflection routing, SAFARI Technical Report TR-2014-002: http:// safari.ece.cmu.edu/tr.html (2014).*

5.1 Node router operation

At each node on a local ring, we place a single node router, shown in Fig. 6. A node router is very simple: it passes through circulating traffic, allows new traffic to enter the ring through a MUX, and allows traffic to leave the ring when it arrives at its destination. Each router contains one pipeline register for the router stage, and one pipeline register for link traversal, so the router latency is exactly one cycle and the per-hop latency is two cycles. Such a design is very common in ring-based and ring-like designs (e.g., [79]).

As flits enter the router on the ring, they first travel to the ejector. Because we use bidirectional rings, each node router has two ejectors, one per direction.[d] Note that the flits constituting a packet may arrive out-of-order and at widely separated times. Re-assembly into packets is thus necessary. Packets are re-assembled and reassembly buffers are managed using the Retransmit-Once scheme, borrowed from the CHIPPER bufferless router design [4]. With this scheme, receivers reassemble packets in-place in MSHRs [77], eliminating the need for separate reassembly buffers. The key idea in Retransmit-Once is to avoid ejection backpressure-induced deadlocks by ensuring that all arriving flits are consumed immediately at their receiver nodes. When a flit from a new packet arrives, it allocates a new reassembly buffer slot if available. If no slot is available, the receiver drops the flit and sets a bit in a retransmit queue which corresponds to the sender and transaction ID of the dropped flit. Eventually, when a buffer slot becomes available at the receiver, the receiver reserves the slot for a

---

[d] For simplicity, we assume that up to two ejected flits can be accepted by the processor or reassembly buffers in a single cycle. For a fair comparison, we also implement two-flit-per-cycle ejection in our baselines.

sender/transaction ID in its retransmit queue and requests a retransmit from the sender. Thus, all traffic arriving at a node is consumed (or dropped) immediately, so ejection never places backpressure on the ring. Retransmit-Once hence avoids protocol-level deadlock [4]. Furthermore, it ensures that a ring full of flits always drains, thus ensuring forward progress (as we will describe more fully in [23,24]).

After locally-destined traffic is removed from the ring, the remaining traffic travels to the injection stage. At this stage, the router looks for "empty slots," or cycles where no flit is present on the ring, and injects new flits into the ring whenever they are queued for injection. The injector is even simpler than the ejector, because it only needs to find cycles where no flit is present and insert new flits in these slots. Note that we implement two separate injection buffers (FIFOs), one per ring direction; thus, two flits can be injected into the network in a single cycle. A flit enqueues for injection in the direction that yields a shorter traversal toward its destination.

5.2 Bridge routers

The *bridge routers* connect a local ring and a global ring, or a global ring with a higher-level global ring (if there are more than two levels of hierarchy). A high-level block diagram of a bridge router is shown in Fig. 7. A bridge router resembles two node routers, one on each of two rings, connected by FIFO buffers in both directions. When a flit arrives on one ring that requires a transfer to the other ring (according to the routing function described below in Section 5.3), it can leave its current ring and wait in a FIFO as long as there is space available. These *transfer FIFOs* exist so that a transferring flit's arrival need not be perfectly aligned with a free slot on the destination ring. However, this transfer FIFO will sometimes fill. In that case, if any flit arrives that requires a transfer, the bridge router simply does

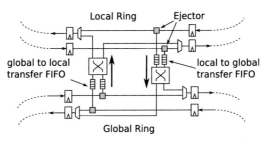

Fig. 7 Bridge router. *Reproduced from R. Ausavarungnirun, et al., Improving energy effciency of hierarchical rings via deflection routing, SAFARI Technical Report TR-2014-002: http://safari.ece.cmu.edu/tr.html (2014).*

not remove the flit from its current ring; the flit will continue to travel around the ring, and will eventually come back to the bridge router, at which point there may be an open slot available in the transfer FIFO. This is analogous to a *deflection* in hot-potato routing [29], also known as deflection routing, and has been used in recent on-chip mesh interconnect designs to resolve contention [2–4,8,9,57]. Note that to ensure that flits are *eventually* delivered, despite any deflections that may occur, we introduce two *guarantee mechanisms* in [23,24]. Finally, note that deflections may cause flits to arrive out-of-order (this is fundamental to any non-minimal adaptively-routed network). Because we use Retransmit-Once [4], packet reassembly works despite out-of-order arrival.

The bridge router uses *crossbars* to allow a flit ejecting from either ring direction in a bidirectional ring to enqueue for injection in either direction in the adjoining ring. When a flit transfers, it picks the ring direction that gives a shorter distance, as in a node router. However, these crossbars actually allow for a more general case: the bridge router can actually join several rings together by using larger crossbars. For our network topology, we use hierarchical rings. We use wider global rings than local rings (analogous to a *fat tree* [53]) for performance reasons. These wider rings perform logically as separate rings as wide as one flit. Although not shown in the figure for simplicity, the bridge router in such a case uses a larger crossbar and has one ring interface (including transfer FIFO) per ring-lane in the wide global ring. The bridge router then load-balances flits between rings when multiple lanes are available. (The crossbar and transfer FIFOs are fully modeled in our evaluations.)

When building a two-level design, there are many different arrangements of global rings and bridge routers that can efficiently link the local rings together. Fig. 8A shows three designs denoted by the number of bridge routers in total: 4-bridge, 8-bridge, and 16-bridge. We assume an 8-bridge design for the remainder of this paper. Also, note that the hierarchical structure that we propose can be extended to more than two levels. We use a 3-level hierarchy, illustrated in Fig. 8B, to build a 64-node network.

Finally, in order to address a potential deadlock case (which will be explained more in [23,24,52]), bridge routers implement a special *Swap Rule*. The Swap Rule states that when the flit that just arrived on each ring requires a transfer to the other ring, the flits can be *swapped*, bypassing the transfer FIFOs altogether. This requires a bypass datapath (which is fully modeled in our hardware evaluations). It ensures correct operation in the case when transfer FIFOs in both directions are full. Only one swap needs

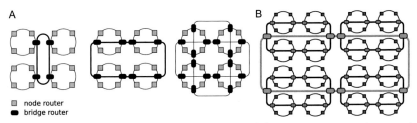

Fig. 8 Hierarchical ring design of HiRD. *Reproduced from R. Ausavarungnirun, et al., Improving energy effciency of hierarchical rings via deflection routing, SAFARI Technical Report TR-2014-002: http://safari.ece.cmu.edu/tr.html (2014).*

to occur in any given cycle, even when the bridge router connects to a wide global ring. Note that because the swap rule requires this bypass path, the behavior is always active (it would be more difficult to definitively identify a deadlock and enable the behavior only in that special case). The Swap Rule may cause flits to arrive out-of-order when some are bypassed in this way, but the network already delivers flits out-of-order, so correctness is not compromised.

5.3 Routing

Finally, we briefly address routing. Because a hierarchical ring design is fundamentally a *tree*, routing is very simple: when a flit is destined for a node in another part of the hierarchy, it first travels *up* the tree (to more global levels) until it reaches a common ancestor of its source and its destination, and then it travels *down* the tree to its destination. Concretely, each node's address can be written as a series of parts, or digits, corresponding to each level of the hierarchy (these trivially could be bitfields in a node ID). A ring can be identified by the common prefix of all routers on that ring; the root global ring has a null (empty) prefix, and local rings have prefixes consisting of all digits but the last one. If a flit's destination does not match the prefix of the ring it is on, it takes any bridge router to a more global ring. If a flit's destination does match the prefix of the ring it is on (meaning that it is traveling down to more local levels), it takes any bridge router which connects to the next level, until it finally reaches the local ring of its destination and ejects at the node with a full address match.

5.4 Guaranteed delivery: Correctness in hierarchical ring interconnects

In order for the system to operate correctly, the inter-connect must guarantee that every flit is eventually delivered to its destination. HiRD ensures

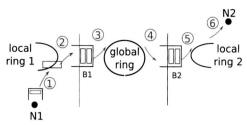

Fig. 9 The need for the injection and transfer guarantees: contention experienced by a flit during its journey. *Reproduced from R. Ausavarungnirun, et al., Improving energy effciency of hierarchical rings via deflection routing, SAFARI Technical Report TR-2014-002: http://safari.ece.cmu.edu/tr.html (2014).*

correct operation through two mechanisms that provide two guarantees: the *injection guarantee* and the *transfer guarantee*. The injection guarantee ensures that any flit waiting to inject into a ring will eventually be able to enter that ring. The transfer guarantee ensures that any flit waiting to enter a bridge router's transfer queue will eventually be granted a slot in that queue.

To understand the need for each guarantee, let us consider an example, shown in Fig. 9. A flit is enqueued for network injection at node N1 on the leftmost local ring. This flit is destined for node N2 on the rightmost local ring; hence, it must traverse the leftmost local ring, then the global ring in the center of the figure, followed by the rightmost local ring. The flit traverses rings twice, at the two bridge routers B1 and B2 shown in the figure. The figure also indicates the six points (labeled as ① to ⑥) at which the flit moves from a queue to a ring or vice-versa: the flit first enters N1's injection queue, transfers to the leftmost local ring ①, the bridge router B1 ②, the global ring ③, the bridge router B2 ④, the rightmost local ring ⑤, and finally the destination node N2 ⑥.

In the worst case, when the network is heavily contended, the flit could wait for an unbounded amount of time at ① to ⑤. First, recall that to enter any ring, a flit must wait for an empty slot on that ring (because the traffic on the ring continues along the ring once it has entered, and thus has higher priority than any new traffic). Because of this, the flit traveling from node N1 to N2 could wait for an arbitrarily long time at ①, ③, and ⑤ if no other mechanism intercedes. This first problem is one of *injection starvation*, and we address it with the *injection guarantee* mechanism described below. Second, recall that a flit that needs to transfer from one ring to another via a bridge router enters that bridge router's queue, but if the bridge router's queue is full, then the transferring flit must make another trip around its current ring and try again when it next encounters a bridge router. Because of this rule, the flit traveling from N1 to N2 could be *deflected* an arbitrarily large number of times at ② and ④ (at entry to bridge routers B1 and B2) if no other mechanism intercedes. This second problem is one of *transfer starvation*, and we address it with the *transfer guarantee* mechanism described below.

Our goal in this section is to demonstrate that HiRD provides both the injection guarantee (Section 5.5) and the transfer guarantee (Section 5.5.1) mechanisms. We show correctness in Section 5.5.2, and quantitatively evaluate both mechanisms in Section 5.10 and in [52].

5.5 Preventing injection starvation: Injection guarantee

The *injection guarantee* ensures that every router on a ring can eventually inject a flit. This guarantee is provided by a very simple throttling-based mechanism: when any node is starved (cannot inject a flit) past a threshold number of cycles, it asserts a signal to a global controller, which then throttles injection from every other node. No new traffic will enter the network while this throttling state is active. All existing flits in the network will eventually drain, and the starved node will be able to finally inject its new flit. At that time, the starved node de-asserts its throttling request signal to the global controller, and the global controller subsequently allows all other nodes to resume normal operation.

Note that this injection guarantee can be implemented in a hierarchical manner to improve scalability. In the hierarchical implementation, each individual local ring in the network monitors only its own injection and throttles injection locally if any node in it is starved. After a threshold number of cycles,[e] if any node in the ring still cannot inject, the bridge routers connected to that ring start sending throttling signals to any other ring in the next level of the ring hierarchy they are connected to. In the worst case, every local ring stops accepting flits and all the flits in the network drain and eliminate any potential livelock or deadlock. Designing the delivery guarantee this way requires two wires in each ring and small design overhead at the bridge router to propagate the throttling signal across hierarchy levels. In our evaluation, we faithfully model this hierarchical design.

5.5.1 Ensuring ring transfers: Transfer guarantee

The *transfer guarantee* ensures that any flit waiting to transfer from its current ring to another ring via a bridge router will eventually be able to enter that bridge router's queue. Such a guarantee is non-trivial because the bridge router's queue is finite, and when the destination ring is congested, a slot

---

[e] In our evaluation, we set this threshold to be 100 cycles.

may become available in the queue only infrequently. In the worst case, a flit in one ring may circulate indefinitely, finding a bridge router to its destination ring with a completely full queue each time it arrives at the bridge router. The transfer guarantee ensures that any such circulating flit will eventually be granted an open slot in the bridge router's transfer queue. Note in particular that this guarantee is *separate from* the injection guarantee: while the injection guarantee ensures that the bridge router will be able to inject flits from its transfer queue into the destination ring (and hence, have open slots in its transfer queue eventually), these open transfer slots may not be distributed *fairly* to flits circulating on a ring waiting to transfer through the bridge router. In other words, some flit may always be "unlucky" and never enter the bridge router if slots open at the wrong time. The transfer guarantee addresses this problem.

In order to ensure that any flit waiting to transfer out of a ring eventually enters its required bridge router, each bridge router *observes a particular slot on its source ring* and monitors for flits that are "stuck" for more than a threshold number of retries. (To observe one "slot," the bridge router simply examines the flit in its ring pipeline register once every N cycles, where N is the latency for a flit to travel around the ring once.) If any flit circulates in its ring more than this threshold number of times, the bridge router reserves the next open available entry in its transfer queue for this flit (in other words, it will refuse to accept other flits for transfer until the "stuck" flit enters the queue). Because of the injection guarantee, the head of the transfer queue must inject into the destination ring eventually, hence an entry must become available eventually, and the stuck flit will then take the entry in the transfer queue the next time it arrives at the bridge router. Finally, the slot which the bridge router observes rotates around its source ring: whenever the bridge router observes a slot the second time, if the flit that occupied the slot on first observation is no longer present (i.e., successfully transferred out of the ring or ejected at its destination), then the bridge router begins to observe the next slot (the slot that arrives in the *next* cycle). In this way, every slot in the ring is observed eventually, and any stuck flit will thus eventually be granted a transfer.

5.5.2 Putting it together: Guaranteed delivery

Before we prove the correctness of these mechanisms in detail, it is helpful to summarize the basic operation of the network once more. A flit can inject into a ring whenever a free slot is present in the ring at the injecting router (except when the injecting router is throttled by the injection guarantee mechanism). A flit can eject at its destination whenever it arrives, and destinations always consume flits as soon as they arrive (which is ensured despite finite reassembly buffers using the Retransmit-Once mechanism [4], as already described in Section 5.1). A flit transfers between rings via a transfer queue in a bridge router, first leaving its source ring to wait in the queue and then injecting into its destination ring when at the head of the queue, and can enter a transfer queue whenever there is a free entry in that transfer queue (except when the entry is reserved for another flit by the transfer guarantee mechanism). Finally, when two flits at opposite ends of a bridge router each desire to traverse through the bridge router, the *Swap Rule* allows these flits to exchange places directly, bypassing the queues (and ensuring forward progress).

Our proof is structured as follows: we first argue that if no new flits enter the network, then the network will drain in finite time. The injection guarantee ensures that any flit can enter the network. Then, using the injection guarantee, transfer guarantee, the swap rule, and the fact that the network is hierarchical, any flit in the network can eventually reach any ring in the network (and hence, its final destination ring). Because all flits in a ring continue to circulate that ring, and any node on a ring must consume any flits that are destined for that node, final delivery is ensured once a flit reaches its final destination ring.

5.5.2.1 Network drains in finite time

Assume no new flits enter the network (for now). A flit could only be stuck in the network indefinitely if transferring flits create a cyclic dependence between completely full rings. Otherwise, if there are no dependence cycles, then if one ring is full and cannot accept new flits because other rings will not accept its flits, then eventually there must be some ring which depends on no other ring (e.g., a local ring with all locally-destined flits), and this ring will drain first, followed by the others feeding into it. However, because the network is hierarchical (i.e., a tree), the only cyclic dependences possible are between rings that are immediate parent and child (e.g., global ring and local ring, in a two-level hierarchy). The *Swap Rule* ensures that when a parent and child ring are each full of flits that require transfer to the other ring, then transfer is always possible, and forward progress will be ensured. Note in particular that we do not require the injection or transfer guarantee for the network to drain. Only the *Swap Rule* is necessary to ensure that no deadlock will occur.

5.5.2.2 Any node can inject

Now that we have shown that the network will drain if no new flits are injected, it is easy to see that the injection guarantee ensures that any node can eventually inject a flit: if any node is starved, then all nodes are throttled, no new flit enters the network, and the network must eventually drain (as we just showed), at which point the starved node will encounter a completely empty network into which to inject its flit. It likely will be able to inject before the network is completely empty, but in the worst case, the guarantee is ensured in this way.

5.5.2.3 All flits can transfer rings and reach their destination rings

With the injection guarantee in place, the transfer guarantee can be shown to provide its stated guarantee as follows: because of the injection guarantee, a transfer queue in a bridge router will always inject its head flit in finite time, hence will have an open entry to accept a new transferring flit in finite time. All that is necessary to ensure that *all* transferring flits eventually succeed in their transfers is that *any* flit stuck for long enough gets an available entry in the transfer queue. The transfer guarantee does exactly this by observing ring slots in sequence and reserving a transfer queue entry when a flit becomes stuck in a ring. Because the mechanism will eventually observe every slot in the ring, all flits will be allowed to make their transfers eventually. Hence, all flits can continue to transfer rings until reaching their destination rings (and thus, their final destinations).

5.5.3 Hardware cost

Our injection and transfer guarantee mechanisms have low hardware overhead. To implement the injection guarantee, one counter is required for each injection point. This counter tracks how many cycles have elapsed while injection is starved, and is reset whenever a flit is successfully injected. Routers communicate with the throttling arbitration logic with only two wires, one to signal blocked injection and one control line that throttles the router. The wiring is done hierarchically instead of globally to minimize the wiring cost (Section 5.5). Because the correctness of the algorithm does not depend on the delay of these wires, and the injection guarantee mechanism is activated only rarely (in fact, *never* for our evaluated realistic workloads), the signaling and central coordinator need not be optimized for speed. To provide the transfer guarantee, each bridge router implements "observer" functionality for each of the two rings it sits on, and the observer consists only of three small counters (to track the current timeslot being observed, the current timeslot at the ring pipeline register in this router, and the number of times the observed flit has circled the ring) and a small amount of control logic. Importantly, note that neither mechanism impacts the router critical path nor affects the router datapath (which dominates energy and area).

5.6 HiRD: Evaluation methodology

We perform our evaluations using a cycle-accurate simulator of a CMP system with 1.6GHz interconnect to provide application-level performance results [80]. Our simulator is publicly available and includes the source code of all mechanisms we evaluated [80]. Tables 2 and 3 provide the configuration parameters of our simulated systems.

Our methodology ensures a rigorous and isolated evaluation of NoC capacity for especially cache-resident workloads, and has also been used in other studies [2–4,8,9]. Instruction traces for the simulator are taken using a Pintool [81] on representative portions of SPEC CPU2006 workloads.

We mainly compare to a single bidirectional ring and a state-of-the-art buffered hierarchical ring [46]. Also, note that while there are many possible ways to optimize each baseline (such as congestion control [5,8,9], adaptive routing schemes, and careful parameter tuning), we assume a fairly typical aggressive configuration for each.

Table 2 Simulation and system configuration parameters.

Parameter	Setting
System topology	CPU core and shared cache slice at every node
Core model	Out-of-order, 128-entry ROB, 16 MSHRs (maximum simultaneous outstanding requests)
Private L1 cache	64KB, 4-way associative, 32-byte block size
Shared L2 cache	Perfect (always hits) to stress the network and penalize our reduced-capacity deflection-based design; cache-block-interleaved
Cache coherence	Directory-based protocol (based on SGI Origin [178]), directory entries co-located with shared cache blocks
Simulation length	5 M-instruction warm-up, 25 M-instruction active execution per node [2–5]

Table 3 Network parameters for HiRD evaluation.

Parameter	Network	Setting
Interconnect Links	Single Ring	Bidirectional, **4×4:** 64-bit and 128-bit width, **8×8:** 128-bit and 256-bit width
	Buffered HRing	Bidirectional, **4×4:** 3-cycle per-hop latency (link + router); 64-bit local and 128-bit global rings, **8×8:** three-level hierarchy, 4×4 parameters, with second-level rings connected by a 256-bit third-level ring
	HiRD	**4×4:** 2-cycle (local), 3-cycle (global) per-hop latency (link + router); 64-bit local ring, 128-bit global ring; **8×8:** 4×4 parameters, with second-level rings connected by a 256-bit third-level ring
Router	Single Ring	1-cycle per-hop latency (as in [43])
	Buffered HRing	Node (NIC) and bridge (IRI) routers based on [46];4-flit in-ring and transfer FIFOs. Bidirectional links of dual-flit width (for fair comparison with our design). Bubble flow control [179] for deadlock freedom.
	HiRD	Local-to-global buffer depth of 1, global-to-local buffer depth of 4

Reproduced from R. Ausavarungnirun, C. Fallin, X. Yu, K. K.-W. Chang, G. Nazario, R. Das, G. H. Loh, O. Mutlu, A Case for Hierarchical Rings with Deflection Routing, Parallel Comput. (2016) 54 29–45.

5.6.1 Data mapping

We map data in a cache-blockinterleaved way to different shared L2 cache slices. This mapping is agnostic to the underlying locality. As a result, it does not exploit the low-latency data access in the local ring. One can design systematically better mapping in order to keep frequently used data in the local ring as in [82,83]. However, such a mapping mechanism is orthogonal to our proposal and can be applied in all ring-based network designs.

5.6.2 Application and synthetic workloads

The system is run with a set of 60 multiprogrammed workloads. Each workload consists of one single-threaded instance of a SPEC CPU2006 benchmark on each core, for a total of either 16 (4 × 4) or 64 (8 × 8) benchmark instances per workload. These multiprogrammed workloads are representative of many common workloads for large CMPs. Workloads are constructed at varying network intensities as follows: first, benchmarks are split into three classes (Low, Medium and High) by L1 cache miss intensity (which correlates directly with network injection rate), such that benchmarks with less than 5 misses per thousand instructions (MPKI) are "Low-intensity," between 5 and 50 are "Medium-intensity," and above 50 MPKI are "High-intensity." Workloads are then constructed by randomly selecting a certain number of benchmarks from each category. We form workload sets with four intensity mixes: High (H), Medium (M), Medium-Low (ML), and Low (L), with 15 workloads in each (the average network injection rates for each category are 0.47, 0.32, 0.18, and 0.03 flits/node/cycle, respectively).

5.6.3 Multithreaded workloads

We use the GraphChi implementation of the GraphLab framework [84,85]. The implementation we use is designed to run efficiently on multi-core systems. The workload consists of Twitter Community Detection (CD), Twitter Page Rank (PR), Twitter Connected Components (CC), Twitter Triangle Counting (TC) [86], and Graph500 Breadth First Search (BFS). We simulated the representative portion of each workload and each workload has a working set size of greater than 151.3 MB. On every simulation of these multithreaded workloads, we warm up the cache with the first 5 million instructions, then we run the remaining code of the representative portion.

5.6.4 Energy and area

We measure the energy and area of routers and links by individually modeling the crossbar, pipeline registers, buffers, control logic, and other datapath components. For links, buffers and datapath elements, we use DSENT 0.91 [76]. Control logic is modeled in Verilog RTL. Both energy and area are calculated based on a 45 nm technology. The link lengths we assume are based on the floorplan of our designs, which we describe in the next paragraph.

We assume the area of each core to be 2.5×2.5 mm. We assume a 2.5 mm link length for single-ring designs. For the hierarchical ring design, we assume 1 mm links between local-ring routers, because the four routers on a local ring can be placed at four corners that meet in a tiled design. Global-ring links are assumed to be 5.0 mm (i.e., five times as long as local links), because they span across two tiles on average if local rings are placed in the center of each four-tile quadrant. Third-level global ring links are assumed to be 10 mm (i.e., 10 times as long as local links) in the 8×8 evaluations. This floorplan is illustrated in more detail in Fig. 10 for the 3-level

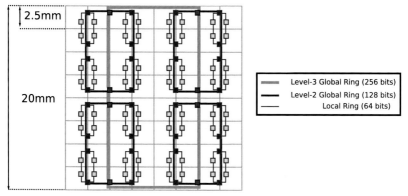

Fig. 10 Assumed floorplan for HiRD 3-level (64-node) network. Two-level (16-node) network consists of one quadrant of this floorplan. *Reproduced from R. Ausavarungnirun, et al., Improving energy effciency of hierarchical rings via deflection routing, SAFARI Technical Report TR-2014-002: http://safari.ece.cmu.edu/tr.html (2014).*

(64-node) HiRD network. Note that one quadrant of the floorplan of Fig. 10 corresponds to the floorplan of the 2-level (16-node) HiRD network. We faithfully take into account all link lengths in our energy and area estimates for all designs.

5.6.5 Application evaluation metrics
For multiprogrammed workloads, we present application performance results using the commonly-used Weighted Speedup metric [87,88]. We use the maximum slowdown metric to measure unfairness [15–17,89–101].

5.7 Performance, energy efficiency and scalability of HiRD
We provide a comprehensive evaluation of our proposed mechanism against other ring baselines. Since our goal is to provide a better ring design, our main comparisons are to ring networks. However, we also provide sensitivity analyses and comparisons to other network designs as well.

5.8 Ring-based network designs
5.8.1 Multiprogrammed workloads
Fig. 11 shows performance (weighted speedup normalized per node), power (total network power normalized per node), and energy-efficiency (perf./power) for 16-node and 64-node HiRD and buffered hierarchical rings in [46], using identical topologies, as well as a single ring (with different bisection bandwidths).

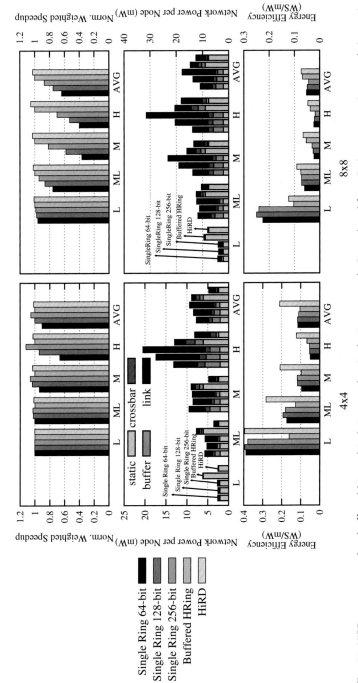

Fig. 11 HiRD as compared to buffered hierarchical rings and a single-ring network. Reproduced from R. Ausavarungnirun, et al., Improving energy effeciency of hierarchical rings via deflection routing, SAFARI Technical Report TR-2014-002: http://safari.ece.cmu.edu/tr.html (2014).

1. A hierarchical topology yields significant performance advantages over a single ring (i) when network load is high and/or (ii) when the network scales to many nodes. As shown, the buffered hierarchical ring improves performance by 7% (and HiRD by 10%) in high-load workloads at 16 nodes compared to a single ring with 128-bit links. The hierarchical design also reduces power because hop count is reduced. Therefore, link power reduces significantly with respect to a single ring. On average, in the 8×8 configuration, the buffered hierarchical ring network obtains 15.6% better application performance than the single ring with 256-bit links, while HiRD attains 18.2% higher performance.

2. Compared to the buffered hierarchical ring, HiRD has significantly lower network power and better performance. On average, HiRD reduces total network power (links and routers) by 46.5% (4×4) and 14.7% (8×8) relative to this baseline. This reduction in turn yields significantly better energy efficiency (lower energy consumption for buffers and slightly higher for links).[f] Overall, HiRD is the most energy-efficient of the ring-based designs evaluated in this paper for both 4×4 and 8×8 network sizes. HiRD also performs better than Buffered HRing due to the reasons explained in the next section (Section 5.9).

3. While scaling the link bandwidth increases the performance of a single ring network, the network power increases 25.9% when the link bandwidth increases from 64-bit to 128-bit and 15.7% when the link bandwidth increases from 128-bit to 256-bit because of higher dynamic energy due to wider links. In addition, scaling the link bandwidth is not a scalable solution as a single ring network performs worse than the bufferred hierarchical ring baseline even when a 256-bit link is used.

We conclude that HiRD is effective in simplifying the design of the hierarchical ring and making it more energy efficient, as we intended to as our design goal. We show that HiRD provides competitive performance compared to the baseline buffered hierarchical ring design with equal or better energy efficiency.

5.8.2 Multithreaded workloads
Fig. 12 shows the performance and power of HiRD on multithreaded applications compared to a buffered hierarchical ring and a single-ring network for both 16-node and 64-node systems. On average, HiRD performs 0.1%

---

[f] Note that the low intensity workloads in the 8x8 network is an exception. HiRD reduces energy efficiency for these as the static link power becomes dominant for them.

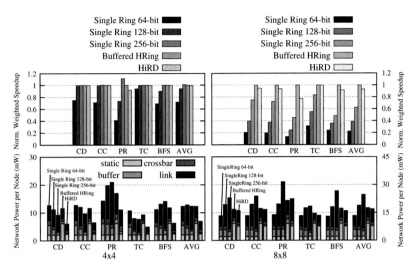

Fig. 12 HiRD as compared to buffered hierarchical rings and a single-ring network on multithreaded workloads. *Reproduced from R. Ausavarungnirun, et al., Improving energy effciency of hierarchical rings via deflection routing, SAFARI Technical Report TR-2014-002: http://safari.ece.cmu.edu/tr.html (2014).*

(4 × 4) and 0.73% (8 × 8) worse than the buffered hierarchical ring. However, on average, HiRD consumes 43.8% (4 × 4) and 3.1% (8 × 8) less power, leading to higher energy efficiency. This large reduction in energy comes from the elimination of most buffers in HiRD.

Both the buffered hierarchical ring and HiRD outperform single ring networks, and the performance improvement increases as we scale the size of the network.

Even though HiRD performs competitively with a buffered hierarchical ring network in most cases, HiRD performs poorly on the Page Ranking application. We observe that Page Ranking generates more non-local network traffic than other applications. As HiRD is beneficial mainly at lowering the local-ring latency, it is unable to speed up such non-local traffic, and is thus unable to help Page Ranking. In addition, Page Ranking also has higher network traffic, causing more congestion in the network (we observe 17.3% higher average network latency for HiRD in an 8 × 8 network), and resulting in a performance drop for HiRD. However, it is possible to use a different number of bridge routers as illustrated in Fig. 8A, to improve the performance of HiRD, which we will analyze in Section 5.15. Additionally, it is possible to apply a locality-aware cache mapping technique [82,83] in order to take advantage of lower local-ring latency in HiRD.

We conclude that HiRD is effective in improving evergy efficiency significantly for both multiprogrammed and multithreaded applications.

5.9 Synthetic-traffic network behavior

Fig. 13 shows the average packet latency as a function of injection rate for buffered and bufferless mesh routers, a single-ring design, the buffered hierarchical ring, and HiRD in 16 and 64-node systems. We show uniform random, transpose and bit complement traffic patterns [1]. Sweeps on injection rate terminate at network saturation. The buffered hierarchical ring saturates at a similar point to HiRD but maintains a slightly lower average latency because it avoids transfer deflections. In contrast to these high-capacity designs, the 256-bit single ring saturates at a lower injection rate.

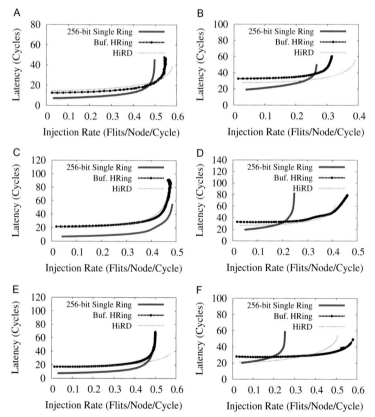

Fig. 13 Synthetic-traffic evaluations for 4×4 and 8×8 networks. *Reproduced from R. Ausavarungnirun, et al., Improving energy effciency of hierarchical rings via deflection routing, SAFARI Technical Report TR-2014-002: http://safari.ece.cmu.edu/tr.html (2014).*

As network size scales to 8×8, HiRD performs significantly better than the 256-bit single ring, because the hierarchy reduces the cross-chip latency while preserving bisection bandwidth. HiRD also performs better than Buffered HRing because of two reasons. First, HiRD is able to allow higher peak utilization (91%) than Buffered HRing (71%) on the global rings. We observed that when flits have equal distance in a clock-wise and counter clock-wise direction, Buffered HRing has to send flits to one direction in order to avoid deadlock while deflections in HiRD allow flits to travel in both directions, leading to better overall network utilization. Second, at high injection rates, the transfer guarantee [23] starts throttling the network, disallowing future flits to be injected into the network until the existing flits arrive at their destinations. This reduces congestion in the network and allows HiRD to saturate at a higher injection rate than the buffered hierarchical ring design.

5.10 Injection and transfer guarantees

In this subsection, we study HiRD's behavior under a worst-case synthetic traffic pattern that triggers the injection and transfer guarantees and demonstrates that they are necessary for correct operation, and that they work as designed.

5.10.1 Traffic pattern

In the worst-case traffic pattern, all nodes on three rings in a two-level (16-node) hierarchy inject traffic (we call these rings Ring A, Ring B, and Ring C). Rings A, B, and C have bridge routers adjacent to each other, in that order, on the single global ring. All nodes in Ring A continuously inject flits to nodes in Ring C, and all nodes in Ring C likewise inject flits to nodes in Ring A. This creates heavy traffic on the global ring across the point at which Ring B's bridge router connects. All nodes on Ring B continuously inject flits (whenever they are able) addressed to another ring elsewhere in the network. However, because Rings A and C continuously inject flits, Ring B's bridge router will not be able to transfer any flits to the global ring in the steady state (unless another mechanism such as the throttling mechanism in [23] intercedes).

5.10.2 Results

Table 4 shows three pertinent metrics on the network running the described traffic pattern: average network throughput (flits/node/cycle) for nodes on Rings A, B, and C, the maximum time (in cycles) spent by any one flit at the

Table 4 Results of worst-case traffic pattern without and with injection/transfer guarantees enabled.

Configuration	Network Throughput (flits/node/cycle)			Transfer FIFO Wait (cycles)	Deflections/ Retries
	Ring A	Ring B	RingC	avg/max	avg/max
Without Guarantees	0.164	0.000	0.163	2.5/299670	6.0/49983
With Guarantees	0.133	0.084	0.121	1./66	2.8/18

Data reproduced from R. Ausavarungnirun, et al., Improving energy effciency of hierarchical rings via deflection routing, SAFARI Technical Report TR-2014-002: http://safari.ece.cmu.edu/tr.html (2014).

head of a transfer FIFO, and the maximum number of times any flit is deflected and has to circle a ring to try again. These metrics are reported with the injection and transfer guarantee mechanisms disabled and enabled. The experiment is run with the synthetic traffic pattern for 300 K cycles.

The results show that *without* the injection and transfer guarantees, Ring B is completely starved and cannot transfer any flits onto the global ring. This is confirmed by the maximum transfer FIFO wait time, which is almost the entire length of the simulation. In other words, once steady state is reached, no flit ever transfers out of Ring B. Once the transfer FIFO in Ring B's bridge router fills, the local ring fills with more flits awaiting a transfer, and these flits are continuously deflected. Hence, the maximum deflection count is very high. Without the injection or transfer guarantees, the network does not ensure forward progress for these flits. In contrast, when the injection and transfer guarantees are enabled, (i) Ring B's bridge router is able to inject flits into the global ring and (ii) Ring B's bridge router fairly picks flits from its local ring to place into its transfer FIFO. The maximum transfer FIFO wait time and maximum deflection count are now bounded, and nodes on all rings receive network throughput. Thus, the guarantees are both necessary and sufficient to ensure deterministic forward progress for all flits in the network.

5.10.3 Real applications

Table 5 shows the effect of the transfer guarantee mechanism on real applications in a 4 × 4 network. Average transfer FIFO wait time shows the average number of cycles that a flit waits in the transfer FIFO across all 60 workloads. Maximum transfer FIFO wait time shows the maximum observed flit wait time in the same FIFO across all workloads. As illustrated in Table 5, some number of flits can experience very high wait times when

Table 5 Effect of transfer guarantee mechanism on real workloads.

Configuration	Transfer FIFO Wait time (cycles)	Deflections/Retries
	(avg/max)	(avg/max)
Without guarantees	3.3/169	3.7/19
With guarantees	0.76/72	0.7/8

Data reproduced from R. Ausavarungnirun, et al., Improving energy effciency of hierarchical rings via deflection routing, SAFARI Technical Report TR-2014-002: http://safari.ece.cmu.edu/tr.html (2014).

there is no transfer guarantee. Our transfer guarantee mechanism reduces both average and maximum FIFO wait times.[g] In addition, we observe that our transfer guarantee mechanism not only provides livelock- and deadlock-freedom but also provides lower maximum wait time in the transfer FIFO for each flit because the guarantee provides a form of throttling when the network is congested. A similar observation has been made in many previous network-on-chip works that use source throttling to improve the performance of the network [5,8,9,102,103].

We conclude that our transfer guarantee mechanism is effective in eliminating livelock and deadlock as well as reducing packet queuing delays in real workloads.

5.11 Network latency and latency distribution

Fig. 14 shows average network latency for our three evaluated configurations: 256-bit single ring, buffered hierarchical ring and HiRD. This plot shows that our proposal can reduce the network latency by having a faster local-ring hop latency compared to other ring-based designs. Additionally, we found that, for all real workloads, the number of deflections we observed is always less than 3% of the total number of flits. Therefore, the benefit of our deflection based router design outweighs the extra cost of deflections compared to other ring-based router designs. Finally, in the case of small networks such as a 4 × 4 network, a 1-cycle hop latency of a single ring provides significant latency reduction compared to the buffered hierarchical design. However, a faster local-ring hop latency in HiRD helps to reduce the network latency of a hierarchical design and provides a competitive network latency compared to a single ring design in small networks.

---

[g] As the network scales to 64 nodes, we observe that the average wait time in the transfer FIFO does not affect the overall performance significantly (adding 1.5 cycles per flit).

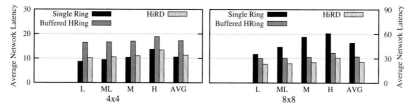

Fig. 14 Average network latency for 4×4 and 8×8 networks. *Reproduced from R. Ausavarungnirun, et al., Design and Evaluation of Hierarchical Rings with Deflection Routing, SBAC-PAD, 2014.*

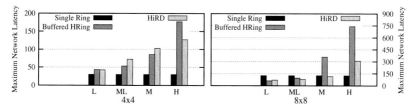

Fig. 15 Maximum network latency for 4×4 and 8×8 networks. *Reproduced from R. Ausavarungnirun, et al., Design and Evaluation of Hierarchical Rings with Deflection Routing, SBAC-PAD, 2014.*

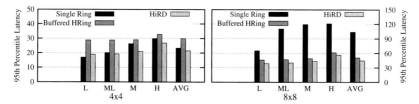

Fig. 16 95th percentile latency for 4×4 and 8×8 networks. *Reproduced from R. Ausavarungnirun, et al., Improving energy effciency of hierarchical rings via deflection routing, SAFARI Technical Report TR-2014-002: http://safari.ece.cmu.edu/tr.html (2014).*

In addition, Fig. 15 shows the maximum latency and Fig. 16 shows the 95th percentile latency for each network design. The 95th percentile latency shows the behavior of the network without extreme outliers. These two figures provide quantitative evidence that the network is deadlock-free and livelock-free. Several observations are in order:

1. HiRD provides lower latency at the 95th percentile and the lowest average latency observed in the network. This lower latency comes from our transfer guarantee mechanism, which is triggered when flits spend more than 100 cycles in each local ring, draining all flits in the network to their destination. This also means that HiRD improves the worst-case latency that a flit can experience because none of the flits are severely delayed.

2. While both HiRD and the buffered hierarchical ring have higher 95th percentile and maximum flit latency compared to a 64-bit single ring network, both hierarchical designs have 60.1% (buffered hierarchical ring) and 53.9% (HiRD) lower average network latency in an 8 × 8 network because a hierarchical design provides better scalability on average.

3. Maximum latency in the single ring is low because contention happens only at injection and ejection, as opposed to hierarchical designs where contention can also happen when flits travel through different level of the hierarchy.

4. The transfer guarantee in HiRD also helps to significantly reduce the maximum latency observed by some flits compared to a buffered design because the guarantee enables the throttling of the network, thereby alleviating congestion. Reduced congestion leads to reduced maximum latency. This observation is confirmed by our synthetic traffic results shown in Section 5.9.

5.12 Fairness

Fig. 17 shows the fairness, measured by the maximum slowdown metric, for our three evaluated configurations. Compared to a buffered hierarchical ring design HiRD, is 8.3% (5.1%) more fair on a 4 × 4 (8 × 8) network. Compared to a single ring design, HiRD is 40.0% (296.4%) more fair on a 4 × 4 (8 × 8) network. In addition, we provide several observations:

1. HiRD is the most fair design compared to the buffered hierarchical ring and the single ring designs. Compared to a single ring design, hierarchical designs are more fair because the global ring in the hierarchical designs allows flits to arrive at the destination faster. Compared to the buffered hierarchical ring design, HiRD is more fair because HiRD has lower

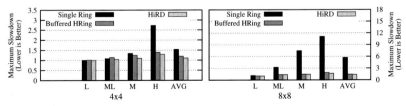

Fig. 17 Unfairness for 4 × 4 and 8 × 8 networks. *Reproduced from R. Ausavarungnirun, et al., Improving energy effciency of hierarchical rings via deflection routing, SAFARI Technical Report TR-2014-002: http://safari.ece.cmu.edu/tr.html (2014).*

average network latency. HiRD is much more fair for medium and high intensity workloads, where the throttling mechanism in HiRD lowers average network latency.

2. Global rings allow both hierarchical designs to provide better fairness compared to the single ring design as the size of the network gets bigger from 4×4 to 8×8.

3. We conclude that HiRD is the most fair ring design among all evaluated designs due to its overall lower packet latencies and reduced congestion across all applications.

5.13 Router area and timing

We show both critical path length and normalized die area for single-ring, buffered hierarchical ring, and HiRD, in Table 6. Area results are normalized to the buffered hierarchical ring baseline, and are reported for all routers required by a 16-node network (e.g., for HiRD, 16 node routers and 8 bridge routers).

Two observations are in order. First, HiRD reduces area relative to the buffered hierarchical ring routers, because the node router required at each network node is much simpler and does not require complex flow control logic. HiRD reduces total router area by 50.3% vs. the buffered hierarchical ring. Its area is higher than a single ring router because it contains buffers in bridge routers. However, the energy efficiency of HiRD and its performance at high load make up for this shortcoming. Second, the buffered hierarchical ring router's critical path is 42.6% longer than HiRD because its control logic must also handle flow control (it must check whether credits are available for a downstream buffer). The single-ring network has a higher operating frequency than HiRD because it does not need to accommodate ring transfers (but recall that this simplicity comes at the cost of poor performance at high load for the single ring).

Table 6 Total router area (16-node network) and critical path.

Metric	Single-Ring	Buffered HRing	HiRD
Critical path (ns)	0.33	0.87	0.61
Normalized area	0.281	1	0.497

Data reproduced from R. Ausavarungnirun, et al., Improving energy effciency of hierarchical rings via deflection routing, SAFARI Technical Report TR-2014-002: http://safari.ece.cmu.edu/tr.html (2014).

5.14 Sensitivity to link bandwidth

The bandwidth of each link also has an effect on the performance of different network designs. We evaluate the effect of different link bandwidths on several ring-based networks by using 32-, 64- and 128-bit links on all network designs. Fig. 18 shows the performance and power consumption of each network design. As links get wider, the performance of each design increases. According to the evaluation results, HiRD performs slightly better than a buffered hierarchical ring design for almost all link bandwidths while maintaining much lower power consumption on a 4×4 network, and slightly lower power consumption on an 8×8 network.

Additionally, we observe that increasing link bandwidth can decrease the network power in a hierarchical design because lower link bandwidth causes more congestion in the network and leads to more dynamic buffer, crossbar and link power consumption due to additional deflections at the buffers. As the link bandwidth increases, congestion reduces, lowering dynamic power. However, we observe that past a certain link bandwidth (e.g., 128 bits for buffered hierarchical ring and HiRD), congestion no longer reduces, because deflections at the buffers become the bottleneck instead. This leads to diminishing returns in performance yet increased dynamic power.

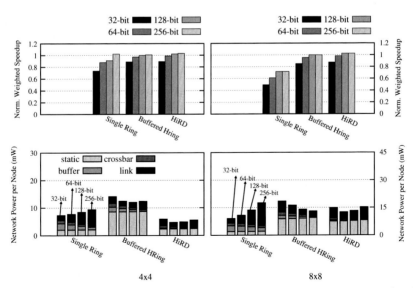

4x4   8x8

Fig. 18 Sensitivity to different link bandwidth for 4×4 and 8×8 networks. *Reproduced from R. Ausavarungnirun, et al., Improving energy effciency of hierarchical rings via deflection routing, SAFARI Technical Report TR-2014-002: http://safari.ece.cmu.edu/tr.html (2014).*

5.15 Sensitivity to configuration parameters

5.15.1 Bridge router organization

The number of bridge routers connecting the global ring(s) to the local rings has an important effect on system performance because the connection between local and global rings can limit bisection bandwidth. In Fig. 8A, we showed three alternative arrangements for a 16-node network, with 4, 8, and 16 bridge routers. So far, we have assumed an 8-bridge design in 4×4-node systems, and a system with 8 bridge routers at each level in 8×8-node networks (Fig. 8B). In Fig. 19A, we show average performance across all workloads for a 4×4-node system with 4, 8, and 16 bridge routers. Buffer capacity is held constant. As shown, significant performance is lost if only 4 bridge routers are used (10.4% on average). However, doubling from 8 to 16 bridge routers gains only 1.4% performance on average. Thus, the 8-bridge design provides the best tradeoff of performance and network cost (power and area) overall in our evaluations.

5.15.2 Bridge router buffer size

The size of the FIFO queues used to transfer flits between local and global rings can have an impact on performance if they are too small (and hence are often full, leading them to deflect transferring flits) or too large (and hence increase bridge router power and die area). We show the effect of

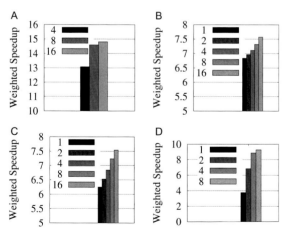

Fig. 19 Performance sensitivity to buffer sizes and the global ring bandwidth in a 4 × 4 network. *Reproduced from R. Ausavarungnirun, et al., Improving energy effciency of hierarchical rings via deflection routing, SAFARI Technical Report TR-2014-002: http://safari.ece. cmu.edu/tr.html (2014).*

local-to-global and global-to-local FIFO sizes in Fig. 19B and C, respectively, for the 8-bridge 4 × 4-node design. In both cases, increased buffer size leads to increased performance. However, performance is more sensitive to global-to-local buffer size (20.7% gain from 1-flit to 16-flit buffer size) than to local-to-global size (10.7% performance gain from 1 to 16 flits), because in the 8-bridge configuration, the whole-loop latency around the global ring is slightly higher than the loop latency in each of the local ring, making a global-to-local transfer retry more expensive than a local-to-global one.

For our evaluations, we use a 4-flit global-to-local and 1-flit local-to-global buffer per bridge router, which results in transfer deflection rates of 28.2% (global-to-local) and 34% (local-to-global) on average for multi-programmed workloads. These deflection rates are less than 1% for all of our multithreaded workloads. The deflection rate is much lower in multi-threaded workloads because these workloads are less memory-intensive and hence the contention in the on-chip interconnect is low for them.

5.15.3 Global ring bandwidth

Previous work on hierarchical ring designs does not examine the impact of global ring bandwidth on performance but instead assume equal bandwidth in local and global rings [49]. In Fig. 19D, we examine the sensitivity of system performance to global ring bandwidth relative to local ring bandwidth, for the all-High category of workloads (in order to stress check bisection bandwidth). Each point in the plot is described by this global-to-local ring bandwidth ratio. The local ring design is held constant while the width of the global ring is adjusted. If a ratio of 1:1 is assumed (leftmost bar), performance is significantly worse than the best possible design. Our main evaluations in 4 × 4 networks use a ratio of 2:1 (global: local) in order to provide equivalent bisection bandwidth to a 4 × 4 mesh baseline. Performance increases by 81.3% from a 1:1 ratio to the 2:1 ratio that we use. After a certain point, the global ring becomes less of a bottleneck, and further global-ring bandwidth increases have massively smaller effects.

5.15.4 Delivery guarantee parameters

We introduced injection guarantee and ejection guarantee mechanisms to ensure every flit is eventually delivered to its destination. These guarantees are clearly described in detail in our original work [23]. The injection guarantee mechanism takes a threshold parameter that specifies how long an injection can be blocked before action is taken. Setting this parameter too low can have an adverse impact on performance, because the system throttles nodes too aggressively and thus underutilizes the network.

Our main evaluations use a 100-cycle threshold. For high-intensity work-loads, performance drops by 21.3% when using an aggressive threshold of only 1 cycle. From 10 cycles upward, variation in performance is at most 0.6%: the mechanism is invoked rarely enough that the exact threshold does not matter, only that it is finite (is required for correctness guarantees). In fact, for a 100-cycle threshold, the injection guarantee mechanism is *never* triggered in our real applications. Hence, the mechanism is necessary only for corner-case correctness. In addition, we evaluate the impact of commu-nication latency between routers and the coordinator. We find less than 0.1% variation in performance for latencies ranging from 1 to 30 cycles (when parameters are set so that the mechanism becomes active); thus, slow, low-cost wires may be used for this mechanism.

The ejection guarantee takes a single threshold parameter: the number of times a flit is allowed to circle around a ring before action is taken. We find less than 0.4% variation in performance when sweeping the threshold from 1 to 16. Thus, the mechanism provides correctness in corner cases but is unimportant for performance in the common case.

5.16 Comparison against other ring configurations

Fig. 20 highlights the energy-efficiency comparison of different ring-based design configurations by showing weighted speedup (Y axis) against power (X axis) for all evaluated 4×4 networks. HiRD is shown with the three dif-ferent bridge-router configurations (described in Section 5.2). Every ring

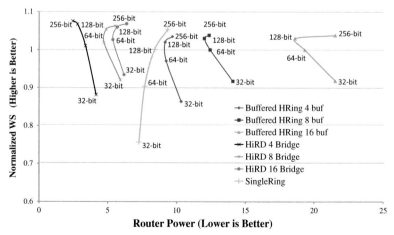

Fig. 20 Weighted speedup (Y) vs. power (X) for 4×4 networks. *Reproduced from R. Ausavarungnirun, et al., Improving energy effciency of hierarchical rings via deflection routing, SAFARI Technical Report TR-2014-002: http://safari.ece.cmu.edu/tr.html (2014).*

design is evaluated at various link bandwidths (32-, 64-, 128- and 256-bit links). The top-left is the ideal corner (high performance, low power). As the results show, at the same link bandwidth, all three configurations of HiRD are more energy efficient than the evaluated buffered hierarchical ring baseline designs at this network size.

We also observe that increasing link bandwidth can sometimes decrease router power as it reduces deflections in HiRD or lowers contention at the buffers in a buffered hierarchical ring design. However, once links are wide enough, this benefit diminishes for two reasons: (1) links and crossbars consume more energy, (2) packets arrive at the destination faster, leading to higher power as more energy is consumed in less time.

5.17 Comparison against other network designs

For completeness, Table 7 compares HiRD against several other network designs on 4×4 and 8×8 networks using the multiprogrammed workloads described in Section 5.6.

We compare our mechanism against a buffered mesh design with buffer bypassing [27,28]. We configure the buffered mesh to have 4 virtual channels (VCs) per port with 8 buffers per VC. We also compare our mechanism against CHIPPER [4], a low-complexity bufferless mesh network. We use 128-bit links for both designs. Additionally, we compare our mechanism against a flattened butterfly [18] with 4 VCs per output port, 8 buffers per VC, and 64-bit links. Our main conclusions are as follows:

1. Against designs using the mesh topology, we observe that HiRD performs very closely to the buffered mesh design both for 4×4 and 8×8 network sizes, while a buffered hierarchical ring design performs slightly worse compared to HiRD and buffered mesh designs. Additionally, HiRD performs better than CHIPPER in both 4×4 and 8×8 networks, though CHIPPER consumes less power in an 8×8 design as there is no buffer in CHIPPER.

2. Compared to a flattened butterfly design, we observe that HiRD performs competitively with a flattened butterfly in a 4×4 network, but consumes lower router power. In an 8×8 network, HiRD does not scale as well as a flattened butterfly network and performs 11% worse than a flattened butterfly network; however, HiRD consumes 59% less power than the flattened butterfly design.

3. Overall, we conclude that HiRD is competitive in performance with the highest performing designs while having much lower power consumption.

Table 7 Evaluation for 4×4 and 8×8 networks against different network designs.

Topologies	4×4		8×8	
	Norm. WS	Power (mWatts)	Norm. WS	Power (mWatts)
Single Ring	0.904	7.696	0.782	13.603
Buffered HRing	1	12.433	1	16.188
Buffered Mesh	1.025	11.947	1.091	13.454
CHIPPER	0.986	4.631	1.013	7.275
Flattened Butterfly	1.037	10.760	1.211	30.434
HiRD	1.020	4.746	1.066	12.480

Data reproduced from R. Ausavarungnirun, et al., Improving energy effciency of hierarchical rings via deflection routing, SAFARI Technical Report TR-2014-002: http://safari.ece.cmu.edu/tr.html (2014).

6. Other methods to improve NoC scalability

In this Section, we now discuss other approaches that are designed to improve scalability of NoCs.

6.1 Ring-based NoCs

Hierarchical ring-based interconnect was proposed in a previous line of work [44–51,104]. We have already extensively compared to past hierarchical ring proposals qualitatively and quantitatively. The major difference between HiRD and these previous approaches is that HiRD uses deflection-based bridge routers with minimal buffering, and node routers with no buffering. In contrast, all of these previous works use routers with in-ring buffering, wormhole switching and flow control. Kim et al. propose tNoCs, hybrid packet-flit credit-based flow control [104] and Clumsy Flow Control (CFC) [64]. However, these two designs add additional complexity because tNoCs [104] requires an additional credit network to guarantee forward progress while CFC requires coordination between cores and memory controllers. Flow control in HiRD is different from that in these works due to HiRD's simplicity (with deflection based flow control, the Retransmit-Once mechanism, and simpler local-to-global and global-to-local buffers). Additionally, throttling decisions in HiRD can be made locally in each local ring as opposed to global decisions in CFC [64] and tNoCs [104].

Udipi et al. propose a hierarchical topology using global and local buses [74]. Using buses limits scalability in favor of simplicity. In contrast, HiRD design has more favorable scaling, in exchange for using more complex flit-switching routers. Das et al. [75] examine several hierarchical designs, including a concentrated mesh (one mesh router shared by several nearby nodes).

A previous system, SCI (Scalable Coherent Interface) [105], also uses rings, and can be configured in many topologies (including hierarchical rings). However, to handle buffer-full conditions, SCI NACKs and subsequently retransmits packets, whereas HiRD deflects only single flits (within a ring), and does not require the sender to retransmit its flits. SCI was designed for off-chip interconnect, where tradeoffs in power and performance are very different from those in on-chip interconnects. The KSR (Kendall Square Research) machine [106] uses a hierarchical ring design that resembles HiRD, yet these techniques are not disclosed in detail and, to our knowledge, have not been publicly evaluated in terms of energy efficiency.

6.2 Scalable topology design

While low-radix topologies (e.g., ring, tori [107], meshes [108], Express Cubes [10] and Kilo-NoC [12,13,109]) offer low area and power consumptoin, high-radix topologies [18,22,110] provide scalable alternatives to large scale systems. A flattened butterfly topology provides a scalable design that allows routers to send flits using two hops [18]. A HyperX network [110] extends the hypercube [61] design to minimize cost and lowering the latency. A SlimNoC [22] network provides a low-diameter, high-radix that further reduces the power and area through a SlimFly topology [111].

6.3 Low cost router designs

Kim [79] proposes a low-cost router design that is superficially similar to HiRD's node router design where routers convey traffic along rows and columns in a *mesh* without making use of crossbars, only pipeline registers and MUXes. Once traffic enters a row or column, it continues until it reaches its destination, as in a ring. Traffic also transfers from a row to a column analogously to a ring transfer in our design, using a "turn buffer." However, because a turn is possible at any node in a mesh, every router requires such a buffer [112,113]; in contrast, HiRD require similar transfer buffers only at bridge routers, and their cost is paid for by all nodes. Additionally, this design does not use deflections when there is contention.

Mullins et al. [114] propose a buffered mesh router with single-cycle arbitration. Abad et al. [115] propose the Rotary Router that consists of two independent rings that join the router's ports and perform packet arbitration similar to standalone ring-based networks. Both the Rotary Router and HiRD allow a packet to circle a ring again in a "deflection" if an ejection (ring transfer or router exit) is unsuccessful. Nicopoulos et al. [14] propose a buffer structure that allows the network to dynamically regulate the number of virtual channels. Kodi et al. [116] propose an orthogonal mechanism that reduces buffering by using links as buffer space when necessary. Multidrop Express Channels [10] also provides a low cost mechanism to connect multiple nodes using a multidrop bus without expensive router changes.

7. Conclusion and future outlook

Scalability and energy are two major concerns as core counts increase in commercial processors. To provide a design that is area-efficient and energy efficient without sacrificing performance, this chapter first presents MinBD [3,72,73]. MinBD is a minimally-buffered deflection router design. It combines deflection routing with a small buffer, such that some network traffic that would have been deflected is placed in the buffer instead. By using the buffer for only a fraction of network traffic, MinBD makes more efficient use of a given buffer size than a conventional input-buffered router. Its average network power is also greatly reduced: relative to an input-buffered router, buffer power is much lower, because buffers are smaller. Relative to a bufferless deflection router, dynamic power is lower, because deflection rate is reduced with the use of a small energy-conscious buffer.

To further improve scalability, this chapter discusses *HiRD* [23,24,52], a simple hierarchical ring-based NoC design that employs deflection routing. Past work has shown that a hierarchical ring design yields good performance and scalability relative to both a single ring and a mesh. HiRD has two new contributions: (1) a simple router design that enables ring transfers *without in-ring buffering or flow control*, instead using limited *deflections* (retries) when a flit cannot transfer to another ring, and (2) two *guarantee mechanisms* that ensure deterministically-guaranteed forward progress despite deflections. The evaluations show that HiRD enables a simple and low-cost implementation of a hierarchical ring network. Our HiRD evaluations also show that HiRD is more energy-efficient than several other topologies while providing competitive performance.

Despite the extensive design space for low-power NoC we considered so far, a number of key challenges remain to enable truly scalable and energy-efficient interconnection networks for modern systems and workloads. We believe that low-cost, energy-efficient network-on-chip design is an important challenge in scaling modern architectures beyond traditional CMPs. For example, heterogeneous architectures in modern SoCs can stress the interconnect through their imbalanced loads that are a consequence of largely different demands across many different types of applications and accelerators. Relatively new technology such as chiplets [117,118] or new types of memory [119–121] can create demands for efficient interconnection network designs that connect multiple memory nodes together. Especially processing-inmemory systems [122–177] can require well-connected memory arrays via efficient interconnects to tightly couple computation and communication. A fundamentally low-cost and energy-efficient interconnection network can further push the boundaries of computing systems, leading to significant improvements in performance and energy, and potentially enabling new applications and computing platforms.

Acknowledgments

This chapter incorporates revised material from another earlier article published in Parallel Computing in 2016 [24], the proceedings of the International Symposium on Computer Architecture and High Performance Computing in 2014 [23], the proceedings of the International Symposium on Networks-on-Chip in 2012 [3] and the proceedings of the International Symposium on High Performance Computer Architecture in 2011 [4].

This chapter is based on research done over the course of the past 13 years in the SAFARI Research Group on the topic of Network-on-Chips (NoC). We thank all of the members of the SAFARI Research Group, and our collaborators at Carnegie Mellon, ETH Zürich, and other universities, who have contributed to the various works we describe in this paper. Thanks also goes to our research group's industrial sponsors over the past 13 years, especially Alibaba, AMD, ASML, Google, Huawei, Intel, Microsoft, NVIDIA, Samsung, Seagate, and VMware. This work was also partially supported by the Intel Science and Technology Center for Cloud Computing, the Semiconductor Research Corporation, the Data Storage Systems Center at Carnegie Mellon University, various NSF and NIH grants, and various awards, including the NSF CAREER Award, the Intel Faculty Honor Program Award, a number of Google and IBM Faculty Research Awards to Onur Mutlu, and the Royal Thai Scholarship to Rachata Ausavarungnirun.

References

[1] W. Dally, B. Towles, Principles and Practices of Interconnection Networks, Morgan Kaufmann, 2004.

[2] T. Moscibroda, O. Mutlu, A Case for Bufferless Routing in On-Chip Networks, ISCA, 2009.

[3] C. Fallin, et al., MinBD: Minimally-Buffered Deflection Routing for Energy-Effcient Interconnect, NOCS, 2012.
[4] C. Fallin, et al., CHIPPER: A Low-Complexity Bufferless Deflection Router, HPCA, 2011.
[5] K. K.-W. Chang, R. Ausavarungnirun, C. Fallin, O. Mutlu, HAT: Heterogeneous Adaptive Throttling for On-Chip Networks, in: SBAC-PAD, 2012.
[6] X. Xiang, W. Shi, S. Ghose, L. Peng, O. Mutlu, N.-F. Tzeng, Carpool: A Bufferless On-Chip Network Supporting Adaptive Multicast and Hotspot Alleviation, ICS, 2017.
[7] X. Xiang, S. Ghose, O. Mutlu, N.-F. Tzeng, A Model for Application Slowdown Estimation in On-Chip Networks and Its Use for Improving System Fairness and Performance, ICCD, 2016.
[8] G.P. Nychis, et al., Next Generation On-Chip Networks: What Kind of Congestion Control Do We Need?, Hotnets, 2010.
[9] G.P. Nychis, et al., On-Chip Networks From a Networking Perspective: Congestion and Scalability in Many-Core Interconnects, SIGCOMM, 2012.
[10] B. Grot, et al., Express Cube Topologies for On-Chip Interconnects, HPCA, 2009.
[11] B. Grot, et al., Preemptive virtual clock: a flexible, Effcient, and cost-effective QOS scheme for networks-on-Chip, MICRO, 2009.
[12] B. Grot, et al., Topology-Aware Quality-of-Service Support in Highly Integrated Chip Multiprocessors, WIOSCA, 2010.
[13] B. Grot, et al., Kilo-NOC: A Heterogeneous Network-On-Chip Architecture for Scalability and Service Guarantees, ISCA, 2011.
[14] C. Nicopoulos, et al., ViChaR: A Dynamic Virtual Channel Regulator for On-Chip Networks, MICRO, 2006.
[15] R. Das, et al., Application-Aware Prioritization Mechanisms for On-Chip Networks, MICRO, 2009.
[16] R. Das, et al., Aérgia: Exploiting Packet Latency Slack in On-Chip Networks, ISCA, 2010.
[17] R. Das, et al., Application-to-Core Mapping Policies to Reduce Memory System Interference in Multi-Core Systems, HPCA, 2013.
[18] J. Kim, W. Dally, Flattened Butterfly: A Cost-Effcient Topology for High-Radix Networks, ISCA, 2007.
[19] R. Alverson, et al., The Tera Computer System, ICS, 1990.
[20] M. Taylor, J. Kim, J. Miller, D. Wentzlaff, The raw microprocessor: a computational fabric for software circuits and general-purpose programs, IEEE Micro 22 (2) (2002).
[21] Intel Corporation, Single-chip cloud computer, http://techresearch.intel.com/articles/Tera-Scale/1826.htm.
[22] M. Besta, S.M. Hassan, S. Yalamanchili, R. Ausavarungnirun, O. Mutlu, T. Hoefler, Slim NoC: a Low-diameter on-Chip network topology for high energy Effciency and scalability, ASPLOS, 2018.
[23] R. Ausavarungnirun, et al., Design and Evaluation of Hierarchical Rings with Deflection Routing, SBAC-PAD, 2014.
[24] R. Ausavarungnirun, C. Fallin, X. Yu, K.K.-W. Chang, G. Nazario, R. Das, G.H. Loh, O. Mutlu, A Case for Hierarchical Rings with Deflection Routing, Parallel Comput. 54 (2016) 29–45.
[25] M. Fattah, A. Airola, R. Ausavarungnirun, N. Mirzaei, P. Liljeberg, J. Plosila, S. Mohammadi, T. Pahikkala, O. Mutlu, H. Tenhunen, A. Low-Overhead, Fully-Distributed, Guaranteed-Delivery Routing Algorithm for Faulty Network-On-Chips, NOCS, 2015.
[26] P. Gratz, C. Kim, R. McDonald, S.W. Keckler, D. Burger, Implementation and Evaluation of On-Chip Network Architectures, ICCD, 2006.

[27] G. Michelogiannakis, et al., Evaluating Bufferless Flow-Control for On-Chip Networks, NOCS, 2010.

[28] H. Wang, et al., Power-Driven Design of Router Microarchitectures in On-Chip Networks, MICRO, 2003.

[29] P. Baran, On distributed communications networks, IEEE Trans. Commun. 12 (1) (1964) 1–9.

[30] S. Borkar, NoCs: What's The Point? NSF Workshop on Emerging Tech. for Interconnects (WETI), 2012.

[31] Y. Hoskote, et al., A 5-GHz mesh interconnect for a teraflops processor, IEEE Micro 27 (5) (2007) 51–61.

[32] P. Gratz, C. Kim, R. McDonald, S. Keckler, Implementation and evaluation of on-chip network architectures, ICCD, 2006.

[33] D. Wentzlaff, et al., On-chip interconnection architecture of the tile processor, IEEE Micro (2007) 27 (5) (2007) 15–31.

[34] Tilera Corporation, Tilera announces the world's first 100-core processor with the new TILE-Gx family, http://www.tilera.com/news_&_events/press_release_091026.php.

[35] 6th Generation Intel Core Processor Family Datasheet, http://www.intel.com/content/www/us/en/processors/core/desktop-6th-gen-core-family-datasheet-vol-1.html.

[36] Second Generation Intel Xeon Scalable Datasheet, https://www.intel.com/content/www/us/en/products/docs/processors/xeon/2nd-gen-xeon-scalable-datasheet-vol-1.html.

[37] 10th Generation Intel Core Processor Families Datasheet, https://www.intel.com/content/dam/www/public/us/en/documents/datasheets/10th-gen-core-families-datasheet-vol-1-datasheet.pdf.

[38] L. Seiler, et al., Larrabee: A Many-Core x86 Architecture for Visual Computing, SIGGRAPH, 2008.

[39] D. Pham, et al., Overview of the Architecture, Circuit Design, and Physical Implementation of a First-Generation CELL Processor, JSSC, 2006.

[40] Intel Corporation, Intel Details 2011 Processor Features, 2011, http://newsroom.intel.com/community/intel_newsroom/blog/2010/09/13/intel-details-2011-processor-features-offers-stunning-visuals-built-in.

[41] 8th and 9th Generation Intel Core Processor Family and Intel Xeon E Processor Families Datasheet, https://www.intel.com/content/dam/www/public/us/en/documents/datasheets/8th-gen-core-family-datasheet-vol-1.pdf.

[42] NVIDIA DGX-1 With Tesla V100 System Architecture, https://images.nvidia.com/content/pdf/dgx1-v100-system-architecture-whitepaper.pdf.

[43] J. Kim, H. Kim, Router Microarchitecture and Scalability of Ring Topology in On-Chip Networks, NoCArc, 2009.

[44] K. Farkas, Z. Vranesic, M. Stumm, Cache Consistency in Hierarchical-Ring-Based Multiprocessors, SC, 1992.

[45] Z. G. Vranesic, M. Stumm, D. M. Lewis, R. White, Hector: A hierarchically structured shared-memory multiprocessor, Computer 1991) 24 (1) 72–79, https://doi.org/10.1109/2.67196.

[46] G. Ravindran, M. Stumm, A Performance Comparison of Hierarchical Ring- and Mesh-Connected Multiprocessor Networks, HPCA, 1997.

[47] X. Zhang, Y. Yan, Comparative Modeling and Evaluation of CCNUMA and COMA On Hierarchical Ring Architectures, IEEE TPDS, 1995.

[48] V.C. Hamacher, H. Jiang, Hierarchical ring network configuration and performance modeling, IEEE Trans. Comput. 50 (1) (2001) 1–12.

[49] G. Ravindran, M. Stumm, On Topology and Bisection Bandwidth for Hierarchical-Ring Networks for Shared Memory Multiprocessors, HPCA, 1998.

[50] R. Grindley, et al., The NUMAchine multiprocessor, ICPP, 2000.

[51] M. Holliday, M. Stumm, Performance evaluation of hierarchical ring-based shared memory multiprocessors, IEEE Trans. Comput. 43 (1) (1994) 52–67, https://doi.org/10.1109/12.250609.

[52] R. Ausavarungnirun, et al., Improving Energy Effciency of Hierarchical Rings Via Deflection Routing, SAFARI Technical Report TR-2014-002, 2014, http://safari.ece.cmu.edu/tr.html.

[53] W. Hillis, The Connection Machine, MIT Press, 1989.

[54] B. Smith, Architecture and Applications of the HEP Multiprocessor Computer System, SPIE, 1981.

[55] M. Hayenga, et al., SCARAB: A Single Cycle Adaptive Routing and Bufferless Network, MICRO, 2009.

[56] C. Gómez, et al., Reducing Packet Dropping in a Bufferless NoC, EuroPar, 2008.

[57] S. Tota, et al., Implementation analysis of NoC: a MPSoC Tracedriven Approach, GLSVLSI, 2006.

[58] Z. Lu, M. Zhong, A. Jantsch, Evaluation of On-Chip Networks Using Deflection Routing, GLSVLSI, 2006.

[59] Y. Cai, K. Mai, O. Mutlu, Comparative evaluation of FPGA and ASIC implementations of bufferless and buffered routing algorithms for on-chip networks, in: ISQED, 2015, pp. 475–484, https://doi.org/10.1109/ISQED.2015.7085472.

[60] T. Chich, P. Fraigniaud, J. Cohen, Unslotted deflection routing: a practical and effcient protocol for multihop optical networks, IEEE/ACM Trans. Netw. 9 (1) (2001) 47–59.

[61] A. Greenberg, B. Hajek, Deflection routing in hypercube networks, IEEE Trans. Commun. 40 (6) (1992) 1070–1081.

[62] A. Bar-Noy, P. Raghavan, B. Schieber, Fast Deflection Routing for Packets and Worms, PODC-12, 1993.

[63] C. Busch, M. Magdon-Ismail, M. Mavronicolas, Effcient bufferless packet switching on trees and leveled networks, J. Parallel Distrib. Comput. 67 (11) (2007) 1168–1186.

[64] H. Kim, et al., Clumsy flow control for high-throughput bufferless on-chip networks, IEEE Comput. Archit. Lett. 12 (2) (2013) 47–50.

[65] C. Gómez, et al., BPS: A Bufferless Switching Technique for NoCs, WINA, 2008.

[66] C. Busch, M. Herlihy, R. Wattenhofer, Hard-Potato Routing, STOC, 2000.

[67] S. Konstantinidou, L. Snyder, Chaos Router: Architecture and Performance, ISCA, 1991.

[68] D. Kroft, Lockup-Free Instruction Fetch/Prefetch Cache Organization, in: ISCA, 1981.

[69] K.I. Farkas, N.P. Jouppi, Complexity/Performance tradeoffs with non-blocking loads, in: ISCA, 1994.

[70] J. Tuck, L. Ceze, J. Torrellas, Scalable Cache Miss Handling for High Memory-Level Parallelism, in: MICRO, 2006.

[71] S. Jafri, et al., Adaptive Flow Control for Robust Performance and Energy, MICRO, 2010.

[72] C. Fallin, et al., MinBD: Minimally-Buffered Deflection Routing for Energy-Effcient Interconnect, SAFARI Technical Report TR-2011-008, 2011, http://safari.ece.cmu.edu/tr.html.

[73] C. Fallin, et al., Bufferless and Minimally-Buffered Deflection Routing, in Routing Algorithms in Networks-on-Chip, Springer, New York, NY, 2014, pp. 241–275.

[74] A.N. Udipi, et al., Towards Scalable, Energy-Effcient, Bus-Based On-Chip Networks, HPCA, 2010.

[75] R. Das, et al., Design and Evaluation of Hierarchical On-Chip Network Topologies for Next Generation CMPs, HPCA, 2009.

[76] C. Sun, et al., DSENT - A Tool Connecting Emerging Photonics With Electronics for Opto-Electronic Networks-On-Chip Modeling, NOCS, 2012.

[77] D. Kroft, Lockup-Free Instruction Fetch/Prefetch Cache Organization, ISCA, 1981.

[78] A. Hansson, K. Goossens, A. Radulescu, Avoiding Message-Dependent Deadlock in Network-Based Systems-On-Chip, VLSI Design, 2007.

[79] J. Kim, Low-Cost Router Microarchitecture for On-Chip Networks, MICRO, 2009.

[80] R. Ausavarungnirun, et al., NOCulator, in: https://github.com/CMU-SAFARI/NOCulator.

[81] C.-K. Luk, et al., Pin: Building Customized Program Analysis Tools With Dynamic Instrumentation, PLDI, 2005.

[82] H. Lee, et al., Cloudcache: Expanding and Shrinking Private Caches, HPCA, 2011.

[83] C. Craik, O. Mutlu, Investigating the Viability of Bufferless NoCs in Modern Chip Multi-Processor Systems, SAFARI Technical Report TR-2011-004, 2011, http://safari.ece.cmu.edu/tr.html.

[84] A. Kyrola, et al., GraphChi: Large-Scale Graph Computation on Just a PC, OSDI, 2012.

[85] Y. Low, et al., GraphLab: A New Parallel Framework for Machine Learning, UAI, 2010.

[86] H. Kwak, et al., What Is Twitter, a Social Network or a News Media?, WWW, 2010.

[87] A. Snavely, D.M. Tullsen, Symbiotic Jobscheduling for a Simultaneous Multithreaded Processor, ASPLOS, 2000.

[88] S. Eyerman, L. Eeckhout, System-level performance metrics for multiprogram workloads, IEEE Micro 28 (3) (2008) 42–53.

[89] Y. Kim, et al., ATLAS: a Scalable and High-Performance Scheduling Algorithm for Multiple Memory Controllers, HPCA, 2010.

[90] Y. Kim, et al., Thread Cluster Memory Scheduling: Exploiting Differences in Memory Access Behavior, MICRO, 2010.

[91] H. Vandierendonck, A. Seznec, Fairness metrics for multithreaded processors, IEEE Comput. Archit. Lett. 10 (1) (2011) 4–7.

[92] E. Ebrahimi, et al., Fairness Via Source Throttling: A Configurable and High-Performance Fairness Substrate for Multi-Core Memory Systems, ASPLOS, 2010.

[93] E. Ebrahimi, et al., Prefetch-Aware Shared Resource Management for Multi-Core Systems, ISCA, 2011.

[94] E. Ebrahimi, et al., Coordinated Control of Multiple Prefetchers in Multi-Core Systems, MICRO, 2009.

[95] S.P. Muralidhara, et al., Reducing Memory Interference in Multicore Systems Via Application-Aware Memory Channel Partitioning, MICRO, 2011.

[96] L. Subramanian, et al., The Application Slowdown Model: Quantifying And Controlling The Impact OF Inter-Application Interference at Shared Caches and Main Memory, MICRO, 2015.

[97] L. Subramanian, et al., MISE: Providing Performance Predictability and Improving Fairness in Shared Main Memory Systems, HPCA, 2013.

[98] H. Usui, et al., DASH: Deadline-Aware High-Performance Memory Scheduler for Heterogeneous Systems with Hardware Accelerators, ACM TACO, 2016.

[99] R. Ausavarungnirun, et al., Staged Memory Scheduling: Achieving High Performance and Scalability in Heterogeneous Systems, ISCA, 2012.

[100] L. Subramanian, et al., The Blacklisting Memory Scheduler: Achieving High Performance and Fairness at Low Cost, ICCD, 2014.

[101] L. Subramanian, et al., The Blacklisting Memory Scheduler: Balancing Performance, Fairness and Complexity, TPDS, 2016.

[102] M. Thottethodi, et al., Self-Tuned Congestion Control for Multiprocessor Networks, HPCA, 2001.

[103] E. Baydal, et al., A family of mechanisms for congestion control in wormhole networks, IEEE Trans. Parallel Distrib. Syst. (2005) 16.

[104] H. Kim, et al., Transportation-Network-Inspired Network-On-Chip, HPCA, 2014.

[105] D. Gustavson, The scalable coherent interface and related standards projects, IEEE Micro 12 (1) (1992) 10–22.

[106] T.H. Dunigan, Kendall square multiprocessor: early experiences and performance, in: Of the Intel Paragon, ORNL/TM-12194, 1994.

[107] R. Alverson, D. Roweth, L. Kaplan, The Gemini System Interconnect, HOTL, 2010.

[108] J. Balfour, W.J. Dally, Design Tradeoffs for Tiled CMP On-Chip Networks, ICS, 2006.

[109] B. Grot, J. Hestness, S. Keckler, O. Mutlu, A QoS-enabled on-die interconnect fabric for kilo-node chips, IEEE Micro (2012) 32 (3) (2012) 17–25.

[110] J.H. Ahn, N. Binkert, A. Davis, M. McLaren, R.S. Schreiber, HyperX: Topology, Routing, and Packaging of Effcient Large-Scale Networks, SC, 2009.

[111] M. Besta, T. Hoefler, Slim Fly: A Cost Effective Low-Diameter Network Topology, SC, 2014.

[112] C.J. Glass, L.M. Ni, The Turn Model for Adaptive Routing, ISCA, 1992.

[113] L. Ni, Retrospective: The Turn Model for Adaptive Routing, ISCA, 1998.

[114] R. Mullins, et al., Low-Latency Virtual-Channel Routers for On-Chip Networks, ISCA, 2004.

[115] P. Abad, et al., Rotary Router: An Effcient Architecture for CMP Interconnection Networks, ISCA, 2007.

[116] A. Kodi, et al., iDEAL: Inter-Router Dual-Function Energy and Area-Effcient Links for Network-on-Chip (NoC) Architectures, ISCA, 2008.

[117] J. Yin, Z. Lin, O. Kayiran, M. Poremba, M.S.B. Altaf, N.E. Jerger, G.H. Loh, Modular Routing Design for Chiplet-Based Systems, ISCA, 2018.

[118] R. Hwang, T. Kim, Y. Kwon, M. Rhu, Centaur: A Chiplet-Based, Hybrid Sparse-Dense Accelerator for Personalized Recommendations, ISCA, 2020.

[119] S.H.S. Rezaei, M. Modarressi, R. Ausavarungnirun, M. Sadrosadati, O. Mutlu, M. Daneshtalab, NoM: Networkon-memory for inter-Bank data transfer in highly-banked memories, IEEE Comput. Archit. Lett. 19 (1) (2020) 80–83.

[120] M. Ogleari, Y. Yu, C. Qian, E. Miller, J. Zhao, String Figure: A Scalable and Elastic Memory Network Architecture, HPCA, 2019.

[121] Y. Kwon, M. Rhu, Beyond the Memory Wall: A Case for Memory-Centric HPC System for Deep Learning, MICRO, 2018.

[122] J. Ahn, S. Hong, S. Yoo, O. Mutlu, K. Choi, A Scalable Processing-in-Memory Accelerator for Parallel Graph Processing, ISCA, 2015.

[123] M. Zhang, Y. Zhuo, C. Wang, M. Gao, Y. Wu, K. Chen, C. Kozyrakis, X. Qian, GraphP: Reducing Communication for PIM-Based Graph Processing with Effcient Data Partition, HPCA, 2018.

[124] O. Mutlu, S. Ghose, J. Gómez-Luna, R. Ausavarungnirun, A modern primer on processing in memory, arXiv arXiv:2012. 03112 (2020).

[125] O. Mutlu, et al., Processing data where it makes sense: enabling in-memory computation, Microprocess. Microsyst. 67 (2019) 28–41.

[126] Q. Zhu, T. Graf, H.E. Sumbul, L. Pileggi, F. Franchetti, Accelerating Sparse Matrix-Matrix Multiplication with 3D-Stacked Logic-in-Memory Hardware, HPEC, 2013.

[127] S.H. Pugsley, J. Jestes, H. Zhang, R. Balasubramanian, V. Srinivasan, A. Buyuktosunoglu, A. Davis, F. Li, NDC: Analyzing the Impact of 3D-Stacked Memory + Logic Devices on MapReduce Workloads, ISPASS, 2014.

[128] D.P. Zhang, N. Jayasena, A. Lyashevsky, J.L. Greathouse, L. Xu, M. Ignatowski, TOP-PIM: Throughput-Oriented Programmable Processing in Memory, HPDC, 2014.

[129] A. Farmahini-Farahani, J.H. Ahn, K. Morrow, N.S. Kim, NDA: Near-DRAM Acceleration Architecture Leveraging Commodity DRAM Devices and Standard Memory Modules, HPCA, 2015.

[130] D.S. Cali, G.S. Kalsi, Z. Bingöl, C. Firtina, L. Subramanian, J.S. Kim, R. Ausavarungnirun, M. Alser, J. Gomez-Luna, A. Boroumand, et al., GenASM: A High-Performance, Low-Power Approximate String Matching Acceleration Framework for Genome Sequence Analysis, MICRO, 2020.

[131] J. Ahn, S. Yoo, O. Mutlu, K. Choi, PIM-Enabled Instructions: A Low-Overhead, Locality-Aware Processing-in-Memory Architecture, ISCA, 2015.

[132] G.H. Loh, N. Jayasena, M. Oskin, M. Nutter, D. Roberts, M. Meswani, D.P. Zhang, M. Ignatowski, A Processing in Memory Taxonomy and a Case for Studying Fixed-Function PIM, WoNDP, 2013.

[133] K. Hsieh, E. Ebrahimi, G. Kim, N. Chatterjee, M. O'Conner, N. Vijaykumar, O. Mutlu, S. Keckler, Transparent Offloading and Mapping (TOM): Enabling Programmer-Transparent Near-Data Processing in GPU Systems, ISCA, 2016.

[134] K. Hsieh, S. Khan, N. Vijaykumar, K.K. Chang, A. Boroumand, S. Ghose, O. Mutlu, Accelerating Pointer Chasing in 3DStacked Memory: Challenges, Mechanisms, Evaluation, ICCD, 2016.

[135] O.O. Babarinsa, S. Idreos, JAFAR: Near-Data Processing for Databases, SIGMOD, 2015.

[136] J.H. Lee, J. Sim, H. Kim, BSSync: Processing Near Memory for Machine Learning Workloads with Bounded Staleness Consistency Models, PACT, 2015.

[137] M. Gao, C. Kozyrakis, HRL: Effcient and Flexible Reconfigurable Logic for Near-Data Processing, HPCA, 2016.

[138] P. Chi, S. Li, C. Xu, T. Zhang, J. Zhao, Y. Liu, Y. Wang, Y. Xie, PRIME: A Novel Processing-In-Memory Architecture for Neural Network Computation In ReRAM-Based Main Memory, ISCA, 2016.

[139] B. Gu, A.S. Yoon, D.-H. Bae, I. Jo, J. Lee, J. Yoon, J.-U. Kang, M. Kwon, C. Yoon, S. Cho, J. Jeong, D. Chang, Biscuit: A Framework for Near-Data Processing of Big Data Workloads, ISCA, 2016.

[140] D. Kim, J. Kung, S. Chai, S. Yalamanchili, S. Mukhopadhyay, Neurocube: A Programmable Digital Neuromorphic Architecture with High-Density 3D Memory, ISCA, 2016.

[141] H. Asghari-Moghaddam, Y.H. Son, J.H. Ahn, N.S. Kim, Chameleon: Versatile and Practical Near-DRAM Acceleration Architecture for Large Memory Systems, MICRO, 2016.

[142] A. Boroumand, S. Ghose, M. Patel, H. Hassan, B. Lucia, K. Hsieh, K.T. Malladi, H. Zheng, O. Mutlu, LazyPIM: An Effcient Cache Coherence Mechanism for Processing-in-Memory, CAL, 2016.

[143] M. Hashemi, E. Khubaib, O. Ebrahimi, Y.N.P. Mutlu, Accelerating Dependent Cache Misses with an Enhanced Memory Controller, ISCA, 2016.

[144] M. Gao, G. Ayers, C. Kozyrakis, Practical Near-Data Processing for In-Memory Analytics Frameworks, PACT, 2015.

[145] Q. Guo, N. Alachiotis, B. Akin, F. Sadi, G. Xu, T.M. Low, L. Pileggi, J.C. Hoe, F. Franchetti, 3D-Stacked Memory-Side Acceleration: Accelerator and System Design, WoNDP, 2014.

[146] Z. Sura, A. Jacob, T. Chen, B. Rosenburg, O. Sallenave, C. Bertolli, S. Antao, J. Brunheroto, Y. Park, K. O'Brien, R. Nair, Data Access Optimization in a Processing-in-Memory System, CF, 2015.

[147] A. Morad, L. Yavits, R. Ginosar, GP-SIMD Processing-in-Memory, ACM TACO, 2015.

[148] S.M. Hassan, S. Yalamanchili, S. Mukhopadhyay, Near Data Processing: Impact and Optimization of 3D Memory System Architecture on the Uncore, MEMSYS, 2015.

[149] S. Li, C. Xu, Q. Zou, J. Zhao, Y. Lu, Y. Xie, Pinatubo: A Processing-in-Memory Architecture for Bulk Bitwise Operations in Emerging Non-Volatile Memories, DAC, 2016.

[150] M. Kang, M.-S. Keel, N.R. Shanbhag, S. Eilert, K. Curewitz, An Energy-Effcient VLSI Architecture for Pattern Recognition via Deep Embedding of Computation in SRAM, ICASSP, 2014.

[151] S. Aga, S. Jeloka, A. Subramaniyan, S. Narayanasamy, D. Blaauw, R. Das, Compute Caches, HPCA, 2017.

[152] A. Shafiee, A. Nag, N. Muralimanohar, et al., ISAAC: A Convolutional Neural Network Accelerator with In-Situ Analog Arithmetic in Crossbars, ISCA, 2016.

[153] L. Nai, R. Hadidi, J. Sim, H. Kim, P. Kumar, H. Kim, GraphPIM: Enabling Instruction-Level PIM Offoading in Graph Computing Frameworks, HPCA, 2017.

[154] J.S. Kim, D. Senol, H. Xin, D. Lee, S. Ghose, M. Alser, H. Hassan, O. Ergin, C. Alkan, O. Mutlu, GRIM-Filter: Fast Seed Filtering in Read Mapping Using Emerging Memory Technologies, arXiv:1708.04329 [q-bio.GN], 2017.

[155] J.S. Kim, D. Senol, H. Xin, D. Lee, S. Ghose, M. Alser, H. Hassan, O. Ergin, C. Alkan, O. Mutlu, GRIM-filter: fast seed location filtering in DNA read mapping using processing-in-memory technologies, BMC Genom. 19 (2018), 89.

[156] S. Li, D. Niu, K.T. Malladi, H. Zheng, B. Brennan, Y. Xie, DRISA: A DRAM-Based Reconfigurable In-Situ Accelerator, MICRO, 2017.

[157] G. Kim, N. Chatterjee, M. O'Connor, K. Hsieh, Toward Standardized Near-Data Processing with Unrestricted Data Placement for GPUs, SC, 2017.

[158] A. Boroumand, S. Ghose, Y. Kim, R. Ausavarungnirun, E. Shiu, R. Thakur, D. Kim, A. Kuusela, A. Knies, P. Ranganathan, O. Mutlu, GoogleWorkloads for Consumer Devices: Mitigating Data Movement Bottlenecks, ASPLOS, 2018.

[159] I. Fernandez, R. Quislant, C. Giannoula, M. Alser, J. Gomez-Luna, E. Gutierrez, O. Plata, O. Mutlu, NATSA: A Near-Data Processing Accelerator for Time Series Analysis, ICCD, 2020.

[160] G. Singh, J. Gomez-Luna, G. Mariani, G.F. Oliveira, S. Corda, S. Stujik, O. Mutlu, H. Corporaal, NAPEL: Near-memory Computing Application Performance Prediction via Ensemble Learning, DAC, 2019.

[161] S.H.S. Rezaei, M. Modarressi, R. Ausavarungnirun, M. Sadrosadati, O. Mutlu, M. Daneshtalab, NoM: Networkon-Memory for Inter-Bank Data Transfer in Highly-Banked Memories, CAL (2020).

[162] M. Besta, R. Kanakagiri, G. Kwasniewski, R. Ausavarungnirun, J. Beránek, K. Kanellopoulos, K. Janda, Z. Vonarburg-Shmaria, L. Gianinazzi, I. Stefan, J.G. Luna, J. Golinowski, M. Copik, L. Kapp-Schwoerer, S. Di Girolamo, N. Blach, M. Konieczny, O. Mutlu, T. Hoefler, SISA: Set-Centric Instruction Set Architecture for Graph Mining on Processing-in-Memory Systems, MICRO, 2021.

[163] G. Singh, D. Diamantopoulos, C. Hagleitner, J. Gomez-Luna, S. Stujk, O. Mutlu, H. Corporaal, NERO: A Near High-Bandwidth Memory Stencil Accelerator for Weather Prediction Modeling, FPL, 2020.

[164] Y. Wang, L. Orosa, X. Peng, Y. Guo, S. Ghose, M. Patel, J.S. Kim, J.G. Luna, M. Sadrosadati, N.M. Ghiasi, et al., FIGARO: Improving System Performance via Fine-Grained In-DRAM Data Relocation and Caching, MICRO, 2020.

[165] C. Giannoula, N. Vijaykumar, N. Papadopoulou, V. Karakostas, I. Fernandez, J. Gómez-Luna, L. Orosa, N. Koziris, G. Goumas, O. Mutlu, SynCron: Effcient Synchronization Support for Near-Data-Processing Architectures, HPCA, 2021.

[166] J. Gómez-Luna, I.E. Hajj, I. Fernandez, C. Giannoula, G.F. Oliveira, O. Mutlu, Benchmarking a new paradigm: An experimental analysis of a real processing-in-memory architecture, 2021, arXiv:2105.03814.

[167] A. Boroumand, S. Ghose, B. Akin, R. Narayanaswami, G.F. Oliveira, X. Ma, E. Shiu, O. Mutlu, Google Neural Network Models for Edge Devices: Analyzing and Mitigating Machine Learning Inference Bottlenecks, PACT, 2021.

[168] G.F. Oliveira, J. Gómez-Luna, L. Orosa, S. Ghose, N. Vijaykumar, I. Fernandez, M. Sadrosadati, O. Mutlu, DAMOV: A New Methodology and Benchmark Suite for Evaluating Data Movement Bottlenecks, IEEE Access, 2021.

[169] G. Singh, M. Alser, D. Cali, D. Diamantopoulos, J. Gomez-Luna, H. Corporaal, O. Mutlu, Fpga-based near-memory acceleration of modern data-intensive applications, IEEE Micro 41 (04) (2021) 39–48.

[170] O. Mutlu, Intelligent Architectures for Intelligent Computing Systems, DATE, 2021.

[171] Y.-C. Kwon, S.H. Lee, J. Lee, S.-H. Kwon, J.M. Ryu, J.-P. Son, O. Seongil, H.-S. Yu, H. Lee, S.Y. Kim, Y. Cho, J.G. Kim, J. Choi, H.-S. Shin, J. Kim, B. Phuah, H. Kim, M.J. Song, A. Choi, D. Kim, S. Kim, E.-B. Kim, D. Wang, S. Kang, Y. Ro, S. Seo, J. Song, J. Youn, K. Sohn, N.S. Kim, 25.4 A 20nm 6GB Function-In-Memory DRAM, Based on HBM2 with a 1.2TFLOPS Programmable Computing Unit Using Bank-Level Parallelism, for Machine Learning Applications, ISSCC, 2021.

[172] S. Lee, S.-h. Kang, J. Lee, H. Kim, E. Lee, S. Seo, H. Yoon, S. Lee, K. Lim, H. Shin, J. Kim, O. Seongil, A. Iyer, D. Wang, K. Sohn, N.S. Kim, Hardware Architecture and Software Stack for PIM Based on Commercial DRAM Technology: Industrial Product, ISCA, 2021.

[173] J.G. Luna, I.E. Hajj, I. Fernandez, C. Giannoula, G.F. Oliveira, O. Mutlu, Understanding a modern processing-in-memory architecture: Benchmarking and Experimental Characterization, in: HotChips Presentation, 2019.

[174] N. Hajinazar, G.F. Oliveira, S. Gregorio, J.A.D. Ferreira, N.M. Ghiasi, M. Patel, M. Alser, S. Ghose, J. Gómez-Luna, O. Mutlu, SIMDRAM: A Framework for Bit-Serial SIMD Processing sing DRAM, ASPLOS, 2021.

[175] V. Seshadri, K. Hsieh, A. Boroumand, D. Lee, M.A. Kozuch, O. Mutlu, P.B. Gibbons, T.C. Mowry, Fast Bulk Bitwise AND and OR in DRAM, CAL (2015).

[176] K.K. Chang, P.J. Nair, D. Lee, S. Ghose, M.K. Qureshi, O. Mutlu, Low-Cost Inter-Linked Subarrays (LISA): Enabling Fast Inter-Subarray Data Movement in DRAM, HPCA, 2016.

[177] V. Seshadri, D. Lee, T. Mullins, H. Hassan, A. Boroumand, J. Kim, M.A. Kozuch, O. Mutlu, P.B. Gibbons, T.C. Mowry, Ambit: In-Memory Accelerator for Bulk Bitwise Operations Using Commodity DRAM echnology, MICRO, 2017.

[178] J. Laudon, D. Lenoski, The SGI Origin: A ccNUMA Highly Scalable Server, ISCA, 1997.

[179] C. Carrión, et al., A Flow Control Mechanism to Avoid Message Deadlock in K-Ary N-Cube Networks, HiPC, 1997.

About the authors

Rachata Ausavarungnirun is an assistant professor at the Sirindhorn International Thai-German Graduate School of Engineering (TGGS) at King Mongkut's University of Technology North Bangkok. His research spans multiple topics across computer architecture and system software with emphasis on GPU architecture, heterogeneous CPU-GPU architecture, virtual memory, containerization, management of GPUs in the cloud, memory subsystems, memory management, processing-in-memory, non-volatile memory, network-on-chip, and accelerator designs. He received his PhD in Electrical and Computer Engineering from Carnegie Mellon University in 2017.

Onur Mutlu is a professor of Computer Science at ETH Zurich. He is also a faculty member at Carnegie Mellon University, where he previously held the Strecker Early Career Professorship. His current broader research interests are in computer architecture, systems, hardware security, and bioinformatics. A variety of techniques he, along with his group and collaborators, has invented over the years have influenced industry and have been employed in commercial microprocessors and memory/storage systems. He obtained his PhD and MS in ECE from the University of Texas at Austin and BS degrees in Computer Engineering and Psychology from the University of Michigan, Ann Arbor. He started the Computer Architecture Group at Microsoft Research (2006–2009) and held various product and research positions at Intel Corporation, Advanced Micro Devices, VMware, and Google. He received the IEEE High Performance Computer Architecture Test of Time Award, the IEEE Computer Society Edward J. McCluskey

Technical Achievement Award, ACM SIGARCH Maurice Wilkes Award, the inaugural IEEE Computer Society Young Computer Architect Award, the inaugural Intel Early Career Faculty Award, US National Science Foundation CAREER Award, Carnegie Mellon University Ladd Research Award, faculty partnership awards from various companies, and a healthy number of best paper or "Top Pick" paper recognitions at various computer systems, architecture, and security venues. He is an ACM Fellow "for contributions to computer architecture research, especially in memory systems", IEEE Fellow for "contributions to computer architecture research and practice", and an elected member of the Academy of Europe (Academia Europaea). His computer architecture and digital logic design course lectures and materials are freely available on YouTube (https://www.youtube.com/OnurMutluLectures), and his research group makes a wide variety of software and hardware artifacts freely available online (https://safari.ethz.ch/). For more information, please see his webpage at https://people.inf.ethz.ch/omutlu/.

CHAPTER NINE

Power-gating in NoCs

Hossein Farrokhbakht[a], Shaahin Hessabi[b], and N. Enright Jerger[a]

[a]University of Toronto, Toronto, ON, Canada
[b]Sharif University of Technology, Tehran, Iran

Contents

Advances in Computers, Volume 124
ISSN 0065-2458
https://doi.org/10.1016/bs.adcom.2021.11.013

Abstract

Driven by trends such as machine learning, the internet of things and 5G, it is increasingly common to see computer chips with tens to hundreds of processing cores, ranging from general purpose CPUs to application specific accelerators. All processing cores must share data and coordinate their operation, leading to contention and performance penalties. Communication between cores is handled by a network-on-chip (NoC). However, NoCs for modern chips contribute considerably to the chips overall power budget. Thus, the NoC power consumption is critical for ensuring that chips can continue to keep pace with increasing demand. Although NoCs are over-provisioned to handle worst-case scenarios, the NoCs routers are quite often underutilized, making them a promising candidate for power-gating. However, applying power-gating to the on-chip routers has several challenges. In this chapter, we present different power-gating solutions to alleviate the challenges.

1. Introduction

Network-on-Chip (NoC) is a key component in chip multiprocessors since it supports communication between many nodes. As core counts increase, the need for high-performance and efficient NoCs becomes increasingly important. The NoC routers consume a considerable portion of the chip's power Hoskote2007mesh. For instance, SCORPIO's 36-core NoC [1] contributes 19% to tile power and 11% to chip area. Large routers are provisioned to tolerate peak loads, but typically go underutilized for most traffic and consume power even when idle. Also, as transistor size continues to shrink, the contribution of NoC's static power is projected to increase [2,3]. Power analysis for the 32 nm technology node demonstrates that static power consumption increases by 3× compared to the same design in a 65 nm technology [4]. Both the inefficient use of router resources and further transistor size scaling may push routers to have greater static power consumption in future designs.

There are several techniques for reducing static power in digital circuits [5]. Power gating is a widely used technique where idle blocks are powered-off for reducing static power [6]. Although this method can effectively reduce the power consumption in many digital circuits, it has several shortcomings particularly when applied to NoC routers. First, when a packet encounters a powered-off router, it has to wait until the router wakes up. Hence, power gating may impose a significant latency overhead to the NoC. The problem gets worse if the packet encounters further powered-off routers along its path. Second, although the purpose of power gating is to save static power when blocks are idle, power gating itself incurs power cost. In other words, powering on a router leads to a non-negligible power overhead. Thus, consecutive idle periods should be large enough to be able to compensate the imposed wake-up power overhead. While real applications have a relatively low average traffic load, even in these low utilization rates it is hard to find/predict large enough consecutive idle cycles as packets arrive intermittently and do not follow a particular arrival pattern [7,8].

Several attempts have been made to overcome the power gating issues. A common optimization technique to alleviate the latency overhead of power gating is the early wake-up technique. In this technique, wake-up signals are generated multiple hops earlier before a packet arrives at the powered-off router. Thus, it can partially [9] or almost completely [2] hide the wake-up latency. However, in addition to its area overhead, this technique only addresses the wake-up latency issue associated with power gating; it still suffers from intermittent packet arrival. Other recent efforts in effective power-gating methods for routers [4,9,10] are either complex and impose significant area and performance overhead, or do not scale. As an example, NoRD [10] avoids waking up a powered-off router by using a ring to maintain connectivity of the network. Nevertheless, the long bypass ring makes NoRD scale poorly, as a packet may traverse a ring with the size of the network to avoid encountering powered-off routers.

In this chapter, we present several power-gating solutions applied to NoCs routers: **TooT** [11] reduces the number of wake-ups by leveraging the characteristics of deterministic routing algorithms and mesh topologies. TooT avoids powering a router on when it forward a *straight* packet or ejects a packet, i.e., a router is powered on only when either a packet turns through it or its associated node injects a packet. **SMART** [12] makes power-gating more efficient by reducing wake-up overheads through a novel application-specific mapping and routing technique to regularize traffic patterns. **SPONGE** [13] keeps the majority of the routers in power-gated state by using a deterministic routing algorithm. Packets traverse power-gated routers without waking them up, and can only turn at predetermined powered-on routers. **Muffin** [14] exploits a comprehensive bypass mechanism to minimize the number of wake-ups.

2. Background

Due to shrinking transistor feature sizes and reductions in supply voltage, the contribution of leakage power in the chip's total power has increased. NoCs, as a promising substitute for the old bus structures in today's *Multi-processor Systems-on-Chip* (MPSoCs), are not exceptions to this trend. While the NoC by itself may consume up to 35% of a chip's total power [15], it has been shown that the ratio of static power in a typical 8×8 NoC increases from 43% in 45 nm technology to more than 63% in 22 nm [16], making the NoC's static power a major contributor to chip's power. Therefore, methods to alleviate the role of static power are indispensable.

Power gating is a widely accepted approach to eliminate the static power of unutilized/idle components in digital circuits. To efficiently apply power gating to a digital circuit, it requires idle time of the circuit to be long enough to compensate for the corresponding wake-up energy overhead. While NoCs are provisioned to tolerate high inter-node communications to avoid performance bottleneck, real-world applications typically exhibit low traffic rates, leading to underutilized on-chip resources. Analysis on PARSEC benchmarks on an 8×8 mesh topology reveals that, on average, 92.7% of routers are idle during application runtime [11]. Altogether, power gating is crucial for NoCs due to the contribution of its static power to the total chip power, and seems promising because of abundant idleness in components. Power-gating techniques are classified into *coarse grained* and *fine grained*. In coarse-grained power-gating, the entire router is disconnected from the power supply whereas in fine-grained one only a subset of resources (e.g., a virtual channel) are power-gated.

2.1 Power-gating

Power gating is carried out by placing a switch (e.g., high threshold transistor(s)) between a circuit and its supply voltage. For on-chip routers, the switch is cut-off when all the data-paths, i.e., input buffers, output latches and crossbar of a router, are idle. It can be accomplished by a simple always-on monitoring controller. However, the adjacent routers must be notified of a power-gated router, otherwise they may send packets that will be dropped in the network. Thus, handshaking signals are necessary to inform the neighbors if a router is powered-off. In this case, the neighbor tags the corresponding output port as powered-off, and make it unavailable when allocating the switch. Similarly, the adjacent routers need handshaking signals to inform a powered-off router when they send a packet for it. As illustrated in Fig. 1, the right side router is power gated and notifies its neighbors (only the left router is shown) by asserting the PG signals. Subsequently, the neighbors tag the corresponding output buffer as

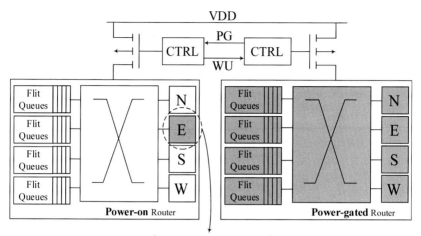

Fig. 1 Power gating of on-chip routers.

powered-off (port E of the left router is tagged as powered-off). The neighboring routers prompt this router to power on via the wake-up (WU) signals.

2.2 Cumulative wake-up latency

A router can remain in power-gated state until confronted with a packet. This happens when either (a) its associated processing node injects a packet, (b) the associated processing node is the destination of a packet (i.e., ejects a packet), or (c) it transmits a packet to any neighbors. Each of the mentioned scenarios causes the router to wake up. Typically, waking up a router takes about 4 ns [9], which corresponds to 8 cycles wake-up latency in 2 GHz frequency. The problem is exacerbated when a packet encounters several power-gated routers within its path. In conventional power gating method, when the network utilization is low, the majority of the routers are power gated, hence there is high probability that such scenario would happen, referred to as cumulative wake-up latency.

2.3 Break-even time

Each time a router is woken up, it imposes a considerable wake-up energy. The term *Break-Even Time* (BET) is defined as the minimum number of consecutive cycles that a gated block (here a router) must remain powered-off to compensate for the wake-up energy overhead. For on-chip routers, the

BET is about 10 cycles [9]. Therefore, power gating the idle intervals for less than 10 cycles imposes energy overhead rather than savings. As mentioned before, routers are idle for 92.7% of time; however, 63.4% of idle intervals violate the BET rule [11], i.e., are shorter than 10 cycles. This imposes significant energy overhead and deteriorates the efficiency of power gating.

Based on the above-mentioned issues, applying efficient power gating on on-chip routers requires an approach to reduce the wake-up latency and energy at the circuit level and/or reduce the number of wake-ups, which subsequently increases the idle intervals as well. While applying power gating in smaller granularities (e.g., buffer and port level) [17] may alleviate the BET issue, these methods are not efficient because of their hardware overhead and complexity. In addition, these methods do not mitigate the wake-up energy overhead and just split it into smaller pieces.

2.4 Related work

Coarse-grained power-gating: *NoRD* [10] eliminates the dependency of a node to its associated router to provide it with the ability to inject and eject packets without powering on the router. To this end, it adds an inport and outport bypass to the *Network Interface* (NI) of the router. As a result, any sent/received packet can inject/eject to/from the network without waking-up the router. Similarly, forwarded packets can bypass a powered-off router by entering through the dedicated inport bypass and passing through its outport bypass. While this solution eliminates the wake-up latency once a packet encounters a powered-off router, it increases the packet latency by detouring it to a neighbor router. The situation becomes worse when a packet faces multiple powered-off routers. Therefore, NoRD is not scalable for large NoCa (e.g., an 8×8 mesh) because a packet may traverse long bypass rings—up to size of the network—to avoid from powering on the router(s), which imposes significant packet latency for large NoC. To mitigate the wake-up latency, Chen *et al.* [2] have proposed *PowerPunch* which aims to power on a powered-off router ahead arrival of the packets, using additional controlling signals. Since PowerPunch does not address the fragmented idle periods of routers and the number of wake-ups, this prevalent power-gating overhead is still an issue in this method. *Router Parking* [4] powers off a subset—based on aggressive or conservative policies—of the routers that are connected to powered-off cores. It is also aware of maintaining the network connectivity and limiting the latency imposed by the packets that detour the powered-off routers.

Based on the information of the network traffic which is collected by a specific component, it chooses the subset of routers to be powered off. It is only able to power-gate the routers connected to powered-off cores, so its efficiency is limited. In addition, Router Parking requires complex design because it updates the routing table once a router is powered-off, and modifies the routing algorithm. Catnap [18] investigates power-gating efficiency in a multi-NoC design. In a multi-NoC, wires and buffers are split-off into several sub-networks wherein the node is connected to all of its associated routers in all the sub-networks. Thus, it is promising for power-gating since a sub-network can be power-gated entirely without affecting the network connectivity. To achieve this goal, they propose a sub-network selection policy that provides it long idleness interval to alleviate the BET problem, and adapts the network bandwidth to application demands.

Fine-grained power-gating: Fine-grained power-gating techniques have been proposed to reduce the power consumption of sub-modules within a router [9,17,19,20]. *FlexiBuffer* has proposed partitioning a buffer into two unequal parts to facilitate power gating [20]. The smaller part is always active to provide routability under light traffics, while the second part is powered-off by default to save energy. When the number of flits in smaller part surpasses a threshold, the second part of the buffer is powered-on. Matsutani et al. [17] propose ultra fine-grained power-gating for NoC routers in which each component (e.g., buffer, crossbar, etc.) can be individually powered off, considering the network flow. To alleviate the wake-up latency, it uses 2-hop look-ahead signals to wake-up the buffers prior to arrival of the packets. While this method reduces the wake-up latency, it does not reduce the number of wake-ups, i.e., idleness intervals of routers are still fragmented that leads to energy overhead due to wake-ups and also loses power-gating opportunity due to idle-detection times. Finally *Panthre* [19] is proposed to provide long sleeping intervals to fine-grained units to facilitate efficient power-gating. This method is based on the observation that only 10% of network traffic flows through the 30% of (low utilization) links, providing a potential for power-gating. Power-gating is done at link-level granularity, i.e., links with utilization lower than an adaptive threshold are powered-off. When any anomaly is detected, all links are woken-up, and after three consecutive successful epochs, the power-gating threshold reduces. In addition to complexity of added components, this method is inefficient since it can exploit only about half of its power-gating opportunity.

3. TooT
3.1 Key observation

TooT is based on a key observation: most of the times a powered-off router needs to be woken up is caused by the arrival of a *straight* packet. Demonstrated in Fig. 2, using the XY routing algorithm, straight packets are the majority of packets that interrupt power-gated routers, which on average contribute to 54.7% of wake-ups. The remaining part is distributed almost evenly between other types of packets, i.e., turn, inject and eject. Notice that the routing algorithm plays a key role here, since in the XY routing algorithm any packet is injected and ejected once, has at most one turn during its path, and mostly traverses straight. As an obvious consequence, if we could avoid waking-up the power-gated routers for straight packets, performance overhead due to wake-up latency would be decreased significantly. In addition, not only will the wake-up energy overhead be mitigated, but also further energy can be saved since router will be in power-gated state instead of being in power-consuming *idle detection* state if conventional power gating method had been applied.

3.2 Main idea

In the conventional power-gating method, a powered-off router is woken up if any packet arrives at the router. On the other hand, as noticed in Section 3.1, more than 54.7% of these packets are straight packets. Accordingly, if we avoid powering on the routers for this large portion of packets, the efficiency of power-gating will be improved by (a) increasing

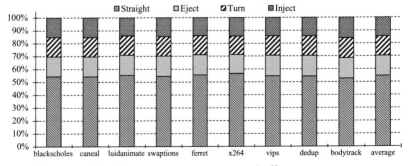

Fig. 2 Breakdown of packets encountering a powered-off router.

the length of intervals a router is powered-off, which not only increases the powered-off periods, but also reduces the energy-overheads incurred while powering-on a router as well, and (b) reducing the accumulated wake-up latency by reducing the cases a packet waits for a power-gated router to be powered-on during its path traversal from the source to the destination.

TooT prevents powering on the power-gated routers by augmenting all the routers with a bypass path. In addition, by appropriately architecting the bypass route, TooT can exploit it for the ejecting packets, as well (details of the bypass will be described in Section 3.3). Thus, it is not necessary anymore to involve a power-gated router datapath to route such packets. Note that the observation of Fig. 2 is central to the proposed method because otherwise, i.e., if all the packet types were distributed evenly, its effectiveness would be diminished, especially taking the potential overheads into account.

3.3 TooT architecture

Fig. 3 demonstrates the proposed TooT architecture. For the sake of simplicity, only the X^+ (the input from the west upstream router) to X^- (the output to the east downstream router) path is represented in this figure. For every bypass path such as X^+ to X^-, one bypass latch and two 2 to 1 multiplexers—all having equal flit widths—are needed. Once a flit arrives at a TooT router, depending on active/power-gated state of the router, different scenarios happen, as detailed next.

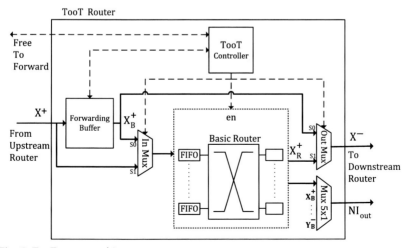

Fig. 3 TooT router architecture.

Router is power-gated: Input flit enters the dedicated bypass latch. Afterward, the controller checks whether the flit turns at the current router. To do that, it compares the destination coordination of the flit with that of the current router. If the flit is supposed to turn at this router, it must wait until the router wakes up (power-on signal is triggered by the controller). Otherwise, the flit is routed to the downstream router using the *S0* port of the *Out Mux*. Its selection signal is also controlled by the controller (represented by dashes in the figure). Note that for the output network interface, the output of the latch (denoted by X_B^+) passes through the *5x1 Mux*. Actually, in the power-gated mode, TooT router acts as a conventional router with one-flit input buffers, capable of merely forwarding straight packets and ejecting. Hence, it uses the *free-to-forward* signals to notify neighbors whether the corresponding bypass latches are empty. Similarly, if the input buffers of the next router (or the bypass latch if it is power-gated) are not empty, the flit remains in bypass latch.

To avoid any conflict or starvation, while the bypass latch is not empty, the controller always grants the priority to this latch, i.e., *S0* port of *Out Mux* is passed to the output, whether the router is power-gated or even active. The latter may happen when one or more flits inside the bypass latches cannot route to the next hops (due to network/router congestion, etc.) and meanwhile the router is triggered by some turning/injecting packets and has been powered-on (or is powering on). In such cases, the controller disables the (input ports of the) bypass latches until the router is completely woken-up, and then grants the priority to flit(s) that may reside in them. Finally, it returns the privilege to the basic router by selecting the port *S1* of the dedicated multiplexers. This procedure is depicted in Fig. 4A, where

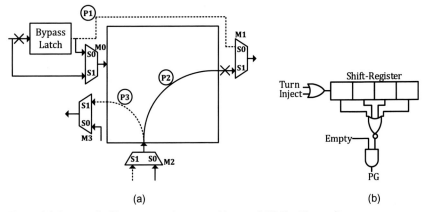

(a)   (b)

Fig. 4 (A) Powered-off to powered-on transition and (B) TooT's predictor.

the router has been powered-on but there is a flit in the shown bypass latch. Both the flits in bypass latch and multiplexer $M2$ (or its corresponding input buffer inside the router) are about to send a flit through the port in $M1$. In this case, as mentioned above, priority is given to the latch (path is denoted by $P1$). However, simultaneous with $P1$, the *non-straight* path $P3$ can also be established since there is no full bypass latch (East to West) that sends a flit through $M3$. Since the router is powered-on (or is powering on) the input of the bypass latch is disabled.

Router is active: This scenario is similar to typical behavior of a router with the delay overhead of *In Mux* and *Out Mux* if the network frequency is the bottleneck. It is worth noting that as discussed above, when the router is active, all input flits enter the input buffers (through $S1$ Port of *In Mux*). However, when there is a flit in a bypass latch, its corresponding TooT router output will be assigned to this latch. For instance, as depicted in Fig. 3, if the bypass latch of X^+ is full, the X_R^+ must wait until the controller grants the $S1$.

3.4 TooT's predictor

In conventional power-gating methods, if a router has been idle for 4 consecutive cycles, it is deduced that it will be idle for the next 10 cycles, as well, providing it with the power-gating opportunity. However, since TooT modifies the wake-up conditions, it necessitates a change in predictor mechanism since it is not anymore necessary to keep a router on when the passing packets are straight/eject. Thus, we aim to power gate a router if no turning or injecting flit encounters it within 4 consecutive cycles. Therefore, as a key advantage of TooT, if frequent/bursty straight packets arrive at a router, the controller can simply power gate it after 4 cycles, which significantly improves the TooT efficiency. This is a substantial benefit to power gate a router during its operation because in on-chip networks, the traffic on a router's port is usually continuous. It is noteworthy that this 4–cycle predictor can be implemented with a simple circuitry. As shown in Fig. 4B, we have implemented the predictor with a (4-bit) shift register and a few logic gates. Based on this figure, in every cycle, if any of the flits traversing through the router is turn or inject, a logical 1 (one) is input into the shift register, which disables the power gating signal ($PG=0$) for the next 4 cycles. On the other hand, if none of the outgoing flits is turn or inject, a 0 (zero) is shifted in. Repeating this procedure for 4 consecutive cycles makes the output of NOR gate logical 1. Then, if all of the router buffers, pipeline, etc. are empty ($Empty=1$), the power-gating signal will be triggered.

4. SMART

4.1 Main idea

Based on the investigation on types of the packets in SPLASH-2 benchmarks, SMART illustrates that 73.5% of all packets encountering power-off routers are *intermediate* packets, i.e., the packets whose sources and destinations are not the current nodes. The remaining proportion is divided between *inject* packets and *eject* packets. Fig. 5 depicts the distribution of the packets encountering a power-gated router in SPLASH-2 benchmarks. Note that this proportionately is valid for all minimal-based routing algorithm. On the other hand, facing power-gated routers is a significant contributor to power consumption in on-chip routers. Therefore, if we could employ a specific deterministic routing algorithm, which decreases the number of encountering power-gated routers, we can dramatically alleviate power-consumption in NoC routers. Hence, we engage a deterministic routing algorithm that can re-use the receiving path for sending in each source-destination pair. It can significantly reduce the number of encountered power-gated routers. Additionally, this routing algorithm can increase some parts' utilization compared to other components if we employ a unique mapping for each communication task graph (CTG). It achieves a network with less irregularity, providing more large idle time intervals. Consequently, SMART achieves greater gains compared to conventional power-gating technique. We also propose an exclusive mapping for each CTG, with a unique and distinctive layout, which is compatible with the proposed routing algorithm. In that case, we can minimize the number of hops for the packets, leads us to attain a more efficient power-gating technique.

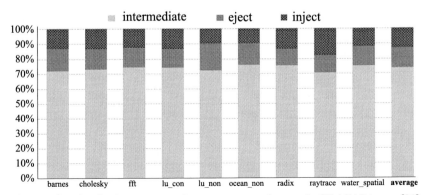

Fig. 5 Packet distribution encountering a power-gated router in SPLASH-2 traces [21].

4.2 XYX routing algorithm

XY routing uses different paths for each source-destination pair while source and destination are not in the same row and column. Fig. 6A illustrates different routes for some source-destination pairs using XY routing algorithm. SMART exploits XYX routing by which packets use a unique path between each source-destination pair. In other words, if SMART uses XY routing for packet traversal from *node1* to *node2*, it will choose YX routing for traversing from *node2* to *node1*, and vice versa. Fig. 6B demonstrates the unique routes for packet traversal based on the XYX routing. As an instance, according to XYX, the path for packet traversal from (4,5) to (6,7) is XY, and the route for sending packets from (6,7) to (4,5) is YX. Using XYX, SMART re-uses the powered-on routers located between both directions, increasing the routers' utilization in the chosen path. Therefore, XYX can avoid powering routers on unnecessarily, saving power consumption on average. Note that if two nodes are in the same row or column, XY and XYX routes are the same.

XYX is a deterministic routing algorithm whose structure automatically eliminates some turns. This turn elimination results in a unique path between each source-destination pair. The set of paths that emerge result in a cycle-free network. Fig. 7A illustrates all possible routes in XYX in a 3×3 mesh network, which proves no possible cycle in the XYX routing algorithm. Consequently, no additional virtual channel (VC) is needed to handle potential deadlocks. Fig. 7B depicts all possible turns in the XYX routing algorithm.

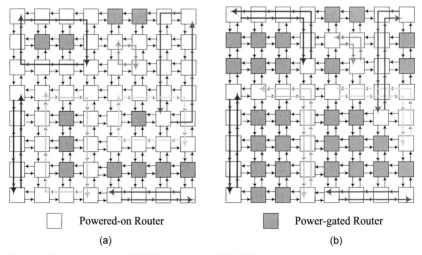

☐ Powered-on Router   ■ Power-gated Router

(a)   (b)

Fig. 6 Utilized routers in (A) XY routing and (B) XYX.

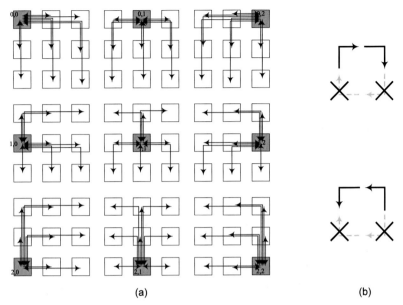

(a)   (b)

Fig. 7 Deadlock avoidance in the XYX routing algorithm. (A) A 3 × 3 mesh network with XYX routing algorithm and (B) prohibited turns in XYX which guarantees deadlock freedom.

4.3 Pseudo-spiral: Exclusive mapping for XYX routing

According to XYX, if SMART engages XY for sending packets from *node1* to *node2*, it will use YX for sending in opposite direction, i.e., form *node2* to *node1*. Thus overall pattern of the routing packets is *L-shape*, as shown in Fig. 8A. Although XYX can avoid unnecessary wake-ups, SMART goes further to increase the number of power-gated routers by employing an exclusive mapping mechanism. XYX, alongside the mapping mechanism, increases utilization in high traffic regions while while lowering traffic in other network regions.

Using a mapping mechanism, we can localize the cores with more communication. This localization can dramatically decrease the number of hops between each source-destination pair. Consequently, decreasing the number of hops between each source-destination reduces the number of *intermediate* routers, reducing the probability of encountering power-gated routers for each packet. Thus engaging a mapping mechanism using offline profiling techniques can considerably improve the SMART efficiency.

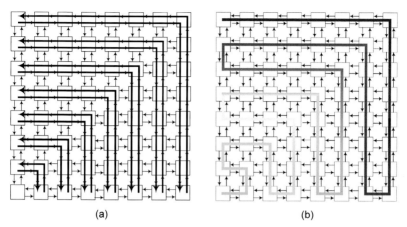

Fig. 8 Reciprocal relationship between XYX and mapping (pseudo-spiral) in a 4 × 4 network: (A) L-shape structure in XYX and (B) L-shape-based mapping (pseudo-spiral).

To increase the mapping efficiency alongside XYX, we must engage an exclusive *L-shape*-based mapping mechanism suited for XYX. Fig. 8B illustrates the overall layout of the mapping, which has a pseudo-spiral structure. Using this pseudo-spiral mapping structure orchestrates the mapping and the routing integration, improving the XYX efficiency and decreasing the number of hops for each packet traversal. To this end, SMART needs offline profiling on CTG. Using offline profiling, SMART makes an extrapolation from the traffic volume between nodes and then arranges the nodes with more communication in a pseudo-spiral structure.

4.4 SMART predictor

The Predictor of a conventional power-gating technique looks for four consecutive idle cycles in a router to infer that the router will be idle in the 10 next cycles. SMART exploits XYX routing algorithm and pseudo-spiral mapping mechanism, changing the traffic load across the routers; the traffic is denser in the upper triangle of the mesh, and the idleness of nodes located in the lower triangle is more probable (Fig. 8B). Thus SMART uses a different approach in deciding when to power gate a router. SMART predictor waits for 10 consecutive idle cycles in the first four L-shapes to predict that the router will be idle in the next 10 cycles. Similar to the conventional power-gating techniques, the number of waiting cycles for other remaining L-shapes is 4.

5. SPONGE

5.1 Main idea

In conventional power-gating methods, the router must be powered on to pass different types of packets, i.e., straight, turn, inject and eject.

SPONGE reveals that using SPLASH-2 benchmarks with XY routing, only 13.28% of packets turn through power-gated routers. In the XY routing, turning can occur in each router. Hence handling straight, inject and eject packets through a bypass mechanism and directing turn packets to specific powered-on routers improves both static power and packet latency. This improvement comes from maximizing the length of idle periods and minimizing the number of wake-ups. Thus in SPONGE first, the middle column of the mesh network is determined as the turning point whose routers are always power-on, and the controllers of the power-gated routers defer the turns to that column using bypass paths. After that, depending on the traffic load, congestion may occur on the middle column, and then the proportion of turning packets may increase. In this case, power-gated columns adjacent to the congested column will be powered on. The recent powered-on columns might experience underutilization in the later cycles; thus, they will be power gated again. Therefore, the window size of powered-on columns can be expanded or shrunk similarly to a *sponge*.

SPONGE uses the following terminology:

- **Major pivot column:** the middle column in the mesh-based network whose routers are always powered on.
- **Boundary pivot column:** a powered-on column that at least one of its adjacent columns are power-gated. SPONGE uses boundary pivot column as the turning point. If all columns except the major pivot column are powered off, the major pivot column is also a boundary pivot column. On the other hand, if there is more than one powered-on column, there will be two boundary pivot columns.
- **pivot columns:** the columns between boundary pivot columns which are powered on.

5.2 Handling turn and straight packets

SPONGE power gates all of the routers in the mesh-based network excluding the major pivot column. For an *even* n in an $n \times n$ mesh, one of the

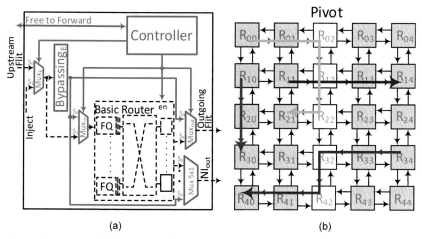

(a)   (b)

Fig. 9 (A) The overall architecture of bypass mechanism for straight, inject and eject flits and (B) different packets traversals scenarios based-on the boundary pivot column.

two middle columns can be selected. Packets traverse straight through power-gated routers using a bypass mechanism. Incoming packets from power-gated routers can only turn through boundary pivot column(s). This restriction saves a significant portion of static power consumption since SPONGE does not power on the routers for turn packets. Fig. 9(b) shows a 5×5 mesh network, whose major pivot is column 2, i.e., $R_{i2}(i=0...4)$. The major pivot column divides the network into two sub-networks. If the source and the destination are located in the same column or the same row, packets traverse straight. Otherwise, packets traverse straight until they encounter a boundary pivot column, and by turning twice (if their destinations are not located in the boundary pivot column) through it, they reach their destinations rows. All packets which are injected to power-gated routers and need to turn, use the routers of boundary pivot column(s); the packets injected to powered-on routers use YX routing. Depending on the location of source-destination pair, these packets can be classified into two types. If these packets are located in different sides of boundary pivot column(s) (e.g., R_{34} to R_{40}), or they are located between boundary pivot columns, the average hop is equal to that of the XY routing. On the other hand, if they are located in the same side of boundary pivot column(s) (e.g., R_{00} to R_{21}), detouring is required which imposes latency. However, traversing packets through the power-gated routers using bypass paths can completely compensate for the detouring-induced hop overhead.

To avoid powering on the power-gated routers, a bypass mechanism is needed to handle straight, inject and eject packets without powering on the routers (details of the bypass mechanism for the inject and eject packets will be described in Section 5.3). Thus straight packets can be handled by adding a buffer of 1-flit depth and three 2x1 multiplexers—all having equal flit widths—for each port as depicted in Fig. 9A. The addition of a 1-flit buffer and multiplexers per port makes the router similar to a switch for straight packets. For simplicity, Fig. 9A demonstrates only one direction. When input flit enters the dedicated bypass buffer, the controller checks whether its destination is located in the row/column of the current router. If the flit is supposed to have a turn to reach its destination, it must be directed toward boundary pivot column for its turn. Otherwise, the flit is routed to its destination which is located in the current row/column. These straight actions use the S_0 port of the Mux_{out}, and the bypass mechanism is accomplished in one cycle. The *free-to-forward* signal in Fig. 9A is used to inform neighbors whether the current bypass buffers are empty.

5.3 Handling inject and eject packets

As shown in Fig. 9A, SPONGE can handle inject and eject packets without powering on the routers using the bypass paths. As shown in Fig. 10A, each direction has its own bypass. Thus, after injection via $Mux_{N,\ S,\ WorE}$, the

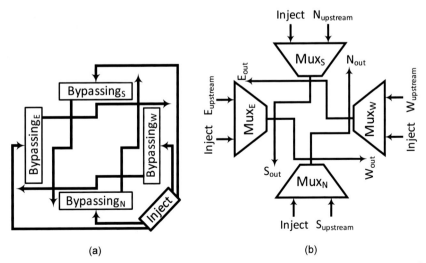

(a)   (b)

Fig. 10 (A) Using bypass paths for sending the inject packets without powering on and (B) simple multiplexers as arbitration between incoming and inject packets.

controller is configured to copy the flit to all bypass buffers, i.e., $bypass_{N,S,W,E}$. The controller checks the copied flits in all bypass buffers. For $bypass_S$ in a power-gated router, if the destination of the copied flit is not located in the current column, *or* the destination is not one of the southern routers in the current column, the flit must be dropped. The conditions for $bypass_N$ in a power-gated router are the same while directions are opposite. The conditions for traversals in a row are different due to the boundary pivot column. For $bypass_E$ in a power-gated router, the controller must know the relative position with the boundary pivot column. Suppose that it is located in the east of the current router, and the destination is a western router in the same row, *or* the destination is located in the same column, the flit must be dropped. On the contrary, for $bypass_W$ in a power-gated router (i.e., the boundary pivot column is located in the east of the current router), if the destination is a eastern router in the same row, *or* the destination is located in the same column, the flit must be dropped. For an inject flit whose destination is not located in the same column and the same row, the flit must be directed toward boundary pivot column for its turn. Considering these conditions allows SPONGE to handle inject packets without waking up the power-gated routers. Note that in Fig. 10B, inject packets take precedence over others in each multiplexer.

It is easier to handle eject packets without waking up the power-gated routers. Since each power-gated router has an bypass buffer for each direction, the controller can check whether the current router is the destination. If the flit is supposed to go to the output network interface, the output of the buffer passes through the *5x1 Mux* as shown in Fig. 9A. In addition, it is possible to have four flits in each port simultaneously, whose destinations are the current router; in this case the controller always grants the priority to N, S, W and E buffers, respectively. Therefore, in the worst-case, the flit passes at most through three multiplexers to reach the ejection port.

5.4 Congestion and deadlock avoidance

The potential challenge in SPONGE is that using boundary pivot column(s) in the network to handle all turning packets can significantly increase the probability of congestion on the boundary pivot column. To avoid this challenge, we need a metric to predict congestion, based on the current information of the boundary pivot column. The metric can be defined as the ratio of the number of grants to the number of refusals of VC allocation requests for each router [22]. The proposed lightweight congestion-aware mechanism is based on this metric. Since each router needs to have only two counters for holding the number of grants and refusals for each router, it can be implemented with negligible area overhead. Thus, by having a congestion-aware metric, the window size of powered-on columns can be expanded or shrunk.

Apart from this metric (i.e., $1 - \frac{No_{granted}}{No_{Request}}$), we need to define two different thresholds for two different circumstances. These thresholds determine the upper and the lower bounds of this metric, which help us estimate the probability of congestion on boundary pivot column. The upper bound threshold (i.e., th_{ON}) implies that we need to utilize other specific column(s) adjacent to the boundary pivot column(s) in the network to avoid congestion. If the value of this metric for only one router exceeds th_{ON}, the adjacent column(s) located at the west and the east of current boundary pivot column must be powered on to avoid probable congestion. Hence, we need a customized wake-up mechanism for the routers in SPONGE. As shown in Fig. 11A, we employ a one bit bus for each column to connect the east and the west adjacent routers of that column. The 1-bit bus can be set to high by any router in the boundary pivot column(s), and this bus informs all adjacent routers to trigger wake-up signal.

On the other hand, the lower bound threshold (i.e., th_{OFF}) implies that the utilization of the current boundary pivot column is too low, and the

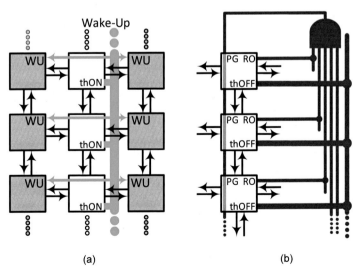

Fig. 11 (A) Wake up (WU) and (B) power gate (PG), in SPONGE.

available active resources are more than required. In a boundary pivot column, if the metric value is less than th_{OFF} for even one router, all of the routers in that column can be power gated. Consequently, we need a customized power-gating technique as well. As shown in Fig. 11A, for an $n \times n$ network, we need a bus whose width is $n + 1$ bits. One bit of this bus can be set to high by any router in the boundary pivot column whose metric value is less than the th_{OFF}. This informs other routers to go to the power-gated state. Thus, they stop receiving packets and process all previous ones to dequeue all packets. Each router has access to one bit of n remaining bus wires, which can be set to high when the corresponding router's buffers are empty. Thus, all routers can be power gated when the output of the AND gate is asserted to high. Since boundary pivot column is responsible for handling turning packets, its routers must be power gated simultaneously, to avoid deadlock for the remaining packets.

After powering on the columns adjacent to the boundary pivot column, we have three powered-on columns. Thus, each column has its own one-bit bus to power on the west and the east adjacent routers depending on the value of metric for its routers. Therefore, depending on the traffic load of the network, the number of columns with powered-on routers can vary. Fig. 12 depicts all possible scenarios in a 5×5 mesh network. The boundary pivot column at the beginning (i.e., S_1) is the middle column. If the metric value of even one of the routers in this column exceeds the upper bound threshold, it can set the one-bit bus to high to power on the adjacent routers (S_2 in Fig. 12). At this time, three columns (i.e., P_{-1}, P_0, and P_1) are powered on, so P_{-1} and P_1 are new boundary pivot columns. Hence, incoming packets from power-gated routers can turn through them. Besides, other packets which are injected to the powered-on routers use YX routing. Based on the network traffic, different circumstances may occur after S_1. In S_3 (S_4), the P_1 (P_{-1}) has been power gated due to its low traffic volume. Due to higher traffic volume and increasing probability of congestion, P_{-2} or P_2 has been powered on, as illustrated in S_7 and S_8. Accordingly, based on the traffic in the network, different number of columns can be powered on. Note that P_0 is always powered on in all states.

Packets traverse straight through power-gated routers without any turns. Besides, packets have up to two turns through boundary pivot column; incoming packets from power-gated (powered-on) to powered-on (power-gated) routers traverse as XY (YX). Also, incoming packets from powered-on to powered-on routers traverse as YX. Therefore, neither $W–N$ nor $W–S$ turns occurs on the left boundary pivot column, and

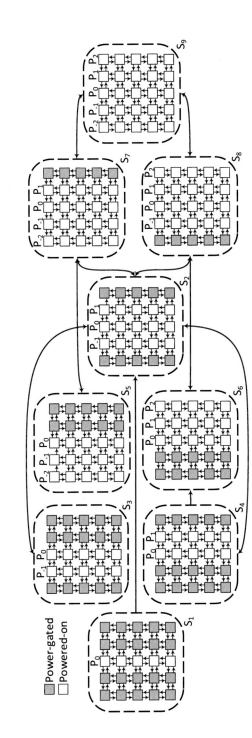

Fig. 12 Different on/off situations based on the congestion.

neither $E–N$ nor $E–S$ turns occurs on the right boundary pivot column. Besides, turn restriction between boundary pivot columns is equal to that of the YX routing. As a result, there is no cycle in SPONGE.

6. MUFFIN

There are *two* major shortcomings in prior power-gating techniques: (1) power overhead of wake-up due to aperiodicity of traffic flow and (2) The delay overhead of power-gating due to detouring or wake-up delay. *Muffin* adds only 5 flit-size buffers and a lightweight controller. These changes enable *Muffin* to efficiently handle a high proportion of flits, regardless of type, i.e., *inject, eject, straight*, and even *turn*, without powering on the router. Since a high proportion of flits can traverse power-gated routers, *Muffin* significantly decreases the number of wake-ups which reduces the delay incurred by blocking flits requesting power-gated routers while extending the length of the power-gated periods for each router. *Muffin* achieves performance benefits; packets using the extra buffers in Muffin, effectively bypass the pipeline stages of the power-gated router to achieve lower latency.

6.1 Enabling bypass for all flit types

To avoid powering on the router for all flit types, *Muffin* requires 5 flit-sized buffers, a small amount of logic and a lightweight power-gating controller in each router. *Muffin* controller handles buffer management using *Free to Forward* signals, i.e., FF_{dir} where *dir* is one of the cardinal directions: N, S, E or W. Fig. 13 depicts the *Muffin* router microarchitecture. The basic router functionality is highlighted in gray with dotted lines and remains unchanged.

When a flit arrives and the router is powered on (i.e., $en\_t = 1$), the flit enters the input buffers of the basic router through the upstream multiplexers, i.e., $MUXUP_{dir}$. When the router is in a power-gated mode (i.e., $en\_t = 0$), the incoming flits from the upstream router and the network interface (NI) are buffered in ByP_{dir} and *interject* buffer, respectively. Thereafter, the *Muffin* controller calculates the control signals for all multiplexers to bypass flits. To do that, *Muffin* controller determines the flit type, i.e., *inject, eject, straight*, and *turn*, using the destination information which is stored in the flit header. The logic for this is very similar to the routing logic; due to its simplicity, it can be completed in a single cycle to enable the proper handling of flits when the router is power-gated.

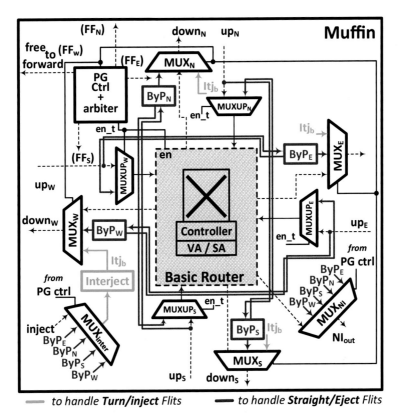

Fig. 13 *Muffin* router micro-architecture.

Straight: Four out of five flit-size buffers are located at each of the four outgoing ports of each side, called ByP_{dir}. These buffers bypass *straight* flits. For example, the flits from upstream link of the E port, UP_E, are directly connected to ByP_W. The *Muffin* controller sets the selectors of MUX_W to bypass this flit to downstream link of port W. Similarly, the remaining three ports handle straight flits using these buffers without powering on the router. Note that one of the inputs of the downstream multiplexers (MUX_{dir}) comes from the basic router when the router is powered on. When the router is powered on, *Muffin* controller selects the output of basic router.

Eject: These four flit-size buffers also handle *eject* flits. Suppose an incoming flit must be ejected to the NI. In this case, the flit will be buffered in corresponding ByP_{dir} and forwarded to the NI. As shown in Fig. 13, we add a *5-to-1* multiplexer (MUX_{NI}) to provide this capability. For periods when the router is powered on, the 5 th input of MUX_{NI}, connects to the NI port of the basic router.

Turn: *Turn* packets require at least two clock cycles to traverse the proposed router. To handle *turn* flits, we add another flit-size buffer, called *interject*. All flits arriving from upstream ports are buffered in bypass buffers. After buffering the incoming flit, the *Muffin* controller determines the outgoing port for this flit. If it is a *turn* flit, the controller sends this flit to the *interject* buffer to turn the flit toward its destination. Accordingly, the select of MUX_{inter} determines which bypass buffer should be stored in the *interject* buffer. After buffering the flit into the *interject* buffer, the *Muffin* controller allocates the link and releases the buffer. Since the downstream router must be checked for *turn* flits before link allocation by the controller, it is not possible to handle turn packets without having the *interject* buffer.

Inject: The *interject* buffer also handles injecting flits without needing to power on the router. The inject input port is one of the inputs of MUX_{inter}. So, *inject* flits will be buffered in *interject* when there is no *turn* flit in it. The *Muffin* controller sets the select of corresponding MUX_{dir} to inject this flit.

6.2 Priority arbitration for different flit types

Since we use shared buffers and additional multiplexers in *Muffin*, its controller must determine the priority of different types of flits. The first arbitration is for the downstream multiplexers (MUX_{dir}). Each MUX has three inputs, the *ByP* buffer that contains a *straight* flit, the *Interject* buffer, and the basic router when it is powered on. Since the third one is only used when the router is powered on, *Muffin* arbitration happens between the first two inputs. The controller grants higher priority to the *ByP* buffer. The *Interject* buffer holding either a *inject* or *turn* flit must wait until there is no *straight* flit in *ByP* buffer. The second arbitration is for the multiplexer of *Interject* buffer (MUX_{inter}). Here, flits stored in *ByP* buffers (*turn* flits), have lower priority than an *inject* flit when they arrive simultaneously. However, if the *Interject* buffer currently holds a *turn* flit, the *inject* flit must wait until the *turn* flit finishes. If multiple *ByP* flits request the *interject* buffer simultaneously, the *Muffin* controller always grants the priority to ByP_N, ByP_S, ByP_E, and ByP_W, respectively. Similarly, for the arbitration of the multiplexer of NI_{out}, the controller ranks the ByP buffers in the aforementioned order. This statically defined order allows us to make the *Muffin* controller simple and lightweight. In Section 6.4, we discuss our congestion and starvation avoidance mechanism. This rank-based arbitration in *Muffin* is easily implementable using cascading multiplexers.

6.3 No restrictions on routing algorithm

Muffin supports both deterministic and adaptive routing algorithms. For XY dimension-order routing (DOR), we only need to support two turns: W to N/S or E to N/S. Hence, we can remove ByP_N and ByP_S wires from MUX_{inter}, and make this multiplexer smaller. To support all possible turns, we connect Itj_b to all downstream multiplexers.

Muffin does not impose any restriction on the capability of the router for handling different scenarios while the router is power-gated compared to the baseline router. Fig. 14 reflects possible scenarios of traffic in a

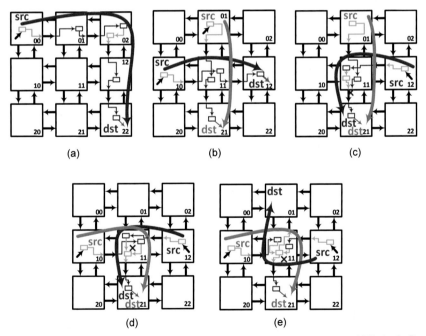

Fig. 14 Example flow control in *Muffin* when basic routers are power gated (A) single flit and no contention, (B) multi flit and no conflict, (C) contention of *Straight* and *Turn*, (D) contention of two *Turns* on same downstream link, and (E) contention of two *Turns* on different downstream links.

Muffin router in a 3×3 mesh network. We assume that all routers are power-gated in these examples to show the efficiency of *Muffin* in handling different scenarios. With no contention (Fig. 14A), flits travel to their destinations with ideal latency. The ideal latency is only *one cycle* for *inject*, *eject*, and *straight* flits, and *two cycles* for *turn* flits. In this example, the flit takes only *6 cycles*. However, for an optimized baseline router with two pipeline stages and no power-gating, it takes *10 cycles*. This gives *Muffin* a performance advantage. *Straight* flits from different dimensions (Fig. 14B) do not affect each other; similar to the first scenario, the latency is ideal. When a *turn* flit and *straight* flit must traverse the same downstream link (Fig. 14C), contention occurs; *straight* flits are prioritized over *turn* flits in downstream multiplexers. For two *turn* flits with the same downstream link (Fig. 14D), prioritization is based on the following order: N, S, E, and W. The winning *turn* flit accesses the *interject* buffer, while the other *turn* flit must wait in its *ByP* buffer. The contention in Fig. 14C and D also occurs in the baseline router; the difference is how the contention is resolved. Since we use a shared buffer for all turns in a router (*interject* buffer), there is a specific type of contention that occurs only in *Muffin* (Fig. 14E). In this scenario, there are two *turn* flits from different directions toward different directions. However, the prioritization for *interject* buffer resolves this contention with a throughput degradation of $1/2$ compared to baseline router. However, as *Muffin* operates with higher throughput in other cases, this throughput degradation is compensated for completely.

6.4 Discussion

Congestion avoidance: In power-gated mode, *Muffin* can handle a high proportion of packets. However, increasing the injection rate increases the probability of congestion in the network when the routers are power-gated. A metric is required to estimate congestion. For powering on routers, we define a time-based metric and a threshold for the added buffers (*interject* and *ByP* buffers). The time-based metric determines the number of cycles that a flit is waiting in one of these buffers. The threshold is the maximum feasible value for this metric to keep the router in power-gated mode. If the number of cycles that a flit is stored in one of these buffers exceeds this threshold, it means that the waiting flit could not access the resource after a long period. This triggers the WU signal by the controller to change the state of this router to powered on. Implementing this threshold only requires a few counters with negligible overhead.

For low injection rates, we need another metric and threshold to determine when router utilization is low while powered-on. At this threshold, it is more efficient to change its state to power-gated. We use a metric from prior work that reflects the ratio of VC grants to refusals for each router, i.e., $1 - \frac{No_{granted}}{No_{Request}}$ [22]. To avoid the cost of division, $No_{Request}$ should be a power of two so that the controller can use a shift register for this calculation. Since only two counters are required, this mechanism has low area overhead. If the metric value is lower than a threshold, the number of refusals is low and the router can be effectively power-gated. When PG signal is triggered in a router, the basic router stops receiving new packets and processes all stored ones prior to using *Muffin*'s buffers.

Both the decision to power on and off are made individually in a decentralized manner at each router, resulting in low overhead and complexity.

Starvation avoidance: To minimize the area overhead, we use a statically defined order for the arbitration of the multiplexer of NI_{out} and MUX_{inter}. Despite this static order, there is no starvation. If a low-priority flit in the arbitration order is being starved by higher priority flits, the power-on threshold will be triggered to wake-up the router and allow all flits to make forward progress. To avoid potential unfairness, the hard-coded arbitration order can be different at each router. For example, one router with priority order: N, S, E, and W, and another router with priority order: E, W, S, and N.

Protocol-level deadlock avoidance: Coherence protocols have several message classes. To avoid resource dependences between messages of different classes, which can lead to deadlock, NoCs separate message classes into different virtual channels. *Muffin* uses single flit buffers at each port; this buffer can mix traffic from different message classes. However, protocol-level deadlock is avoided because congestion due to stalled flits will trigger the power-on threshold in the router which will allow packets to flow into the VCs of their respective message classes, preventing an actual protocol-level deadlock from forming. Flits in the bypass buffers are transferred to the buffers of the current basic routers after powering on. As it can be seen in Fig. 13, four MUXes ($MUXUP_{dir}$) are added in front of input ports of basic router (inputs coming from the upstream routers). The *Muffin* controller selects the *ByP* buffers to transfer into the basic router when powered on. The flit stored in *interject* buffer will be transferred to the next router. Since upstream links are directly connected to bypass flits, the *Muffin* controller sends the flit stored in *interject* buffer to the next router. Then, in the next router, this flit will be stored in these bypass buffers (if the router is power-gated) or in buffers of basic router (if the router is powered on). Note that if the state of the next router is similar to the current one, i.e., it is in the wake-up process, the controller of the current router waits until the next router is powered on, then releases the flit of *interject* router. After releasing the flit of this one-flit buffer, the current router will be powered on.

7. Evaluation

In this section we compare TooT, SMART, SPONGE and Muffin. For the quantitative comparison, we implement the schemes in *Booksim* [23] and run traces gathered from SPLASH-2 benchmarks [21]. We generate traces using *gem5* [24] in *syscall* emulation mode. Table 1 lists all

Table 1 Key simulation parameters.

Core	64 cores, x86 ISA, 2GHz, out-of-order, 8-wide issue
L1 Cache	Private, 16 kb Ins. + 16 kb Data, 4-way set, 3-cycle latency
L2 Cache	Shared, distributed, 256 kb/node, 16-way set, 15-cycle latency
Cache Coherence	MESI, CMP directory, 64-byte block
Memory	1 GB/controller, 1 controller at each mesh corner
Topology	8 × 8 mesh, 3 classes/port, 3 VCs/class, 4 flits/VC
Link Bandwidth	128 bits/cycle
Flow Control	Wormhole
Routing Alg.	XY, Adaptive
Router uArch.	2 stages (No-PG)
	4 stages (No-PG, TooT, PowerPunch, SMART, SPONGE, Muffin)

configuration parameters for our evaluation. We use DSENT [3] to model static and dynamic power for a 45 nm process. Experimental results are compared to the following techniques:

1. **No-PG**: Baseline with no power-gating technique.
2. **Conv-PG**: Simple power-gating technique (immediate power-gating after idle detection).
3. **PG-ConvOpt**: Optimized power-gating endowed with a prediction and early wake-up mechanism [9].
4. **PowerPunch**: A state-of-the-art power-gating method which tries to completely hide wake-up latency [2].

7.1 Qualitative comparison

TooT [11] reduces the number of router wake-ups through a bypass mechanism for straight and eject packets. Routers are only powered-on for turn and inject packets which imposes a delay overhead due to the wake-up process for these packets. This method also has an option to use adaptive routing at the cost of higher latency. SMART [12] reduces static power through a novel mapping and routing technique to regularize traffic patterns to reduce the number of wake-ups. Although it makes power-gating more efficient, the scheme is application specific; the mapping changes from application to application. Moreover, the efficiency of SMART is relatively low for applications with high injection rates. SMART only supports a deterministic routing algorithm. To avoid latency penalties of wake-up, SPONGE [13] keeps a small-set of routers in powered-on state. The main disadvantages of SPONGE are that (1) large networks see high packet latency since turn packets are detoured to the power-on routers, and (2) SPONGE cannot support adaptive routing. In contrast with SPONGE, Muffin [14] virtually eliminates the delay overhead imposed by the wake-up process. This is achieved through a comprehensive bypass mechanism that can efficiently handle all types of flits, i.e., inject, eject, straight, and turn rather than only handling one or two types of flits. Furthermore, Muffin works well with adaptive routing. Major differences between these techniques are given in Table 2.

7.2 Quantitative comparison

Static power: Fig. 15 shows the power dissipation of all techniques normalized to No-PG. Total power consumption includes dynamic power, static

Table 2 Comparison of different power-gating schemes.

Proposed solution	Full path diversity	Intra-routers wiring	Extra buffer	Full bypass mechanism	No misrouting
TooT [11]	✓	None	4 One-flit buffer $\times \# VNs$	Partial	✓
SMART [12]	✗	None	None	None	✓
SPONGE [13]	✗	Bus wires	4 One-flit buffer $\times \# VNs$	Partial	✗
Muffin	✓	None	5 One-flit	✓	✓

power and the overhead of power gating (wake-ups and power-gating controller overhead) which contributes to the total static power. Since Muffin handles all types of flits without powering on the router, it alleviates static power dissipation significantly. Compared to No-PG, Muffin and SPONGE reduce router static power consumption by 97.33% and 86.27% respectively, while PG-ConvOpt, PowerPunch, SMART, TooT save 25.4%, 26.2%, 42%, and 73.1% of static power.

Latency: Fig. 16 shows the average packet latency of different power-gating techniques using XY routing. Since Muffin exploits a comprehensive bypass mechanism, it improves packet latency by 7.5%. This is because that Muffin forward straight, eject, and inject flits in only one cycle, and only turn flits take at least two cycles. However, a 2-stage router pipeline needs 2 cycles for all packet types. Since only ~ 17.5% of flits are turn, Muffin improves the average packet latency compared to a router with a 2-stage bypassing pipeline. Since SPONGE and TooT use partial bypass mechanism, they incur 30.1% and 85.6% delay overhead due to detouring and wake-up delay.

Fig. 17 shows average packet latency for different techniques with an adaptive routing. We exclude SPONGE and SMART from this analysis because they only support deterministic routing. To evaluate the impact of the adaptive routing, we use Duato's protocol and maximize the number of turns for each packet [25]. Although TooT supports adaptive routing, it does not efficiently handle *turn* flits. As the number of turns increase so does a packet's latency. Due to an increase in *turn* flits with adaptive routing, *Muffin* experiences a 5% delay overhead compared to No-PG due to contention for the *interject* buffer.

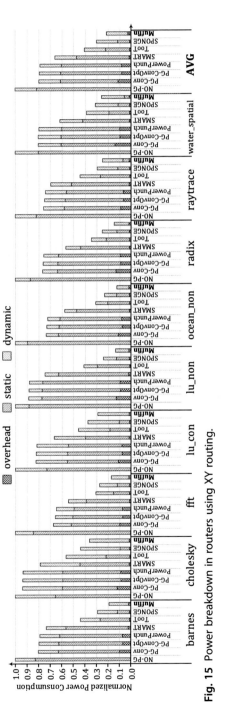

Fig. 15 Power breakdown in routers using XY routing.

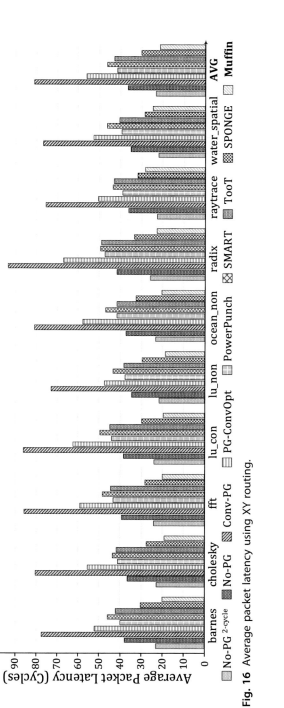

Fig. 16 Average packet latency using XY routing.

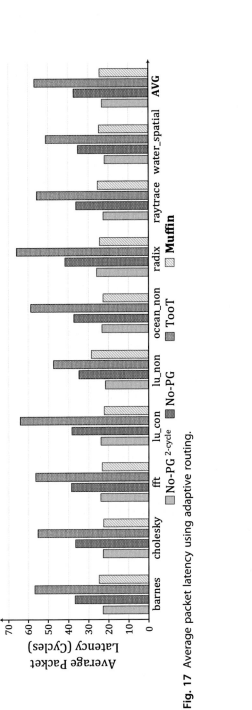

Fig. 17 Average packet latency using adaptive routing.

8. Conclusion

Network-on-Chip (NoC) is a key component in chip multiprocessors (CMPs) as it supports communication between many cores. However, NoCs for modern chips contribute up to 30% of the chips overall power budget. Also, as chips continue to shrink, the contribution of NoC power is projected to increase. Thus, reducing NoC power consumption is critical for ensuring that chips can continue to keep pace with increasing demand. Power-gating is an efficient technique to reduce the static power of underutilized resources in different types of circuits. For NoC, routers are promising candidates for power gating, since they present high idle time. In this chapter, we present TooT, SMART, SPONGE and Muffin. TooT uses bypassing for straight and eject packets; however, routers must be powered on when turn or inject packets arrive incurring wake-up delay. SMART uses an application-specific mapping and routing technique with almost no hardware overhead. However, it is not scalable. SPONGE keeps a small-set of routers in a power-on state. However, its efficiency is degraded in larger networks since turn packets are misrouted to the power-on routers. Muffin exploits a fully bypass mechanism when traffic is low. However, it adds additional buffers, multiplexers and wiring imposing area and power overheads.

Power-gating solutions only save power at low injection rates since the power-gated routers must be powered on when traffic load increases. Thus these solutions are suitable for the low-load applications. To save power regardless of the traffic load, a power-gating technique should take global traffic information into account in deciding when to power gate and when to power on a router (or a part of a router). This might require global coordination between routers imposing overheads. Besides, the ability to support fully adaptive routing is essential to handle irregular topologies and faulty networks. Thus to handle these networks, a power-gating solution should not sacrifice flexibility in routing algorithms.

References

[1] B.K. Daya, C.-H. Chen, S. Subramanian, W.-C. Kwon, S. Park, T. Krishna, J. Holt, A.P. Chandrakasan, L.-S. Peh, Scorpio: a 36-core research chip demonstrating snoopy coherence on a scalable mesh noc with in-network ordering, in: International Symposium on Computer Architecture, 2014, pp. 25–36.
[2] L. Chen, D. Zhu, M. Pedram, T.M. Pinkston, Power punch: towards non-blocking power-gating of NoC routers, in: International Symposium on High Performance Computer Architecture (HPCA), 2015, pp. 378–389.

[3] C. Sun, C.O. Chen, G. Kurian, L. Wei, J. Miller, A. Agarwal, L. Peh, V. Stojanovic, DSENT—a tool connecting emerging photonics with electronics for opto-electronic networks-on-chip modeling, in: International Symposium on Networks-on-Chip, 2012.

[4] A. Samih, R. Wang, A. Krishna, C. Maciocco, C. Tai, Y. Solihin, Energy-efficient interconnect via router parking, in: International Symposium on High Performance Computer Architecture, 2013, pp. 508–519.

[5] N.S. Kim, T. Austin, D. Baauw, T. Mudge, K. Flautner, J.S. Hu, M.J. Irwin, M. Kandemir, V. Narayanan, Leakage current: Moore's law meets static power, IEEE Comput. 36 (12) (2003) 68–75.

[6] M. Keating, D. Flynn, R. Aitken, A. Gibbons, K. Shi, Low Power Methodology Manual: For System-on-Chip Design, Springer Publishing Company, Inc, 2007.

[7] R. Hesse, N. Enright Jerger, Improving DVFS in NoCs with coherence prediction, in: International Symposium on Networks-on-Chip, 2015, pp. 24–32.

[8] M. Badr, N. Enright Jerger, SynFull: synthetic traffic models capturing cache coherent behaviour, in: International Symposium on Computer Architecture, 42, 2014, pp. 109–120.

[9] H. Matsutani, M. Koibuchi, H. Amano, D. Wang, Run-time power gating of on-chip routers using look-ahead routing, in: Asia and South Pacific Design Automation Conference (ASP-DAC), 2008, pp. 55–60.

[10] L. Chen, T. M. Pinkston, Nord: node-router decoupling for effective power-gating of on-chip routers: International Symposium on Microarchitecture, 2012, pp. 270–281.

[11] H. Farrokhbakht, M. Taram, B. Khaleghi, S. Hessabi, Toot: an efficient and scalable power-gating method for NoC routers, in: International Symposium on Networks-on-Chip (NOCS), 2016, pp. 1–8.

[12] H. Farrokhbakht, H.M. Kamali, S. Hessabi, SMART: a scalable mapping and routing technique for power-gating in noc routers, in: International Symposium on Networks-on-Chip (NOCS), 2017, pp. 15:1–15:8.

[13] H. Farrokhbakht, H.M. Kamali, N. Enright Jerger, S. Hessabi, Sponge: a scalable pivot-based on/off gating engine for reducing static power in noc routers, in: International Symposium on Low Power Electronics and Design, 2018, p. 17.

[14] H. Farrokhbakht, H.M. Kamali, N. Enright Jerger, Muffin: minimally-buffered zero-delay power-gating technique in on-chip routers, in: International Symposium on Low Power Electronics and Design, 2019.

[15] J.S. Kim, M.B. Taylor, J. Miller, D. Wentzlaff, Energy characterization of a tiled architecture processor with on-chip networks, in: International Symposium on Low power Electronics and Design, 2003, pp. 424–427.

[16] L. Chen, L. Zhao, R. Wang, T.M. Pinkston, Mp3: minimizing performance penalty for power-gating of clos network-on-chip, in: International Symposium on High Performance Computer Architecture (HPCA), 2014.

[17] H. Matsutani, M. Koibuchi, D. Ikebuchi, K. Usami, H. Nakamura, H. Amano, Ultra fine-grained run-time power gating of on-chip routers for cmps, in: International Symposium on Networks on Chip (NoCS), 2010, pp. 61–68.

[18] R. Das, S. Narayanasamy, S.K. Satpathy, R.G. Dreslinski, Catnap: energy proportional multiple network-on-chip, in: International Symposium on Computer Architecture, ACM, 2013, pp. 320–331.

[19] R. Parikh, R. Das, V. Bertacco, Power-aware nocs through routing and topology reconfiguration, in: Design Automation Conference (DAC), 2014, pp. 1–6.

[20] G. Kim, J. Kim, S. Yoo, Flexibuffer: reducing leakage power in on-chip network routers, in: Design Automation Conference (DAC), 2011, pp. 936–941.

[21] S.C. Woo, M. Ohara, E. Torrie, J.P. Singh, A. Gupta, The SPLASH-2 programs: characterization and methodological considerations, in: International Symposium on Computer Architecture, 1995, pp. 24–36.

[22] A. Mirhosseini, M. Sadrosadati, A. Fakhrzadehgan, M. Modarressi, H. Sarbazi-Azad, An energy-efficient virtual channel power-gating mechanism for on-chip networks, in: Design, Automation Test in Europe Conference Exhibition (DATE), 2015, pp. 1527–1532.
[23] N. Jiang, D.U. Becker, G. Michelogiannakis, J. Balfour, B. Towles, D.E. Shaw, J. Kim, W.J. Dally, A detailed and flexible cycle-accurate network-on-chip simulator, in: International Symposium on Performance Analysis of Systems and Software (ISPASS), 2013, pp. 86–96.
[24] N. Binkert, B. Beckmann, G. Black, S.K. Reinhardt, A. Saidi, A. Basu, J. Hestness, D.R. Hower, T. Krishna, S. Sardashti, R. Sen, K. Sewell, M. Shoaib, N. Vaish, M.D. Hill, D.A. Wood, The gem5 simulator, ACM SIGARCH Comput. Architect. News 39 (2011) 1–7.
[25] J. Duato, A new theory of deadlock-free adaptive routing in wormhole networks, IEEE Trans. Parallel Distrib. Syst. 4 (12) (1993) 1320–1331.

About the authors

Hossein Farrokhbakht received his PhD from the Edward S. Rogers Sr. Department of Electrical and Computer Engineering at the University of Toronto in 2021, under the supervision of Natalie Enright Jerger. His research interests lie in computer architecture, interconnection networks, low-power NoCs, and hardware accelerators.

Shaahin Hessabi received the BS and MS degrees in electrical engineering from Sharif University of Technology, Tehran, Iran, in 1986 and 1990, respectively, and the PhD degree in electrical and computer engineering from the University of Waterloo, Waterloo. He joined Sharif University of Technology in 1996. Since 2007, he has been an associate professor in the Department of Computer Engineering, Sharif University of Technology, Tehran, Iran. He has published more than 130 refereed papers in the related areas. His research interests include cyber-physical systems, reconfigurable and heterogeneous architectures, network-on-chip, and system-on-chip. He has served as the program chair, general chair, and program committee member of various conferences, like DATE, NOCS, NoCArch, and CADS.

N. Enright Jerger is a Professor in the Department of Electrical and Computer Engineering at the University of Toronto. Prior to joining the University of Toronto, she received her PhD from the University of Wisconsin-Madison studying computer architecture, co-advised by Mikko Lipasti and Li-Shiuan Peh. She received her Bachelor of Science degree in 2002 from Purdue University. In 2019, she received the McLean Award from the University of Toronto and was appointed as the Canada Research Chair in Computer Architecture. In 2015, she was awarded an Alfred P. Sloan Research Fellowship and the Borg Early Career Award. She was the recipient of the 2014 Ontario Professional Engineers Young Engineer Medal. She is a Distinguished Member of the ACM and Fellow of the IEEE. In 2017, she co-authored the second edition of On-Chip Networks with Tushar Krishna and Li-Shiuan Peh. Her research focuses on networks-on-chip, approximate computing, IoT devices and hardware acceleration. Beyond research, she is also involved in outreach activities for women in computer architecture, including serving as chair of WICARCH. From 2019–2021, she served as the co-chair of ACM's Council on Diversity and Inclusion.

Printed in the United States
by Baker & Taylor Publisher Services